Water Science and Technology Library

Volume 82

The aim of the Water Science and Technology Library is to provide a forum for dissemination of the state-of-the-art of topics of current interest in the area of water science and technology. This is accomplished through publication of reference books and monographs, authored or edited. Occasionally also proceedings volumes are accepted for publication in the series.

Water Science and Technology Library encompasses a wide range of topics dealing with science as well as socio-economic aspects of water, environment, and ecology. Both the water quantity and quality issues are relevant and are embraced by Water Science and Technology Library. The emphasis may be on either the scientific content, or techniques of solution, or both. There is increasing emphasis these days on processes and Water Science and Technology Library is committed to promoting this emphasis by publishing books emphasizing scientific discussions of physical, chemical, and/or biological aspects of water resources. Likewise, current or emerging solution techniques receive high priority. Interdisciplinary coverage is encouraged. Case studies contributing to our knowledge of water science and technology are also embraced by the series. Innovative ideas and novel techniques are of particular interest.

Comments or suggestions for future volumes are welcomed.

Vijay P. Singh, Department of Biological and Agricultural Engineering & Zachry Department of Civil Engineering, Texas A&M University, USA
Email: vsingh@tamu.edu

More information about this series at http://www.springer.com/series/6689

Vijay P. Singh · Shalini Yadav
Ram Narayan Yadava
Editors

Climate Change Impacts

Select Proceedings of ICWEES-2016

Springer

Editors
Vijay P. Singh
Department of Biological and Agricultural Engineering, and Zachry Department of Civil Engineering
Texas A&M University
College Station, TX
USA

Ram Narayan Yadava
AISECT University
Hazaribagh, Jharkhand
India

Shalini Yadav
Department of Civil Engineering
AISECT University
Bhopal, Madhya Pradesh
India

ISSN 0921-092X ISSN 1872-4663 (electronic)
Water Science and Technology Library
ISBN 978-981-10-5713-7 ISBN 978-981-10-5714-4 (eBook)
https://doi.org/10.1007/978-981-10-5714-4

Library of Congress Control Number: 2017946670

Printed on acid-free paper

This Springer imprint is published by Springer Nature
The registered company is Springer Nature Singapore Pte Ltd.
The registered company address is: 152 Beach Road, #21-01/04 Gateway East, Singapore 189721, Singapore

Preface

Fundamental to sustainable economic development, functioning of healthy ecosystems, reliable agricultural productivity, dependable power generation, maintenance of desirable environmental quality, continuing industrial growth, enjoyment of quality lifestyle, and renewal of land and air resources is water. With growing population, demands for water for agriculture and industry are skyrocketing. On the other hand, freshwater resources per capita are decreasing. There is therefore a need for effective water resources management strategies. These strategies must also consider the nexus between water, energy, environment, food, and society. With these considerations in mind, the International Conference on Water, Environment, Energy and Society (WEES-2016) was organized at AISECT University in Bhopal, MP, India, from March 15–18, 2016. The conference was fifth in the series and had several objectives.

The first objective was to provide a forum to not only engineers, scientists, and researchers, but also practitioners, planners, managers, administrators, and policy makers from around the world for discussion of problems pertaining to water, environment, and energy that are vital for the sustenance and development of society.

Second, the Government of India has embarked upon two large projects one on cleaning of River Ganga and the other on cleaning River Yamuna. Further, it is allocating large funds for irrigation projects with the aim to bring sufficient good quality water to all farmers. These are huge ambitious projects and require consideration of all aspects of water, environment, and energy as well as society, including economics, culture, religion, politics, administration, law, and so on.

Third, when water resources projects are developed, it is important to ensure that these projects achieve their intended objectives without causing deleterious environmental consequences, such as water logging, salinization, loss of wetlands, sedimentation of reservoirs, loss of biodiversity, etc.

Fourth, the combination of rising demand for water and increasing concern for environmental quality compels that water resources projects are planned, designed, executed and managed, keeping changing conditions in mind, especially climate change and social and economic changes.

Fifth, water resources projects are investment intensive and it is therefore important to take a stock of how the built projects have fared and the lessons that can be learnt so that future projects are even better. This requires an open and frank discussion among all sectors and stakeholders.

Sixth, we wanted to reinforce that water, environment, energy, and society constitute a continuum and water is central to this continuum. Water resources projects are therefore inherently interdisciplinary and must be so dealt with.

Seventh, a conference like this offers an opportunity to renew old friendships and make new ones, exchange ideas and experiences, develop collaborations, and enrich ourselves both socially and intellectually. We have much to learn from each other.

Now the question may be: Why India and why Bhopal? India has had a long tradition of excellence spanning several millennia in the construction of water resources projects. Because of her vast size, high climatic variability encompassing six seasons, extreme landscape variability from flat plains to the highest mountains in the world, and large river systems, India offers a rich natural laboratory for water resources investigations.

India is a vast country, full of contrasts. She is diverse yet harmonious, mysterious yet charming, old yet beautiful, ancient yet modern. Nowhere can we find as high mountains as snow-capped Himalayas in the north, the confluence of three seas and large temples in the south, long and fine sand beaches in the east as well as architectural gems in the west. The entire country is dotted with unsurpassable monuments, temples, mosques, palaces, and forts and fortresses that offer a glimpse of India's past and present.

Bhopal is located in almost the centre of India and is situated between Narmada River and Betwa River. It is a capital of Madhya Pradesh and has a rich, several century-long history. It is a fascinating amalgam of scenic beauty, old historic city, and modern urban planning. All things considered, the venue of the conference could not have been better.

We received an overwhelming response to our call for papers. The number of abstracts received exceeded 450. Each abstract was reviewed and about two-thirds of them, deemed appropriate to the theme of the conference, were selected. This led to the submission of about 300 full length papers. The subject matter of the papers was divided into more than 40 topics, encompassing virtually all major aspects of water and environment as well energy. Each topic comprised a number of contributed papers and in some cases state-of-the art papers. These papers provided a natural blend to reflect a coherent body of knowledge on that topic.

The papers contained in this volume, "Climate Change Impacts," represent one part of the conference proceedings. The other parts are embodied in six companion volumes entitled, "Hydrologic Modelling," "Groundwater," "Environmental Pollution," "Water Quality Management," "Energy and Environment," and "Water Resources Management." Arrangement of contributions in these seven books was a natural consequence of the diversity of papers presented at the conference and the topics covered. These books can be treated almost independently, although significant interconnectedness exists amongst them.

This volume contains four parts. The first part deals with some aspects of climatic characteristics ranging from changes in temperature and sunshine hours to downscaling to global climate patterns and effects of ENSO and IOD on extreme rainfall. Part II covers rainfall analysis, including changes in regional rainfall series, analysis of non-stationarity, summer monsoon, and rainfall scenarios. Impacts of climate change are treated in Part III. Change point analysis, greenhouse gas emissions, rainfall variability, water resources variability, and water resources sustainability are discussed in this part. The concluding Part IV is on low flow and drought. It deals with the SPI concept and assessment of drought.

The book will be of interest to researchers and practitioners in the field of water resources, hydrology, environmental resources, agricultural engineering, watershed management, earth sciences, as well as those engaged in natural resources planning and management. Graduate students and those wishing to conduct further research in water and environment and their development and management may find the book to be of value.

WEES-16 attracted a large number of nationally and internationally well-known people who have long been at the forefront of environmental and water resources education, research, teaching, planning, development, management, and practice. It is hoped that long and productive personal associations and friendships will be developed as a result of this conference.

College Station, USA — Vijay P. Singh, Conference Chair
Bhopal, India — Shalini Yadav, Conference Organizing Secretary
Hazaribagh/Bhopal, India — Ram Narayan Yadava, Conference Co-Chair

Acknowledgements

We express our sincere gratitude to Shri Santosh Choubey, Chancellor, and Dr. V.K. Verma, Vice Chancellor, Board of Governing Body, and Board of Management of the AISECT University, Bhopal, India, for providing their continuous guidance and full organizational support in successfully organizing this international conference on Water, Environment, Energy and Society on the AISECT University campus in Bhopal, India.

We are also grateful to the Department of Biological and Agricultural Engineering, and Zachry Department of Civil Engineering, Texas A&M University, College Station, Texas, U.S.A., and International Centre of Excellence in Water Management (ICE WaRM), Australia, for their institutional cooperation and support in organizing the ICWEES-2016.

We wish to take this opportunity to express our sincere appreciation to all the members of the Local Organization Committee for helping with transportation, lodging, food, and a whole host of other logistics. We must express our appreciation to the Members of Advisory Committee, Members of the National and International Technical Committees for sharing their pearls of wisdom with us during the course of the Conference.

Numerous other people contributed to the conference in one way or another, and lack of space does not allow us to list all of them here. We are also immensely grateful to all the invited Keynote Speakers, and Directors/Heads of Institutions for supporting and permitting research scholars, scientists and faculty members from their organizations for delivering keynote lectures and participating in the conference, submitting and presenting technical papers. The success of the conference is the direct result of their collective efforts. The session chairmen and co-chairmen administered the sessions in a positive, constructive, and professional manner. We owe our deep gratitude to all of these individuals and their organizations.

We are thankful to Shri Amitabh Saxena, Pro-Vice Chancellor, Dr. Vijay Singh, Registrar, and Dr. Basant Singh, School of Engineering and Technology, AISECT University, who provided expertise that greatly helped with the conference organization. We are also thankful to all the Heads of other Schools, Faculty Members and Staff of the AISECT University for the highly appreciable assistance in different organizing committees of the conference. We also express our sincere thanks to all the reviewers at national and international levels who reviewed and moderated the papers submitted to the conference. Their constructive evaluation and suggestions improved the manuscripts significantly.

Sponsors and Co-sponsors

The International Conference on Water, Environment, Energy and Society was Jointly organized by the AISECT University, Bhopal (M.P.), India and Texas A&M University, Texas, USA in association with ICE WaRM, Adelaide, Australia. It was partially supported by the International Atomic Energy Agency (IAEA), Vienna, Austria; AISECT University, Bhopal; M.P. Council of Science and Technology (MPCOST); Environmental Planning and Coordination Organization (EPCO), Government of Madhya Pradesh; National Bank for Agriculture and Rural Development (NABARD), Mumbai; Maulana Azad National Institute of Technology (MANIT), Bhopal; and National Thermal Power Corporation (NTPC), Noida, India. We are grateful to all these sponsors for their cooperation and providing partial financial support that led to the grand success to the ICWEES-2016.

Contents

About the Editors

Prof. Vijay P. Singh is a University Distinguished Professor, a Regents Professor, and the inaugural holder of the Caroline and William N. Lehrer Distinguished Chair in Water Engineering in the Department of Biological and Agricultural Engineering and Zachry Department of Civil Engineering at Texas A&M University. He received his B.S., M.S., Ph.D. and D.Sc. degrees in engineering. He is a registered professional engineer, a registered professional hydrologist, and an Honorary Diplomate of American Academy of Water Resources Engineers.

Professor Singh has extensively published the results of an extraordinary range of his scientific pursuits. He has published more than 900 journal articles; 25 textbooks; 60 edited reference books, including the massive Encyclopedia of Snow, Ice and Glaciers and Handbook of Applied Hydrology; 104 book chapters; 314 conference papers; and 72 technical reports in the areas of hydrology, ground water, hydraulics, irrigation engineering, environmental engineering, and water resources.

For his scientific contributions to the development and management of water resources and promoting the cause of their conservation and sustainable use, he has received more than 90 national and international awards and numerous honors, including the Arid Lands Hydraulic Engineering Award, Ven Te Chow Award, Richard R. Torrens Award, Norman Medal, and EWRI Lifetime Achievement Award, all given by American Society of Civil Engineers; Ray K. Linsley Award and Founder's Award, given by American Institute of Hydrology; Crystal Drop Award, given by International Water Resources Association; and Outstanding Distinguished Scientist Award given by Sigma Xi, among others. He has received three honorary doctorates. He is a Distinguished Member of ASCE, and a fellow of EWRI, AWRA, IWRS, ISAE, IASWC, and IE and holds membership in 16 additional professional associations. He is a fellow/member of 10 international science/engineering academies. He has served as President and Senior Vice President of the American Institute of Hydrology (AIH). Currently he is editor-in-chief of two book series and three journals and serves on editorial boards of 20 other journals.

Professor Singh has visited and delivered invited lectures in all most all parts of the world but just a sample: Switzerland, the Czech Republic, Hungary, Austria, India, Italy, France, England, China, Singapore, Brazil, and Australia.

Prof. Shalini Yadav is a Professor and Head of the Department of Civil Engineering, AISECT University, Bhopal, India. Her research interests include Solid and Hazardous Waste Management, Construction Management, Environmental Quality, and Water Resources. She has executed a variety of research projects/consultancy in Environmental and Water Science and Technology and has got rich experience in Planning, formulating, organizing, executing, and management of R&D Programs, Seminars, and Conferences at National and International level. She has got to her credit guiding an appreciable number of M.Tech. and Ph.D. students. She has published more than 10 journal articles and 30 technical reports. Dr. Shalini has also visited and delivered invited lectures at different Institutes/Universities in India and abroad, such as Australia, South Korea, Kenya.

Professor Shalini Yadav graduated with a B.Sc. in Science from the Bhopal University. She earned her M.Sc. in Applied Chemistry with specialization in Environmental Science from Bhopal University and M.Tech. in Civil Engineering with specialization in Environmental Engineering from Malaviya National Institute of Technology, Jaipur, India in 2000. Then she pursued the degree of Ph.D. in Civil Engineering from Rajiv Gandhi Technical University, Bhopal, India in 2011. Also, she is a recipient of national fellowships and awards. She is a reviewer in many International journals. She has been recognized for one and half decades of leadership in research, teaching, and service to the Environmental Engineering Profession.

Dr. Ram Narayan Yadava holds position of Vice Chancellor of the AISECT University, Hazaribag, Jharkhand. His research interests include Solid Mechanics, Environmental Quality and Water Resources, Hydrologic Modelling, Environmental Sciences and R&D Planning and Management. Yadava has executed a variety of research/consultancy projects in the area of Water Resources Planning and Management, Environment, Remote Sensing, Mathematical Modelling, Technology Forecasting, etc.

He has got adequate experience in Establishing Institutes/Organizations, Planning, Formulating, Organizing, Executing and Management of R&D Programs, Seminars, Symposia, Conferences at National and International level. He has got to his credit guiding a number of M.Tech. and Ph.D. students in the area of Mathematical Sciences and Earth Sciences. Dr. Yadava has visited and delivered invited lectures at different Institutes/Universities in India and abroad, such as USA, Canada, United Kingdom, Thailand, Germany, South Korea, Malaysia, Singapore, South Africa, Costa Rica, and Australia.

He earned an M.Sc. in Mathematics with specialization in Special Functions and Relativity from Banaras Hindu University, India in 1970 and a Ph.D. in Mathematics with specialization in Fracture Mechanics from Indian Institute of Technology, Bombay, India in 1975. Also, he is recipient of Raman Research Fellowship and other awards. Dr. Yadava has been recognized for three and half decades of leadership in research and service to the hydrologic and water resources profession. Dr. Yadava's contribution to the state of the art has been significant in many different specialty areas, including water resources management, environmental sciences, irrigation science, soil and water conservation engineering, and mathematical modeling. He has published more than 90 journal articles; four textbooks; seven edited reference books.

Part I
Climatic Characteristics

Trends in Temperature for the Himalayan Environment of Leh (Jammu and Kashmir), India

Rohitashw Kumar, Zeenat Farooq, Deepak Jhajharia and V.P. Singh

Abstract Climate parameters variability affects significantly on water resources, and therefore on the livelihood of the common people, especially in water scarce countries. The aim of this study was to explore changes in the maximum, minimum, and mean temperatures using the monthly data of Leh taking last 15 years from 2000 to 2014, which is situated in the western Indian Himalaya. Trends analyses were performed with nonparametric statistics proposed by Mann-Kendall at different time scales in arid environments of Leh. On monthly basis, a significant falling trend in maximum temperature and minimum temperature has been observed at 5% significance level in the month of July at the rate of 1.7 °C per decade and in the month of August at the rate of 1.3 °C per decade, respectively. However, no trend has been observed in other time scales at 5% level of significance. The observed change in temperature will affect all biochemical reactions of photosynthesis thus in turn will have negative impact on plant growth.

Keywords Trend · Mann-Kendall · Maximum and minimum temperature Himalayan region, Leh

R. Kumar (✉) · Z. Farooq
Division of Agricultural Engineering, Sher-e-Kashmir University of Agricultural Sciences and Technology of Kashmir, Shalimar Campus, Srinagar 190025, Jammu and Kashmir, India
e-mail: rohituhf@rediffmail.com

D. Jhajharia
Department of Soil and Water Engineering, College of Agricultural Engineering and Post-Harvest Technology (Central Agricultural University Imphal), Ranipool, Gangtok 737135, Sikkim, India

V.P. Singh
Department of Biological and Agricultural Engineering, Zachry Department of Civil Engineering, Texas A&M University, College Station, USA

V.P. Singh et al. (eds.), *Climate Change Impacts*, Water Science and Technology Library 82, https://doi.org/10.1007/978-981-10-5714-4_1

Introduction

Climate change has brought in unexpected changes not only in India but all over the regions across the world. Emergence of global warming due to climate change is the new and most talked subject of today's world as it is the most threatening issue for very existence of life on the earth. One of the consequences of climate change is the alteration of rainfall patterns and increase in temperature. According to Intergovernmental Panel on Climate Change (IPCC 2001) reports, the surface temperature of the earth has risen by 0.6 ± 0.2 °C over the twentieth century. Also in the last 50 years, the rise in temperature has been 0.13 ± 0.07 °C per decade. As the warming depends on emissions of greenhouse gases in the atmosphere, the IPCC has projected a warming of about 0.2 °C per decade. Further, surface air temperature could rise by between 1.1 and 6.4 °C over twenty-first century. In case of India, the climate change expected to adversely affect its natural resources, forestry, agriculture, and change in precipitation, temperature, monsoon timing, and extreme events. Due to global warming, precipitation amount, type and timing are changing or expected to change because of increased evaporation, especially in the tropics.

The pattern and amount of rainfall are among the most important factors that affect agricultural production (Jhajharia et al. 2015). Agriculture is vital to India's economy and the livelihood of its people. Agriculture is contributing 21% to the country's GDP, accounting for 115 of total export, employing 56.4% of the total workforce, and supporting 600 million people directly and indirectly (Beena 2010; McVicar et al. 2010, 2012). Temperature drives the hydrological cycle, influencing hydrological processes in a direct or indirect way. A warmer climate leads to intensification of the hydrological cycle, resulting in higher rates of evaporation and increase of liquid precipitation. These processes, in association with a shifting pattern of precipitation, will affect the spatial and temporal distribution of runoff, soil moisture, groundwater reserves, and increase the frequency of droughts and floods. The future climatic change, though, will have its impact globally and will be felt severely in developing countries with agrarian economies, such as India. Surging population and associated demands for freshwater, food, and energy would be areas of concern in the changing climate. Changes in extreme climatic events are of great consequence owing to the high vulnerability of the region to these changes. Parry et al. (2001) have shown that there is a steep rise in the water shortage curve when plotted against rise in temperature. They reported that this is due to large urban populations in China and India being newly exposed to risk. There has not been conducted previous study on behavior of temperature for the Himalayan environment of Leh, this study was carried out to analyze the temperature trend of Leh using Mann-Kendall test for the year 2000–2014.

Materials and Methods

Study Area

The Indian Himalayan Region (IHR) is spreading to 10 states (administrative regions) namely, Jammu & Kashmir, Himachal Pradesh, Uttaranchal, Sikkim, Arunachal Pradesh, Meghalaya, Nagaland, Manipur, Mizoram, Tripura, and hill regions of two states viz. Assam and West Bengal of Indian Republic. It contributes about 16.2% of India's total geographical area, and most of the area is covered by snow-clad peaks, glaciers of higher Himalaya, dense forest cover of mid-Himalaya. The IHR shows a thin and dispersed human population as compared to the national figures due to its physiographic condition and poor infrastructure development, but the growth rate is much higher than the national average.

In this study Himalayan region of Leh was taken into consideration. Mountains dominate the landscape around the Leh as it is at an altitude of 3,500 m. Leh has a cold desert climate with long, harsh winters, with minimum temperatures well below freezing for most of the winter. The city gets occasional snowfall during winter. The weather in the remaining months is generally fine and warm during the day. Average annual rainfall is only 102 mm. The temperature can range from −42 °C in winter to 33 °C in summer. In 2010, the Leh city experienced flash floods which killed more than 100 people. The study area of Himalayan region is shown in Fig. 1.

The data sets used in this study was obtained from High Mountain Research Station Leh, Sher-e-Kashmir University of Agricultural Sciences and Technology of Kashmir for the period of 15 years from 2000 to 2014 of cold desert.

The trend analysis of temperature at monthly, annual and seasonal (winter, spring, summer and autumn) basis was carried out. Trends in data can be identified by using

Fig. 1 Study area of Leh

either parametric or nonparametric methods. In the recent past, both methods have been widely used for the detection of trends (WMO 1997; Mitosek 1992; Chiew and McMahon 1993; Burn and Elnur 2002). Then parametric tests are more suitable for non-normally distributed, censored data, including missing values, which are frequently encountered in hydrological time series (Hirsch et al. 1984).

Mann-Kendall Test

Mann-Kendall test is a statistical test widely used for the analysis of trend in climatologic and in hydrologic time series. There are two advantages of using this test. First, it is a nonparametric test and does not require the data to be normally distributed. Second, the test has low sensitivity to abrupt breaks due to inhomogeneous time series. According to this test, the null hypothesis H_0 assumes that there is no trend (the data is independent and randomly ordered) and this is tested against the alternative hypothesis H_1, which assumes that there is a trend. Mann-Kendall test is a nonparametric test for identifying trends in time series data. This test compares the relative magnitudes of data rather than the data values themselves (Gilbert 1987). This test assumes that there exists only one data value for a time period. When multiple data points exist for a single time period, the median value will be used. The initial value of the Mann-Kendall statistic S is assumed to be 0. If a data value from a later time period is higher than a data value from an earlier time period, S is increased by 1. On the other hand, if the data value from the later time period is lower than a data valued sampled earlier, it is decreased by 1. The net result of increments and decrements yields the final value of S. This method is more suitable for non-normally distributed and censored data, and is less influenced by the presence of outliers in the data (Mann 1945; Kendall 1975).

Let $x_1, x_2, x_3, \ldots, x_n$ represent n data points, then the Mann-Kendall test statistic S is given by

$$S = \sum_{i=1}^{n-1} \sum_{j=i+1}^{n} \operatorname{sgn}(x_j - x_i), \tag{1}$$

where n is the number of observations and x_j is the jth observation and sgn (θ) is the sign function which can be defined as follows:

$$\operatorname{sgn}(\theta) = \begin{bmatrix} 1 & \text{if} & \theta > 0 \\ 0 & \text{if} & \theta = 0 \\ -1 & \text{if} & \theta < 0 \end{bmatrix} \tag{2}$$

Under the assumption that the data are independent and identically distributed, the mean and variance of the S statistic are given by (Kendall 1975)

$$E(S) = 0 \tag{3}$$

$$\frac{n(n-1)(2n+5) - \sum_{i=0}^{m} t_i(t_i-1)(2t_i+5)}{18} = V_S, \tag{4}$$

where m is the number of groups of tied ranks, each with tied observations.

The Z-statistic can be computed as follows:

$$Z = \frac{S-1}{\sqrt{(\mathrm{var}(S))}} \quad \text{IF } S > 0 \tag{5}$$

$$Z = 0 \quad \text{If } S = 0 \tag{6}$$

$$Z = \frac{S+1}{\sqrt{(\mathrm{var}(S))}} \quad \text{if } S<0 \tag{7}$$

Estimation of Magnitude of Trends

The magnitude of the identified trends in the meteorological parameters was obtained through the parametric linear regression test, a commonly used parametric method.

The linear relationship between two variables is represented by a straight line, which is given as

$$y = m \times x + c$$

x denote the time variable
m slope of regression line
c intercept

Results and Discussion

The value of Z statistics with p-value in parenthesis obtained by Mann-Kendall test for all the parameters on monthly and annual time scales are tabulated below. The value test Statistics (z) is summarized in Table 1.

It is witnessed from Table 1 that the statistically significant falling trends are witnessed in maximum temperature in the month of July at the rate of 1.70 °C/decade, at 5% level of significance as the values of Z (test statistics) obtained through the MK

Table 1 The value of test statistics (*z*) obtained through Mann-Kendall test on monthly and annual basis

Month	Maximum temperature	Minimum temperature
January	*0.291 (0.923)*	*0.132 (0.846)*
February	*−0.172 (0.254)*	*−0.357 (0.282)*
March	*0.306 (0.495)*	*0.325 (0.328)*
April	*0.363 (0.282)*	*0.254 (0.298)*
May	*0.038 (0.814)*	*0.166 (0.457)*
June	*0.097 (0.697)*	*0.073 (0.435)*
July	**−2.114 (0.032)**	*0.122 (0.656)*
August	*0.116 (0.137)*	**−1.999 (0.015)**
September	*0.068 (0.657)*	*−0.139 (0.495)*
October	*0.032 (1.000)*	*−0.132 (0.770)*
November	*0.320 (0.804)*	*0.132 (0.626)*
December	*0.052 (0.298)*	*−0.138 (0.499)*
Annual	*0.014 (0.846)*	*0.138 (0.770)*

Note: *Bold values* denote statistically significant at 5% level of significance. *Italic values* are cases of no trends (statistically non-significant even at 10% level of significance)

test are more than −1.96 (Table 1). However, the remaining months witnessed no statistically significant trends in maximum temperature at 5% level of significance as the *Z* values are between +1.96 and −1.96 (or at 10% level of significance as the *Z* values are between +1.65 and −1.65). The monthly trend in maximum temperature during 2000–2014 is shown in Fig. 2a–d.

It is evident from Table 1 that in case of minimum temperature MK test revealed that statistically significant falling trend at 5% level of significance, as the values of *Z* (test statistics) obtained, is more than −1.96 and was witnessed in the month of August at the rate of 1.31 °C/decade (Fig. 3a–d). Statistically significant falling trend was witnessed in the month of December at the rate of 1.74 °C/decade at 10% significance level as the *Z* value is more than 1.65 and less than 1.96. However, the remaining months witnessed no significant trends at 5% level of significance as the *Z* values are between +1.96 and −1.96 (or at 10% level of significance as the *Z* values are between +1.65 and −1.65).

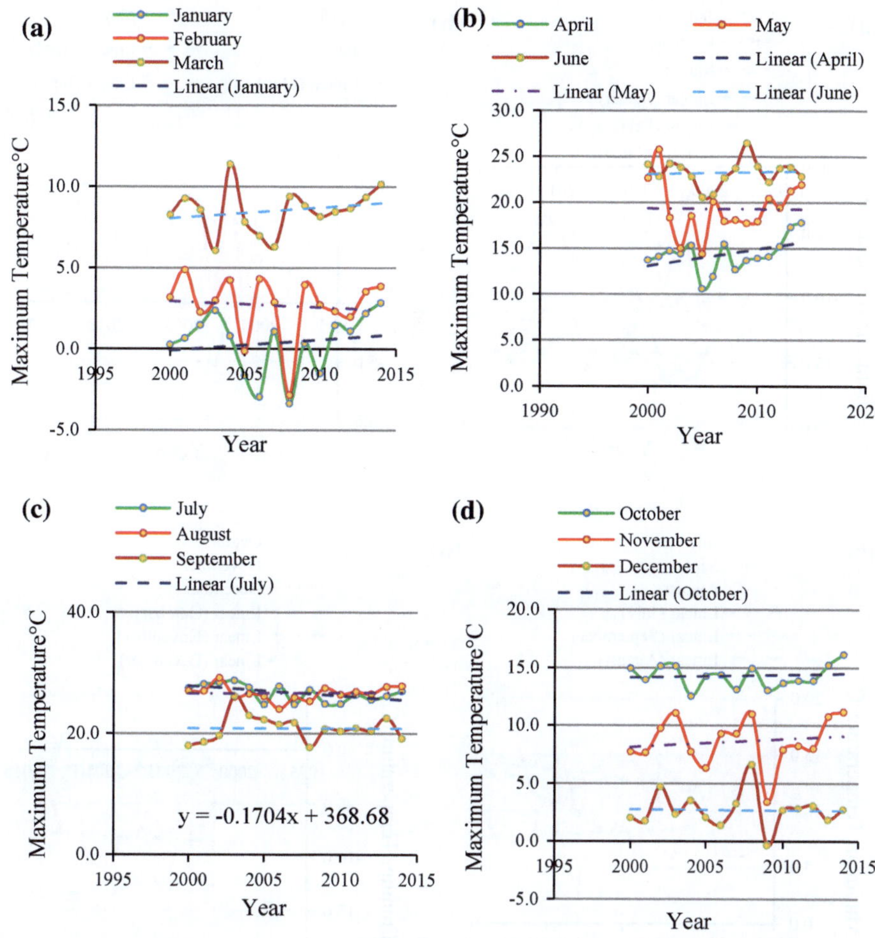

Fig. 2 a–d Time series of maximum temperature on monthly basis with linear trend lines

On annual basis, no significant trend was witnessed in case of maximum and minimum temperature at 5% level of significance as the Z values are between +1.96 and −1.96 (or at 10% level of significance as the Z values are between +1.65 and −1.65). The annual trend of maximum and minimum temperature is shown in Fig. 4a–b.

Test statistics (Z) values with p-value in parenthesis obtained through the Mann-Kendall test on seasonal basis is tabulated and shown in Table 2.

It is evident from Table 2 that on seasonal basis no significant trend was witnessed in case of maximum and minimum temperature at 5% level of significance

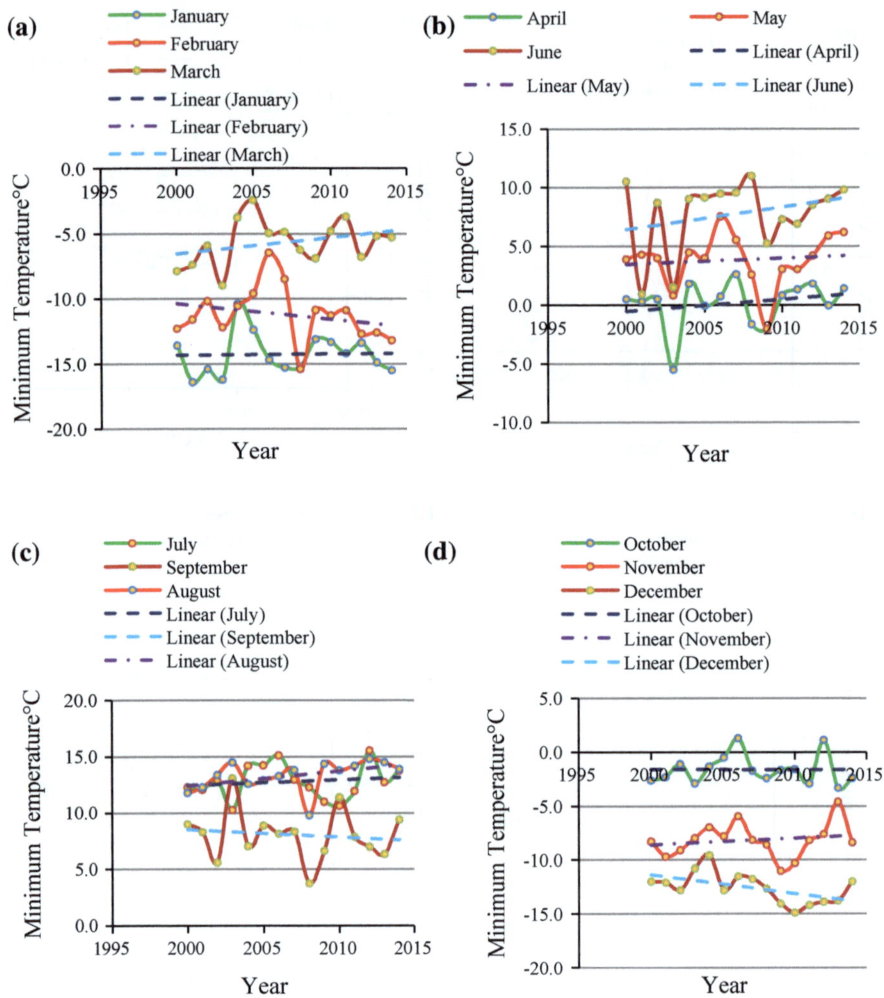

Fig. 3 a–d Time series of minimum temperature on monthly basis with linear trend lines

as the *Z* values are between +1.96 and −1.96 (or at 10% level of significance as the *Z* values are between +1.65 and −1.65). The graphical representation of seasonal basis between maximum temperature is shown in Fig. 5a–b.

The seasonal trend in minimum temperature in different years was also carried. The graphical representation of seasonal basis between minimum temperatures is shown in Fig. 6a–b.

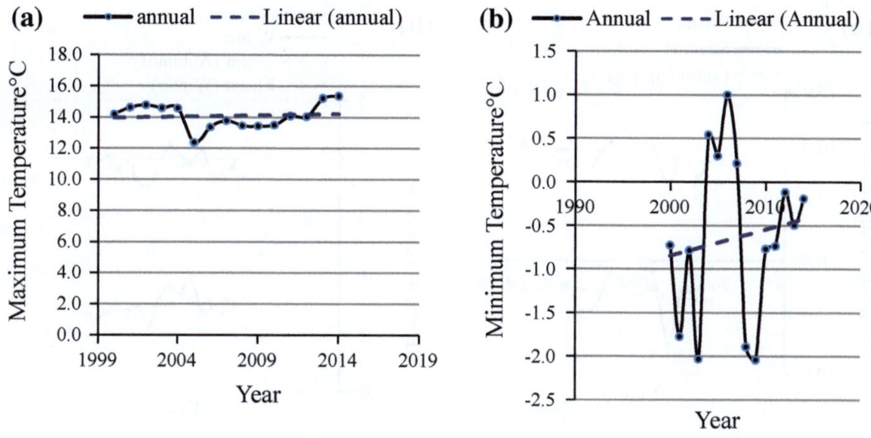

Fig. 4 a–b Time series of different weather parameters on annual basis

Table 2 Value of *Z* statistics with *p*-value in parenthesis on seasonal basis

Season	Maximum temperature	Minimum temperature
Spring	0.306 (0.179)	0.261 (0.667)
Summer	−0.135 (0.667)	0.370 (0.114)
Autumn	0.198 (1.000)	0.058 (0.923)
Winter	−0.017 (0.590)	−0.452 (0.659)

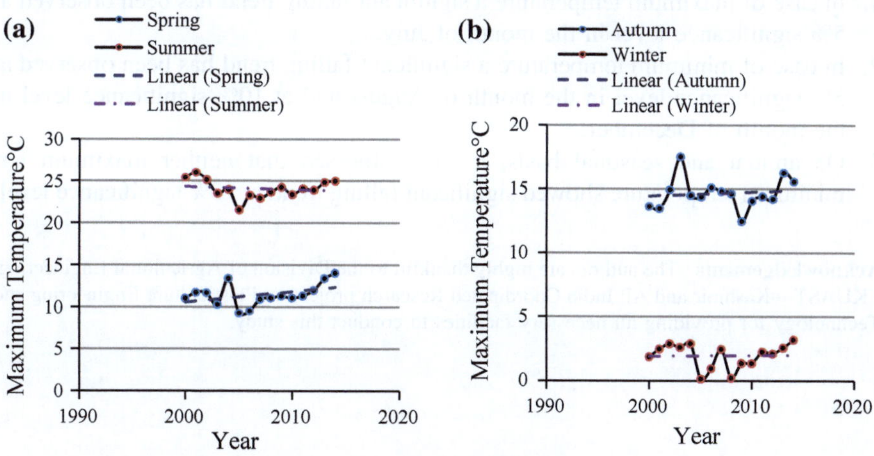

Fig. 5 a–b Maximum temperature on seasonal with linear trend lines

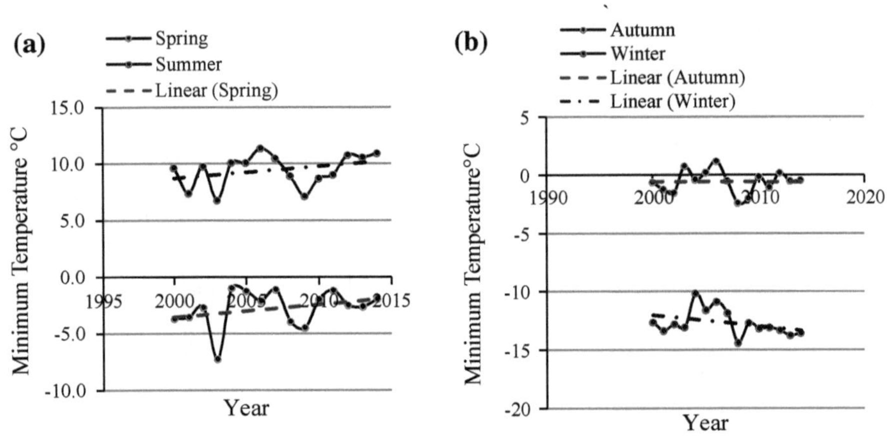

Fig. 6 a–b Minimum Temperature on seasonal with linear trend lines

Conclusions

The weather is a continuous, data-intensive, multidimensional, dynamic and complex process and these properties make weather forecasting a formidable challenge. Thus an attempt has been made in is study to estimate the trends of maximum and minimum temperature, on monthly, seasonal, and annual basis over climatic conditions of Himalayas because of the importance of these parameters in water balance studies, irrigation planning, planning, and operation of reservoirs. The trends in different climatic parameters were investigated using the nonparametric Mann-Kendall (MK) test. The conclusions drawn from the study are summarized as follows:

1. In case of maximum temperature a significant falling trend has been observed at 5% significance level in the month of July.
2. In case of minimum temperature a significant falling trend has been observed at 5% significance level in the month of August and at 10% significance level in the month of December.
3. On annual and seasonal basis, it was witnessed that neither maximum nor minimum temperature showed significant falling trend at 10% significance level.

Acknowledgements The authors are highly thankful to the Division of Agricultural Engineering, SKUAST—Kashmir and All India Coordinated Research project on Plasticulture Engineering and Technology for providing all necessary facilities to conduct this study.

References

Beena S (2010) Global and national concerns on climate change. Univ News 48(24):15–23

Burn DH, Elnur MAH (2002) Detection of hydrologic trends and variability. J Hydrol 255:107–122

Chiew FHS, McMahon TA (1993) Detection of trend or change in annual flow of Australian rivers. Int J Climatol 13:643–653

Gilbert RO (1987) Statistical methods for environmental pollution monitoring. Van Nostrand Reinhold, New York, p 336

Hirsch RM, Slack JR, Smith RA (1984) Techniques of trend analysis for monthly water quality data. Water Resour Res 18:107–121

IPCC (2001) Climate change 2001: synthesis report. In: Watson RT, Core Writing Team (eds) A contribution of working groups I, II and III to the third assessment report of the Intergovernmental Panel on Climate Change. Cambridge University Press, Cambridge

Jhajharia D, Kumar R, Singh VP (2015) Reference evapotranspiration under changing climate over Thar Desert in India. Meteorol Appl 22:425–435. doi:10.1002/met.1471

Kendall MG (1975) Rank correlation methods, 4th edn. London, Charles Griffin

Mann HB (1945) Non-parametric tests against trend. Econometrica 33:245–259

McVicar TR, Michael LR, Donohue RJ, Li LT, Van Niel TG, Thomas A, Grieser J, Jhajharia D, Himri Y, Mahowald NM, Mescherskayai AV, Krugerj AC, Rehman S, Dinpashoh Y (2012) Global review and synthesis of trends in observed terrestrial near-surface wind speeds: implications for evaporation. J Hydrol 416–417:182–205

McVicar TR, Thomas G, Niel Van, Michael LR, Li LT, Mo XG, Zimmermann NE, Dirk RS (2010) Observational evidence from two mountainous regions that near surface wind speeds are declining more rapidly at higher elevations than lower elevations: 1960–2006. Geophys Res Lett 37:L06402. doi:10.1029/2009GL042255

Mitosek HT (1992) Occurrence of climate variability and change within the hydrologic time series: a statistical approach. Report CP-92-05, International Institute for Applied Systems Analysis, Laxemburg, Austria

Parry M, Arnell N, McMichael T, Nicholls R, Martens P, Kovats S, Livermore M, Rosenzweig C, Iglesias A, Fischer G (2001) Millions at risk: defining critical climate change threats and targets. Global Environ Change 11:181–183

WMO (1997) A comprehensive assessment of the freshwater resources of the world. WMO, Geneva

Changes in Sunshine Duration in Humid Environments of Agartala (Tripura), India

D. Jhajharia, P.K. Pandey, Vanita Pandey, P.P. Dabral, R.R. Choudhary, R. Kumar and V.P. Singh

Abstract Study of changes in the global solar radiation is one of the key factors in sustainable agricultural production and management. Therefore, we investigated trends in the solar radiation using sunshine duration as a suitable alternative, based on the recommendation of the FAO Irrigation and Drainage—Paper No. 56, by using the Mann-Kendall (MK) test at different time scales in the humid environments of Northeast India. The average annual bright sunshine hours over Agartala is found to be 6.6 hours (h) with a standard deviation of 0.4 h and coefficient of variation of 6.4%. On annual (seasonal) time scale, statistically significant decreasing trends in bright sunshine duration through the MK test were observed at 5% level of significance at the rate of 0.245 h/decade (0.545 and 0.118 h/decade in winter and monsoon) over Agartala. Similarly, sunshine decreases were observed in the months of January, February, March, May, September, October, and December in the range of 0.237–0.688 h/decade. The observed decreases in sunshine duration

D. Jhajharia
Department of Soil and Water Engineering, College of Agricultural Engineering and Post-Harvest Technology, Ranipool, Gangtok 737135, Sikkim, India
e-mail: jhajharia75@rediffmail.com

P.K. Pandey (✉) · V. Pandey · P.P. Dabral
Department of Agricultural Engineering, North Eastern Regional Institute of Science and Technology, Nirjuli, Itanagar 791109, Arunachal Pradesh, India
e-mail: pandeypk@gmail.com

R.R. Choudhary
Department of Electronic Instrumentation and Control Engineering, Engineering College Bikaner, Karni Industrial Area, Bikaner 334004, Rajasthan, India

R. Kumar
Division of Agricultural Engineering, Sher-e-Kashmir University of Agricultural Sciences and Technology of Kashmir, Shalimar Campus, Srinagar 190025, Jammu and Kashmir, India

V.P. Singh
Department of Biological and Agricultural Engineering, Department of Civil and Environmental Engineering, Texas A&M University, College Station, USA

V.P. Singh et al. (eds.), *Climate Change Impacts*, Water Science and Technology Library 82, https://doi.org/10.1007/978-981-10-5714-4_2

modify the evapotranspiration process and affect the crop water requirements, which, in turn, will negatively influence the agricultural production in humid environments of Northeast India.

Keywords Trend · Mann-Kendall · Sunshine duration · Agartala

Introduction

The anthropogenic-induced climate change has increased greatly in last few decades. Magnus et al. (2011) reported that temperature increased by an estimated 0.73 °C during the 43 years period from 1959 to 2002 globally. The decreasing trends in solar radiation are reported from various parts of the world under different environments. The occurrence of a reduction in global irradiance may lead to the global dimming. Global dimming refers to the decrease in the flux of solar radiation reaching the earth's surface both in the direct solar beam and in the diffuse radiation scattered by the sky and clouds (Stanhill and Cohen 2001). If aerosols and solar radiation would have remained at the 1959 level, then the expected global average temperature would have been 1.09 °C higher (Magnus et al. 2011).

Global dimming created a cooling effect that partially masked the effect of greenhouse gasses on global warming. The sunshine hour is the most important influencing parameter of evapotranspiration process (Pandey et al. 2014, 2016). Also, the global dimming interfered with the hydrological cycle by reducing evaporation, and may have reduced rainfall in some areas. Therefore, solar radiation is one of the most important factors affecting climate. Studying trends in radiation are of much concern in climate change studies and agricultural sustainability assessment. Solar radiation is also responsible for the photosynthesis in plants, which is the main essence in agriculture, and therefore any change in radiation pattern must be analyzed.

The objective of this work is to identify whether there is any sign of solar dimming or brightening over the Northeast India as there is nonavailability of solar radiation measured data not only in this region but around the globe. Usually, the solar radiation is recorded with the help of pyranometers. However, the data of bright sunshine duration are available for a number of sites because of ease in measurements of sunshine duration using the Campbell–Stokes sunshine recorder. Sunshine duration having strong linear relation with global solar radiation could be utilized in place of the solar radiation (Allen et al. 1998). Stanhill (1965) have reported that sunshine duration measurements are the most highly and linearly correlated with the global radiation. In this paper, sunshine duration data were tested for trend identification using the Mann-Kendall (MK) method in monthly, seasonal, and annual time scales in the humid environments of Agartala, Northeast India.

Materials and Methods

Study Area and Meteorological Data

The main ecosystem in the northeastern region of India is a tropical wetland. The Northeast India is the most appropriate region for cultivation of world-class tea, paddy, various forest products, like bamboo, different types of fruit crops, and orchids. The bright sunshine duration data of Agartala needed for this study were obtained from India Meteorological Department (IMD), Pune on a monthly basis from 1969 to 2007. The IMD uses Campbell–Stokes sunshine recorder to measure the bright sunshine hours over different locations in India. The sunshine recorder provides major information on changes in solar irradiance as it automatically records the duration of direct solar beam irradiance above a threshold of 120 W m^{-2} (WMO 1997). The monthly data of 3 years (1993, 1995, and 1997) were missing and were filled with the average values of sunshine of the previous 5 years over Agartala. Figure 1 shows the location of the site in the Tripura state in the northeastern region of India.

Methods of Trend Analysis

Trends in the data can be identified by using parametric or nonparametric methods, and both the methods are widely used. The nonparametric Mann–Kendall (MK) method (Mann 1945; Kendall 1975) is used for identifying the trends in solar

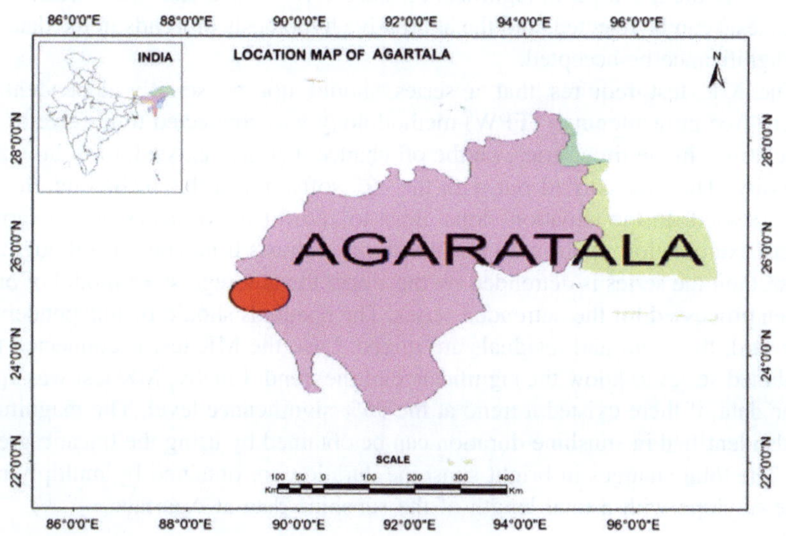

Fig. 1 Location of Agartala site (Tripura), Northeast India

radiation because it is distribution-free and has a higher power than many other commonly used tests (Hess et al. 2001). This method does not require normality of time series and is less sensitive to outliers and missing values. The MK testis based on the test statistic, *S*, defined as follows:

$$S = \sum_{k=1}^{n-1} \sum_{j=k+1}^{n} \operatorname{sgn}(x_j - x_k), \tag{1}$$

where n is the number of observations, and x_j is the jth observation and sgn(·) is the sign function. The mean of the S statistic is zero for the data set assumed to be independent and identically distributed. The variance of the S statistic, under the assumption that the data are independent and identically distributed, is given as:

$$V(S) = \frac{n(n-1)(2n+5) - \sum_{i=1}^{m} t_i(t_i - 1)(2t_i + 5)}{18}, \tag{2}$$

where m is the number of groups of tied ranks, each with t_i tied observations.

The MK statistic, designated by Z, can be computed as

$$Z = \begin{cases} \frac{S-1}{\sqrt{\mathrm{Var}(S)}} & S > 0 \\ 0 & S = 0 \\ \frac{S+1}{\sqrt{\mathrm{Var}(S)}} & S < 0 \end{cases} \tag{3}$$

The values of the Z statistic are calculated and if the value lies within (−) 1.96 and (+) 1.96, i.e., $-Z_{1-\alpha/2} \leq Z \leq Z_{1-\alpha/2}$, then the null hypothesis of no trend can be accepted at the 5% level of significance using a two-tailed test. Otherwise, the null hypothesis can be rejected, and the alternative hypothesis of trends in the data at the 5% significance be accepted.

The MK test requires that a series should not be serially dependent. The pattern-free prewhitening (TFPW) methodology was connected to remove of serial associations in the time series, on the off chance that they existed, to fit in with this necessity. This was carried out with the "R" software. In this technique, the slope was assessed. In the situation slope close to zero, it is not necessary to carry out pattern examination. In the case that the slope contrasts from zero, it is thought to be linear, then the series is detrended by the slope; the autoregressive model of order 1 is then processed for the detrended series. The residuals should be independent, and afterward, the trend and residuals are mixed. Last, the MK test is connected to the combined series to know the significance of the trend. Finally, MK test was applied to the data, if there existed a trend at the 95% significance level. The magnitude of trends identified in sunshine duration can be obtained by using the linear regression test. The total changes in bright sunshine duration are obtained by multiplying the value of slope with a total length of the sunshine data at Agartala.

Results and Discussion

The monthly data obtained from the IMD Pune were used to compute the seasonal and the annual time series of bright sunshine duration over Agartala, Tripura (NE India). The definition of four different seasons in the NE India given by Jhajharia et al. (2009, 2012), are as follows: winter (January–February), pre-monsoon (March–May); monsoon (June–September) and post-monsoon (October–December). Statistical parameters, such as, mean (in hours, h), standard deviation (h), coefficient of variation (%), and maximum (h) and minimum (h) values of sunshine duration are calculated to describe the characteristics of sunshine, and its variability over Agartala (see Table 1). The average annual bright sunshine duration for the considered period over Agartala is found to be 6.6 h with a standard deviation of 0.4 h and coefficient of variation of 6.4%.

The mean sunshine duration in winter and post-monsoon seasons are determined to be about 7.9 h. However, the bright sunshine hours fell drastically to 5.1 in the monsoon season over Agartala due to the presence of dark (rainy) clouds during the months of June to September. The bright sunshine duration in the winter, pre-monsoon and post-monsoon seasons are comparatively higher (mean the sunshine more than 7.3 h) due to the presence of clear skies at Agartala during these three seasons. The C_V of bright sunshine duration in monsoon (10.2%) is found to be the highest, which shows that the bright sunshine intensity has high variability

Table 1 Statistical properties of sunshine duration (h) over Agartala

Time scale	Mean (h)	Max (h)	Max (h)	S (h)	C_v (%)
January	7.69	9.20	6.10	0.83	10.81
February	8.05	9.10	6.40	0.71	8.82
March	7.70	8.80	6.50	0.70	9.09
April	7.49	9.00	5.90	0.68	9.06
May	6.77	9.30	3.10	1.40	20.67
June	4.26	5.90	2.40	1.06	24.81
July	4.02	5.70	2.00	0.71	17.56
August	5.17	6.80	3.00	0.88	16.93
September	5.19	7.80	3.40	1.12	21.50
October	7.04	8.50	5.10	1.03	14.64
November	7.83	9.00	6.90	0.57	7.33
December	8.00	9.10	6.10	0.74	9.25
Yearly	6.60	7.26	5.64	0.40	6.40
Winter	7.87	8.95	6.55	0.69	8.79
Pre-monsoon	7.32	8.77	6.03	0.63	8.60
Monsoon	5.14	6.34	3.98	0.52	10.19
Post-monsoon	7.91	8.65	6.80	0.47	5.95

Note Max, Min, S and C_v denote maximum, minimum, standard deviation, and coefficient of variation, respectively during the months of June to September over Agartala as compared to the C_V (5.95%) of sunshine in post-monsoon season. Similar characteristics of sunshine duration are observed on the monthly time scale at Agartala. The mean bright sunshine duration is found to be the highest (8.05 h) and second largest (8.0 h) during the months of February and December, respectively over Agartala. However, the months of July and June witnessed the lowest (4.02 h) and the second lowest (4.26 h) values of sunshine duration at Agartala, respectively because of intense rainfall activities and cloudy weather in these 2 months. The highest value of the C_V (24.81%) of sunshine duration is witnessed in the month of June that shows comparatively higher variability in bright sunshine in this month at Agartala.

Table 2 shows the trend results obtained through the MK test in different durations: all 12 months; the four seasons; and the yearly time scale in sunshine duration. The slope of the trend lines are computed using the linear regression test, and the magnitude of the trends (in h/decade) are given in Table 2. Dinpashoh et al. (2011) report that the value of the Z statistic lying within the confidence limits presents a value due to random fluctuations, meaning not much in the inferring existence of a trend from the statistical standpoint. On an annual time scale, the statistically significant trend at the 5% level of significance was witnessed as the absolute of Z statistics (3.81) was more than the tabulated value of 1.96. The decreasing trend in sunshine duration at the rate of 0.245 h/decade was witnessed over Agartala in annual time scale. Figure 2 shows the percent deviation of annual bright sunshine hours from the long-term average annual sunshine over Agartala

Table 2 Changes in sunshine duration at Agartala (Tripura)

Time scale	Test statistics (Z) value	Magnitude of trends (h/decade)
January	−3.720	−0.688
February	−3.504	−0.403
March	−2.289	−0.237
April	*−0.933*	*−0.083*
May	−2.421	−0.36
June	*−0.218*	*−0.003*
July	*0.545*	*0.041*
August	*0.182*	*0.069*
September	−1.683	−0.263
October	−3.196	−0.435
November	−1.214	**−0.206**
December	−2.640	−0.373
Yearly	−3.812	−0.245
Winter	−4.091	−0.545
Pre-monsoon	−3.123	−0.227
Monsoon	−1.986	−0.118
Post-monsoon	−2.809	−0.289

The bold number denote the statistically significant trends at 5% level of significance obtained through the t-test

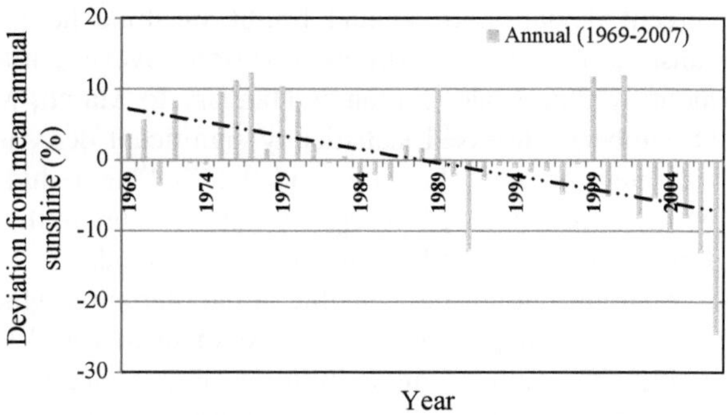

Fig. 2 Deviation of sunshine from average annual sunshine duration (1969–2007)

from 1969–2007. About 65% cases witnessed positive deviation from long-term average values of sunshine during 20 years from 1969 indicating to the phenomenon of brightening happening over the city of Agartala. However, 16 years witnessed negative deviation from long-term average values of sunshine since 1990 at Agartala, which was quite opposite from the recordings of possible sunshine hours before 1990 (see Fig. 2). The decreasing trends witnessed in the sunshine in humid environments of Agartala confirm global dimming since 1984 over the northeastern region of India.

On a seasonal time scale, Agartala witnessed statistically significant decreasing trends in sunshine duration in all the four seasons at 5% level of significance in the range of 0.545 (winter) and 0.118 (monsoon) h/decade. Figures 3a–d show the

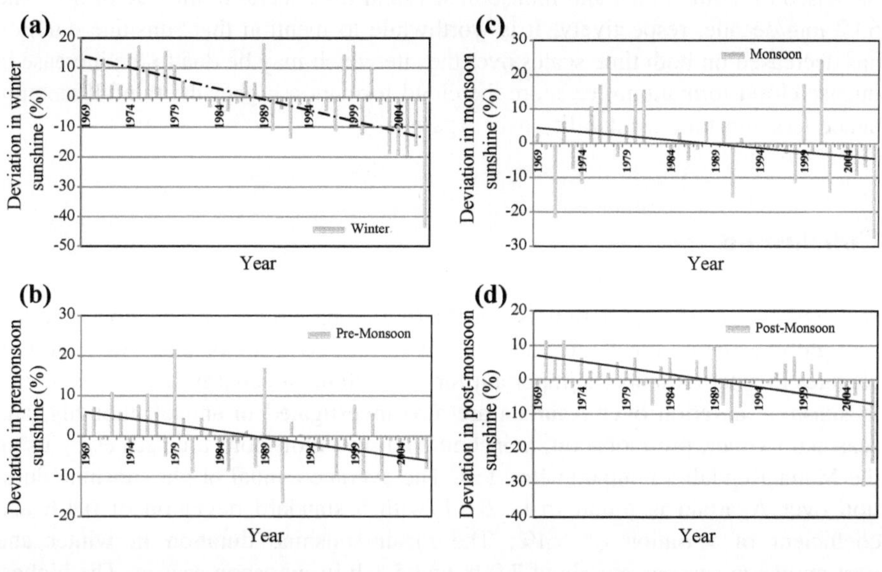

Fig. 3 **a** Deviation of sunshine from average sunshine duration in winter season. **b** Deviation of sunshine from average sunshine duration in pre-monsoon season. **c** Deviation of sunshine from average sunshine duration in monsoon season. **d** Deviation of sunshine from average sunshine duration in post-monsoon season

percent deviation of annual bright sunshine hours from the long-term average sunshine data in four different seasons over Agartala from 1969 to 2007. On a monthly time scale, 7 months (January to March, May, September, October, and December) witnessed statistically significant decreasing trends in the sunshine in the range of 0.69 (January) and 0.237 (March) h/decade. On the other hand, no trends were witnessed in bright sunshine hours in the remaining five months (April, June to August, and November) over Agartala.

Various studies are available in the literature, which reports changes in different meteorological parameters over Agartala in last 30 years or more. These studies confirm the anthropogenic-induced climate change is occurring over humid environments of Agartala. Decreasing trends in annual rainfall were reported at the rate of 2.4 mm/year at Agartala, Tripura (Jhajharia et al. 2007, 2009). The trends in annual rainfall at Agartala were in agreement with the decreasing trends observed in the total cloud amount over Agartala. Decreasing trends in pan evaporation and reference evapotranspiration over different sites of the NE India are also reported in last one decade (Jhajharia et al. 2009, 2012). Significant reduction in pan evaporation and reference evapotranspiration occurred mainly due to stilling (wind speed decreases) and dimming over different parts of Northeast India. Stilling is the process of significant reduction in wind speed at a place in last few years. McVicar et al. (2012) and Jhajharia et al. (2009) reported decreasing trends in wind speed on annual and seasonal time scales over different sites of NE India.

Figure 4a, b depict the relationship between rainfall and sunshine duration in annual time scale and monsoon season. Findings reveal that the total rainfall has increased on both annual and monsoon seasonal time scales at the rate of 30.3 and 63.2 mm/decade, respectively. It is worthwhile to mention that sunshine duration has decreased on both time scales over the site which may be due to the increase in intense cloud formation, i.e., more the cloud formation higher the rainfall amount, hence less sunshine availability over Agartala.

Conclusions

Availability of solar radiation is necessary for agricultural production, because sunlight is essential for photosynthesis and various solar appliances. Any study on changes in solar radiation is important for agriculture and energy sector. The trends in sunshine duration over Agartala are also investigated in annual, monthly, and seasonal (winter, pre-monsoon, monsoon, and post-monsoon) time scales by using the Mann-Kendall's nonparametric test. The average annual bright sunshine duration over Agartala is found to be 6.6 h with a standard deviation of 0.4 h and coefficient of variation of 6.4%. The mean sunshine duration in winter and post-monsoon seasons are about 7.9 h, and 5.1 h in monsoon season. The highest (lowest) mean bright sunshine duration is found to be 8.05 h (4.02 h) in the months of February (July) over Agartala.

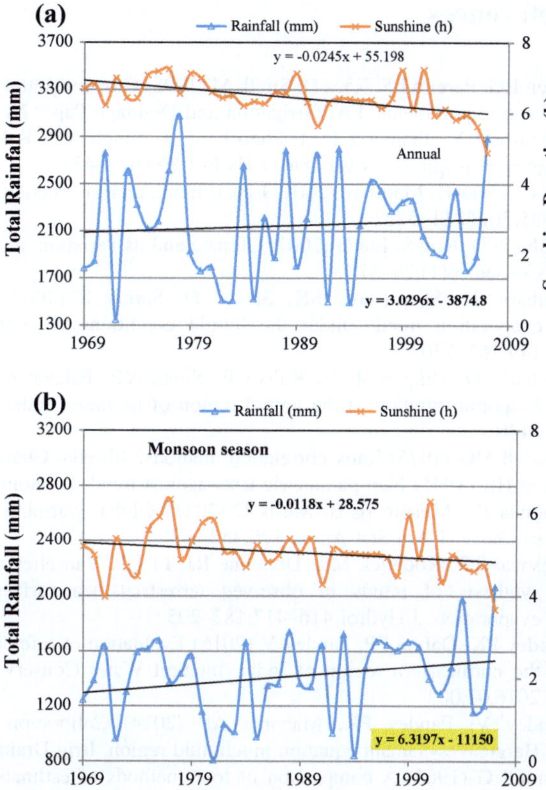

Fig. 4 **a** Annual time series of total rainfall and sunshine duration at Agartala. Note: *Straight line* denotes trend in respective meteorological parameter. **b** Time series of total rainfall and sunshine duration at Agartala in monsoon

Statistically, significant decreasing trends in bright sunshine duration were observed by using the MK test at 5% level of significance at the rate of 0.245 h/decade in annual duration (in the range of 0.545 and 0.118 h/decade in seasonal time scale) over Agartala. On a monthly time scale, Agartala observed sunshine decreases during 7 months in the range of 0.237–0.688 h/decade. Decreasing trends witnessed in sunshine over Agartala support the happening of global dimming over Northeast India in last two decades. Solar brightening phenomenon was observed from 1969 to 1982 over Agartala due to increased bright sunshine duration. However, a solar-dimming phenomenon occurred due to decreased sunshine duration for the remaining period of analysis. Crop water balance and evapotranspiration are closely coupled to solar radiation, and therefore, solar radiation decreases are likely to reduce water use and evapotranspiration.

Acknowledgements The authors thank the India Meteorological Department (Pune) for providing the climatic data used in this study.

References

Allen RG, Pereira LS, Raes D, Smith M (1998) Crop evapotranspiration: guidelines for computing crop requirements. FAO Irrigation and Drainage Paper No. 56, FAO Rome, Italy

Dinpashoh Y, Jhajharia D, Fakheri-Fard A, Singh VP, Kahya E (2011) Trends in reference evapotranspiration over Iran. J Hydrol 399:422–433

Hess A, Iyer H, Malm W (2001) Linear trend analysis: a comparison of methods. Atmos Environ 35(30):5211–5222

Jhajharia D, Roy S, Ete G (2007) Climate and its variation: a case study of Agartala. J Soil Water Conserv 6(1):29–37

Jhajharia D, Shrivastava SK, Sarkar D, Sarkar S (2009) Temporal characteristics of pan evaporation trends under the humid conditions of northeast India. Agric For Meteorol 149:763–770

Jhajharia D, Dinpashoh Y, Kahya E, Singh VP, Fakheri-Fard A (2012) Trends in reference evapotranspiration in the humid region of northeast India. Hydrol Process. doi:10.1002/hyp.8140

Kendall MG (1975) Rank correlation methods, 4th edn. Charles Griffin, London

Mann HB (1945) Non-parametric tests against trend. Econometrica 33:245–259

Magnus JR, Melenberg B, Muris C (2011) Global warming and local dimming: the statistical evidence. J Am Stat Assoc 106:452–464

McVicar TR, Roderick ML, Donohue RJ, Li LT, Van Niel TG et al (2012) Global review and synthesis of trends in observed terrestrial near-surface wind speeds: implications for evaporation. J Hydrol 416–417:182–205

Pandey PK, Dabral PP, Pandey V (2016) Evaluation of reference evapotranspiration methods for the northeastern region of India. Int Soil Water Conserv Res 4:52–63. doi:10.1016/j.iswcr.2016.02.003

Pandey V, Pandey PK, Mahanta AP (2014) Calibration and performance verification of Hargreaves-Samani equation in a humid region. Irrig Drain 63:659–667. doi:10.1002/ird.1874

Stanhill G (1965) A comparison of four methods of estimating solar radiation. In: Eckardt FE (ed) Methodology of plant eco-physiology. In: Proceedings of the Montpellier symposium, Arid Zone Research, XXV. UNESCO, Paris, pp 55–61

Stanhill G, Cohen S (2001) Global dimming: a review of the evidence for a widespread and significant reduction in global radiation with discussion of its probable causes and possible agricultural consequences. Agric For Meteorol 107:255–278

WMO (1997) Measurement of radiation guide to meteorological instruments and methods of observation (Chap. 7), 6th edn. World Meteorological Organization, Geneva

Application of Multiple Linear Regression as Downscaling Methodology for Lower Godavari Basin

Gayam Akshara, K. Srinivasa Raju, Ajit Pratap Singh and A. Vasan

Abstract This paper focused on future precipitation scenarios adopting statistical downscaling approach, namely, Multiple linear regression (MLR) for Lower Godavari basin, India. Global Climate Model (GCM), namely, GFDL-CM3 simulations, are used for downscaling purpose. Five grid points of Lower Godavari basin are considered. Reanalysis data from National Centre for Environmental Prediction (NCEP) of the study area from 1969 to 2005 is used for analysis. Precipitation is chosen as predictand. Representative Concentration Pathways (RCPs) scenarios, 4.5 and 6.0 are used for the study. Projected precipitation from 2006 to 2100 is obtained by the developed MLR model. Downscaled precipitation predictions show that there is increase in precipitation in the future.

Introduction

Climate change is becoming a challenge in the global environment. Continued emissions of greenhouse gases would further increase the existing risks and create new ones for society and ecosystems (IPCC 2014). In this regard, it is required to predict the future climate changes to analyze its possible impacts in a river basin. Global Climate Models (GCM's) are one of the approaches available for assessing change in climate. In addition, statistical downscaling approaches are becoming

G. Akshara (✉) · K. Srinivasa Raju · A. Vasan
Department of Civil Engineering, BITS Pilani, Hyderabad Campus, Hyderabad 500078, India
e-mail: aksharareddygayam@gmail.com

K. Srinivasa Raju
e-mail: ksraju@hyderabad.bits-pilani.ac.in

A. Vasan
e-mail: vasan@hyderabad.bits-pilani.ac.in

A.P. Singh
Department of Civil Engineering, BITS Pilani, Dubai Campus, Dubai, UAE
e-mail: aps@pilani.bits-pilani.ac.in

V.P. Singh et al. (eds.), *Climate Change Impacts*, Water Science and Technology Library 82, https://doi.org/10.1007/978-981-10-5714-4_3

important due to their ability to derive quantitative relationship between local surface variables, i.e., predictands and large-scale atmospheric variables, i.e., predictors (Wilby and Wigley 1997). This study has been taken to synthesize the future precipitation scenarios adopting statistical downscaling approach, namely, Multiple Linear Regression (MLR) for Lower Godavari basin, India using the simulations of GCM, GFDL-CM3. Present paper covers the literature review, study area, methodology, results and discussion followed by summary and conclusions.

Literature Review

Numerous studies have been carried out for assessing the climate change impact adopting various downscaling approaches. Schoof and Pryor (2001) conducted a study to downscale temperature and precipitation using regression and Artificial Neural Networks. They compared the performance of downscaling models and discussed the possible improvements. It was concluded that performance of ANN's was clearly superior to MLR. Tatli et al. (2004) carried out a study to downscale monthly total precipitation over Turkey. They observed that a large-scale analysis is required due to the existence of complex relationships between large-scale processes and local atmospheric conditions. Fowler et al. (2007) presented state-of-the-art review on downscaling techniques for hydrologic modeling applications. They suggested applied research that can handle future uncertainties.

Hashmi et al. (2009) presented a study to deal uncertainty aspects associated with statistical downscaling of monthly precipitation in a watershed by applying a Bayesian framework to develop Weighted Multi-Model Ensemble (WMME). They strongly supported multi-model ensemble downscaling for hydrological impact assessment. Goyal and Ojha (2010) analyzed various linear regression models as a part of downscaling for the case study of Pichola lake region in Rajasthan. The results of the study indicated that direct regression outperformed all other methods.

Similar studies were made by Meenu et al. (2012) for Tungabhadra river basin, India; Latt and Wittenberg (2014) for multi-step forecasting of Chindwin River floods, northern Myanmar; Elhakeem et al. (2015) for United Arab Emirates (UAE).

It is observed from literature review that statistical downscaling approaches are useful and can be used as the basis for adaptation studies. Keeping this in view, it is felt necessary to study the effect of climate change on precipitation for the case study of Lower Godavari Basin, India. Summing up, the following objectives have been carried out:

(a) To study the applicability of MLR as statistical downscaling approach.
(b) To generate future precipitation at the basin level.

Study Area

The study area, Lower Godavari Basin starts from the river's confluence with the Manjira to the mouth which lies between North latitudes 16° 19′–19° 03′ and East longitudes 80° 01′–82° 94′. Godavari basin has a tropical climate. Comparatively the weather is hotter in the western most parts of the basin as compared to the northern, central and eastern regions. Southwest monsoon is the major source of rainfall. Variation of annual rainfall of the basin is from 755 to 1531 mm whereas average annual rainfall is 1096.92 mm over a period of (1971–2005). Variation of annual maximum temperature is from 31 to 33.5 °C over a period of (1969–2004) (Godavari River Basin 2014). Accordingly predictor variables are chosen in the study to arrive at the future climatic scenario in the basin.

Methodology

Selection of Predictors and Predictands

Lower Godavari Basin is divided into 5 grid points for the present study based on the grid resolution available for the India Meteorological Department (IMD) data, i.e., 17.5N–80.5E, 17.5N–81.5E, 18.5N–81.5E, 18.5N–80.5E, 18.5N–82.5E. All the data were processed using following steps (Fig. 1).

Precipitation (Prec) is chosen as predictand in downscaling methodology. In this regard, historical monthly data of Prec for the five grids of Lower Godavari Basin was obtained from the IMD, for the period of 1969–2005 (baseline period). Predictors were established for predictand and presented in Table 1. All the chosen predictors are interdependent which collectively affect the change in precipitation. Hence it is expected to obtain a relationship between the predictands and predictors which would essentially help to evaluate performance of downscaling process (Anandhi et al. 2009; Mujumdar and Nagesh Kumar 2012).

NCEP reanalysis data of the study area have been taken from 1969 to 2005 for the listed predictor variables (Table 1) and used for calibration along with IMD data. Once the model is developed, it can be used for generating future scenarios from GCM data. This data is obtained at a grid resolution of 2.5° × 2.5° (latitude × longitude) on a daily scale and averaged to obtain monthly scale data (http://pcmdi9.llnl.gov/esgf-web-fe).

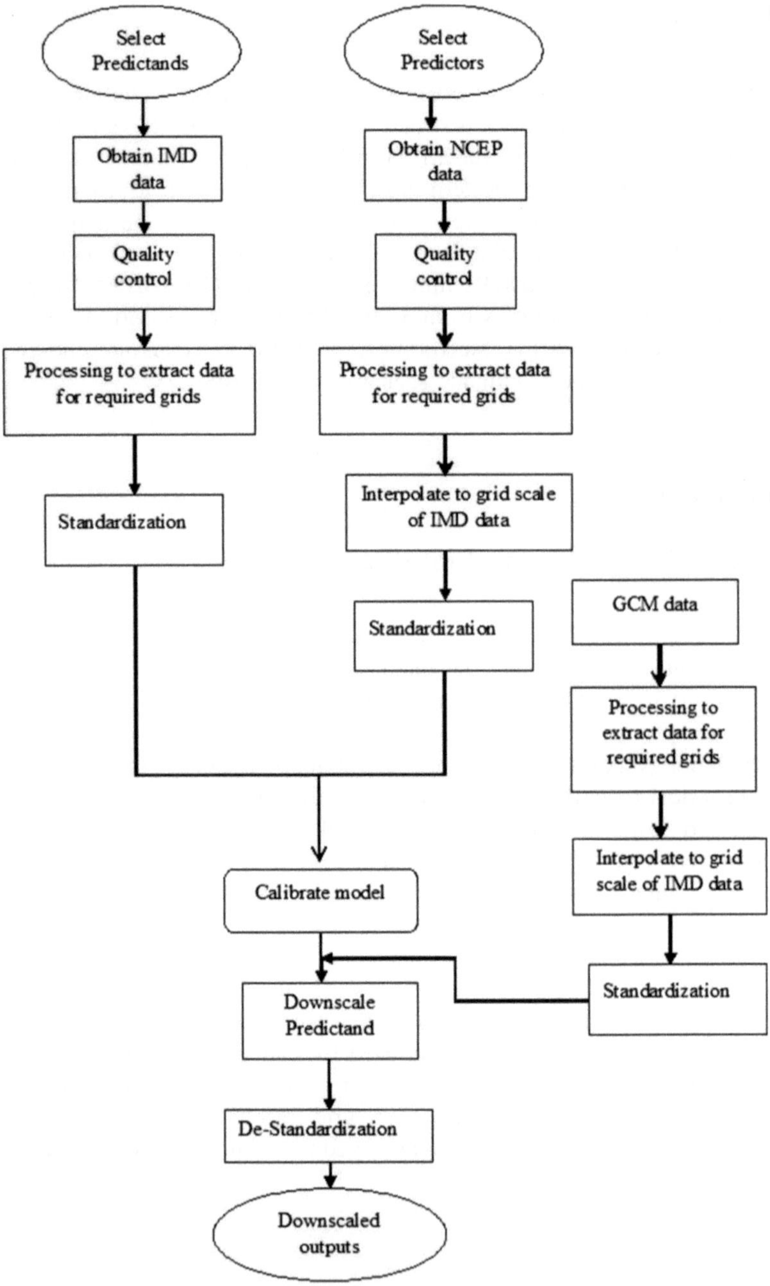

Fig. 1 Schematic view of the present methodology

Table 1 Predictor variables for downscaling precipitation

Predictor variables	Notation
Eastward wind @ 500 hpa	p5_u
Eastward wind @ 850 hpa	p8_u
Northward wind @ 500 hpa	p5_v
Northward wind @ 850 hpa	p8_v
Geopotential height @ 500 hpa	p500
Geopotential height @ 850 hpa	p850
Air pressure @ mean sea level	mslp
Near surface relative humidity	hurr
Near surface specific humidity	hurs

Selection of GCM and Representative Concentration Pathways (RCPs)

Choosing a suitable GCM is a crucial step as it decides the outcome of the study. Chaturvedi et al. (2012) suggested GFDL-CM3 for both temperature and precipitation. Hence, GCM outputs of GFDL-CM3 have been considered from 2006 to 2100 for projecting future scenarios of precipitation (Akshara 2015).

GFDL-CM3 was developed by Geophysical Fluid Dynamics Laboratory of USA. The model has a resolution of 2° × 2.5° (latitude × longitude) covering 90 × 144 global grids. Information about input data is presented in Table 1. The GCM data for RCP 4.5 and RCP 6.0 are used in this study (IPCC 2014). Interpolation technique is used to bring the data at common resolution for comparative purposes.

Data Preparation

The first step in climate modeling is to acquire observed meteorological data relevant to the study area. The future time series has been developed using MLR technique with predictor variables of NCEP and GFDL-CM3 (RCP4.5 and RCP6.0) data sets which are described below:

NCEP 1969–2005: Set contains 37 years of normalized daily observed predictor data, extracted from NCEP reanalysis.

GFDL-CM3 (RCP4.5 and RCP 6.0): 1969–2100: This set contains 132 years of monthly predictor data, derived from the GFDL-CM3, RCP 4.5 and RCP 6.0 experiments, normalized over the 1969–1998 (30 years) period.

Statistical Downscaling

MLR analysis is used as the statistical downscaling approach to find the degree of interrelationship among number of variables. The initial screening of predictor variables has also been performed (Ghosh and Mujumdar 2008; Wilby and Dawson 2013).

Results and Discussion

Regression analysis is performed with respect to each grid and R^2 is evaluated. p5_u, p8_u, p8_v, p500, p850, mslp, hurr, hurs are the chosen predictors after screening for grids 1, 2, 3, and 4 whereas p8_v, p500, p850, mslp, hurs for grid 5. Both the dependent variable and the independent variables are organized grid-wise and are given as input to the linear regression. Five regression equations are formulated with respect to each of the five grids considered in the study area. The calculated R^2 has been found in the range of 0.596–0.696 and grid 4 has the highest R^2 value compared to all other grids. Overall, the regression models are found to be satisfactory. These inferences were based on the outputs related to five grids and innumerable trial and errors conducted on varying combinations of predictors.

Application of Downscaled Scenarios for Future Periods

The standardized GCM predictors for the period 2006–2100 are now introduced in the models developed by calibrating the IMD observed data with NCEP predictor variables for the baseline period (1969–2005) to project the future precipitation for both the scenarios, i.e., RCP 4.5 and RCP 6.0. Three time periods 2020s (2020–2029), 2050s (2050–2059) and 2080s (2081–2089) were considered in order to observe the changes occurring in these periods.

Analysis for Grid 1: 17.5N–80.5E

It is observed from Fig. 2 (for RCP 4.5) that there is minimum precipitation in January and December and maximum in July and August; the trend for future precipitation exhibits intermediate peaks in April; the change in precipitation is significant from baseline period to 2020s as well as 2020s–2050s. It is observed from Fig. 3 (for RCP 6.0) that precipitation is increasing on the whole. Heavy precipitation can be expected in July and August; Precipitation before and after monsoon followed a decreasing pattern with monsoon period as a center in the baseline period. It is not the same with future periods; the maximum precipitation observed in 2020s is 9 mm/month, 2050s is 14 mm/month and in 2080s 19 mm/month.

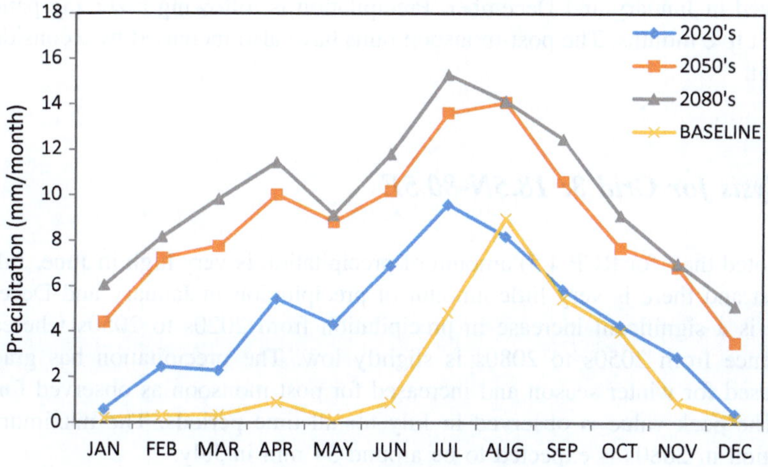

Fig. 2 Prec trend for RCP 4.5 for different time periods for grid 1

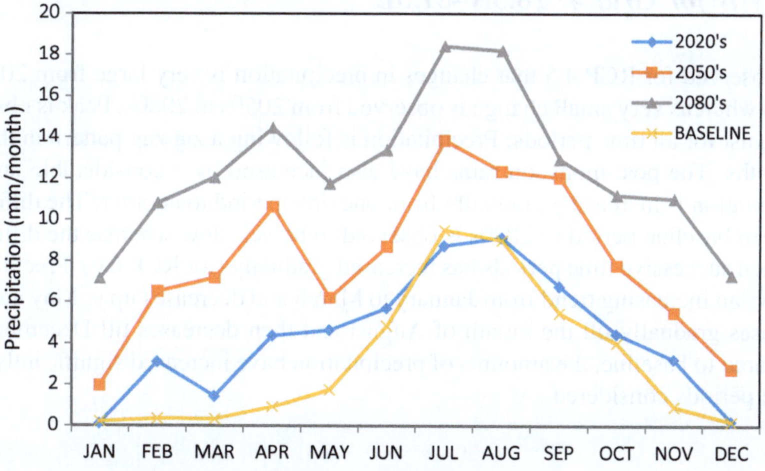

Fig. 3 Prec trend for RCP 6.0 for different time periods for grid 1

Analysis for Grid 2: 17.5N–81.5E

It can be inferred (for RCP 4.5) that precipitation decreases by a small amount for 2020s in the post-monsoon and increases significantly for all months in 2050s and 2080s relative to baseline period; It is noted that precipitation in 2020s in few months coincides with that of baseline period; Significant changes in precipitation are observed from 2020s to 2050s rather than 2050s to 2080s. It is observed (for RCP 6.0) that maximum precipitation occurs in July and August and minimum is observed in January and December. Precipitation is following a zig zag pattern in the first five months. The post-monsoon rains have also increased by a considerable amount.

Analysis for Grid 3: 18.5N-80.5E

It is noted that (for RCP 4.5) amount of precipitation is very high in June, July and August and there is very little amount of precipitation in January and December. There is a significant increase in precipitation from 2020s to 2050s whereas the difference from 2050s to 2080s is slightly low. The precipitation has gradually decreased for winter season and increased for post monsoon as observed for RCP 6.0. The peak value is observed in July for all time periods. The maximum precipitation in 2080s is expected to be around 34 mm in July.

Analysis for Grid 4: 18.5N–81.5E

It is observed for RCP 4.5 that changes in precipitation is very large from 2020s to 2050s whereas very small change is observed from 2050s to 2080s. Peak is observed in August for all time periods; Precipitation is following a zig zag pattern in the first 5 months. The post-monsoon rains have also increased by a considerable amount. Precipitation is increasing gradually from one time period to another. The difference between baseline period to 2020s is observed to be very low whereas the difference between successive time periods has increased gradually (for RCP 6.0). Precipitation follows an increasing trend from January to March and decreases up to May and then increases gradually till the month of August and then decreases till December, but compared to baseline, the amounts of precipitation have increased significantly in all future periods considered.

Analysis for Grid 5: 18.5N–82.5E

There is a gradual decrease in increase of precipitation in pre-monsoon and post-monsoon period and significant increase in monsoon period for RCP 4.5. Precipitation in winter has reduced by a great amount; no significant decrease is observed in precipitation change from 2050s to 2080s. It can be inferred from RCP 6.0 scenario that precipitation increased gradually till April and a decrease in trend is formed in May; whereas June, July and August have the highest amounts of precipitation compared to the remaining months; January and December are reported to be the driest months.

Summary and Conclusions

Linear Regression analysis is applied to Lower Godavari Basin. The predictor variables of NCEP from 1969 to 2005 are used to develop the relationship with observed monthly precipitation obtained from IMD. The relationship obtained by regression analysis is used to predict the monthly precipitation using the surface predictors obtained from GCM outputs. The GFDL-CM3 has been used in present study.

Downscaled precipitation predictions show that there is an increase in the amount of precipitation in the future. Precipitation for all the grids on an average has increased from 8 to 10 mm/month in the baseline period to 20–25 mm/month in 2080s. Considering the RCP's, no noticeable difference between RCP 4.5 and RCP 6.0 is observed except for slight variations in few months. Amount of precipitation will be very high from June to August and very low in January and December.

The present study is based on the chosen GCM, RCPs, and downscaling approach. The results may vary depending on the chosen combinations. However, focus of the present work is to suggest a methodology that can be replicated in due course of time depending on the availability of data and resources.

Acknowledgements The second author is grateful to Council of Scientific and Industrial Research, New Delhi for supporting the present work, through project no. 23(0023)/12/EMR-II dated 15.10.2012 and to Prof D. Nagesh Kumar, IISc, Bangalore for providing valuable inputs while preparing this paper. Authors acknowledge the modeling groups for making their simulations available for analysis, the Program for Climate Model Diagnosis and Intercomparison (PCMDI) for collecting and archiving the CMIP5 model output, and the WCRP's Working Group on Coupled Modeling (WGCM) for organizing the model data analysis activity. First author is thankful to Mr. I Nagababu, Formerly Senior Research Fellow of CSIR project and Ms. Radhika, NRSC Hyderabad for providing data and inputs.

References

Akshara G (2015) Downscaling of climate variables using multiple linear regression—a case study, Lower Godavari Basin, India. M.E Thesis, BITS Pilani Hyderabad Campus, Hyderabad, India

Anandhi A, Srinivas VV, Kumar DN, Nanjundiah RS (2009) Role of predictors in downscaling surface temperature to river basin in India for IPCC SRES scenarios using support vector machine. Int J Clim 29:583–603

Chaturvedi RJ, Joshi J, Jayaraman M, Bala G, Ravindranath NH (2012) Multi-model climate change projections for India under representative concentration pathways. Curr Sci 103:7–10

Elhakeem A, Elshorbagy WE, AlNaser H, Dominguez F (2015) Downscaling global circulation model projections of climate change for the United Arab Emirates. J Water Resour Plan Manage 10:1–15

Fowler HJ, Blenkinsop S, Tebaldi C (2007) Linking climate change modelling to impact studies: recent advances in downscaling techniques for hydrologic modelling. Int J Clim 27:1547–1578

Ghosh S, Mujumdar PP (2008) Statistical downscaling of GCM simulations to streamflow using relevance vector machine. Adv Water Resour 31:132–146

Godavari River basin (2014) India WRIS—water resources information system of India. http://india-wris.nrsc.gov.in. Accessed on 10 May 2015

Goyal MK, Ojha CSP (2010) Evaluation of various linear regression methods for downscaling of mean monthly precipitation in arid pichola watershed. Nat Resour 1:11–18

Hashmi MZ, Shamseldin AY, Melville BV (2009) Statistical downscaling of precipitation: state-of-the-art and application of Bayesian multi-model approach for uncertainty assessment. J Hydrol Earth Syst Sci 6:6535–6579

IPCC (2014) Synthesis report. Contribution of working groups I, II and III to the fifth assessment report of the Intergovernmental Panel on Climate Change

Latt ZZ, Wittenberg H (2014) Improving flood forecasting in a developing country: a comparative study of stepwise multiple linear regression and artificial neural network. Water Resour Mange 28:2109–2128

Meenu R, Rehana S, Mujumdar PP (2012) Assessment of hydrologic impacts of climate change in Tunga-Bhadra river basin, India with HEC-HMS and SDSM. Hydrol Proc 27:1572–1589

Mujumdar PP, Nagesh Kumar D (2012) Floods in a changing climate-hydrologic modelling. Cambridge University Press, UK

Schoof JT, Pryor SC (2001) Downscaling temperature and precipitation: a comparison of regression-based methods and artificial neural networks. Int J Clim 21:773–790

Tatli HA, Zhet DB, Sibel M (2004) A statistical downscaling method for monthly total precipitation over Turkey. Int J Clim 24:161–180

Wilby RL, Wigley TML (1997) Downscaling general circulation model output: a review of methods and limitations. Prog Phys Geogr 21:530–548

Wilby RL, Dawson CW (2013) The statistical downscaling model: insights from one decade of application. Int J Clim 33:1707–1719

Statistical Downscaling of Minimum Temperature of Raipur (C.G.) India

R.K. Jaiswal, H.L. Tiwari, A.K. Lohani and R.N. Yadava

Abstract The future projected climate parameters obtained from using generalized circulation models (GCMs) cannot be used directly on regional or basin scale because of coarse resolution. The dynamic or statistical downscaling procedures are used to convert global scale output to regional scale condition. The statistical downscaling because of its less computational skills is preferably used for generation of future climate and in the present study, minimum temperature of Raipur was forecasted for three future periods using Canadian Global Climate Model (CGCM) predictors for A1B and A2 climate forcing conditions. The statistical downscaling model (SDSM) has been used using *k-fold* validation technique for generation of multitemporal series for periods FP-1 (2020–2035), FP-2 (2046–2064), and FP-3 (2081–2100). The specific humidity at 850 hpa (ncep-s850gl), 500 hpa geopotential height (ncepp500gl), and surface airflow strength (ncep_fgl) were found to be the most appropriate parameters to generate future scenarios. The comparison of mean monthly minimum temperature of generated scenarios with base period confirmed 1.1–11.2% increase of minimum temperature under A1B climate forcing and 2.88–24.44% in summer months will have adverse effect on various demands and human health in future and adaptation measures need to be devised for the region.

Keywords Climate change · Generalized circulation model (GCM)
Regional circulation model (RCM) · Downscaling · Predictor · Predictand

R.K. Jaiswal (✉)
National Institute of Hydrology, RC Bhopal, Madhya Pradesh, India
e-mail: rkjaiswal_sagar@yahoo.co.in

H.L. Tiwari
Department of Civil Engineering, MANIT, Bhopal, Madhya Pradesh, India

A.K. Lohani
National Institute of Hydrology, Jal Vigyan Bhavan, Roorkee, Uttarakhand, India

R.N. Yadava
AISECT University, Bhopal, Madhya Pradesh, India

V.P. Singh et al. (eds.), *Climate Change Impacts*, Water Science and Technology Library 82, https://doi.org/10.1007/978-981-10-5714-4_4

Introduction

The different reports of Intergovernmental Panel on Climate Changes (IPCC 2003, 2007) and other independent researches have confirmed that climate is changing on global and regional scale which is likely to affect availability and supplies of water (Milly et al. 2005; Gleick 1987), health, agriculture, and livestock (McCarthy et al. 2001; Ravindran et al. 2000; FAO 2001; Menzel et al. 2006; Sivakumar et al. 2012) and many more areas of human life. It can be emphasized here that changing climate has intensified probability of extreme events such as floods (Milly et al. 2002, 2005), droughts (Huntington 2006), etc. The temperature among other climatological parameters is the most important and easily detectable parameter to show impact of climate change on water availability and demands, agriculture production, human health, and many more areas of life. The prediction of future climate, its implication, and adaptation measures are keys to cope up the future challenges.

This problem of coarse grid data can be solved by downscaling GCMs to local and basin scale with the help of dynamic or statistical downscaling techniques that bridge the large-scale atmospheric conditions with local scale climatic data (Wilby and Wigley 1997; Xu 1999; Fowler et al. 2007; Tisseuil et al. 2010; etc.). The dynamic downscaling techniques use physically based model run in time slice mode and limited area (Giorgi and Mearns, 1999) having major drawback of dynamic downscaling is its complexity and high computation cost (Anandhi et al. 2008) and propagation of systematic bias from GCM to RCM (Salathe 2003). However, statistical downscaling techniques are reasonably accurate in developing relationships between GCM predictors and regional/station climatic data (Fowler et al. 2007), simple, flexible in adjustment and movement to different regions, less costly, and computationally undemanding in comparison to dynamic downscaling proved its reliability and compatibility in future projections (Hewitson and Crane 2006; Tripathi and Nanjundiah 2006; Lopes 2008; Ethan et al. 2011). In the present study, statistical downscaling model (SDSM 5.2) has been used to predict variability in minimum temperature using Canadian Global Circulation Model (CGCM) weather predictor data for A1B and A2 SRES scenarios.

Statistical Down Scaling Model (SDSM)

The SDSM is user-friendly software developed under sponsorship of A Consortium for Application of Climate Impact Assessments (ACACIA), Canadian Climate Impacts Scenarios (CCIS) Project and Environment Agency of England and Wales. The SDSM can develop multiple, low-cost scenarios of daily surface weather variables using seven key functions namely quality control and data transformation, selection of downscaling predictor variables, model calibration, weather generator, data analysis, graphical analysis, and scenarios generation for the task of daily weather downscaling and forecasting. The quality control function is used to identify the gross error, gaps, statistics, and outliers in the data which is an important step prior to calibration. The spatial and temporal variability in explanatory power of predictors makes selection of appropriate predictors difficult and screen variable operation in SDSM assists examination of seasonal variation, intercorrelation in predictors, and their correlations with predictand. The scatter diagram, correlation, partial correlation, explained variance, etc., can be used to select suitable predictors to develop statistical relationships.

After selecting the most appropriate predictors, the calibrate model tab is used to develop multiple linear regression techniques with efficient dual simplex algorithm (forced entry method) to develop a relationship between predictand and user specified set of predictors in conditional (in case of precipitation) or unconditional (in case of temperature, wind speed, etc.) process. The synthetic series for future periods can be generated using weather generator tab of SDSM software using developed relationships from calibration and CGCMs obtained predictors set from future periods. The SDSM links automatically all required predictors in a regression model developed in calibration process for a user specified period. The data analysis operation in SDSM model is carried out using summary statistics and frequency analysis operation. The frequency analysis tab is useful to compare observed and synthesized series with the help of quantile plot, PDF plot, line plot, and frequency analysis. The time series analysis tool is used to analyze observed and modeled series graphically. The scenario generation can be used to generate ensembles of synthetic daily weather series giving a treatment of percent changes in mean, occurrence or variance or linear, exponential or logistic trend in any series. The detail about the application of SDSM can be found in Goodess et al. (2003), Wilby and Dettinger (2000), Wilby et al. (2001, 2003). The graphical representation of various steps used in SDSM based downscaling can be seen in Fig. 1.

Study Area and Data Used

The study area for the present study is Raipur city, the capital of Chhattisgarh state of India. The Raipur is an important city of eastern India and has large-scale commercial and industrial development since its inception of capital of Chhattisgarh state in 2000. The river Seonath, an important tributary of river Mahanadi, passes through the city and is used to supply water for industrial and domestic demands of district. The map of the study area is presented in Fig 2. The long-term series of observed daily minimum temperature from 1971 to 2003 of Raipur city, the capital city lying in upper Mahanadi basin, has been used for calibration and validation of statistical model. The NCEP reanalyzed predictors from 1971 to 2003 and SRES A1B and B2 data of Canadian Global Circulation Model CGCM 3.1/T47 from 2001 to 2100 were used to generate future scenarios.

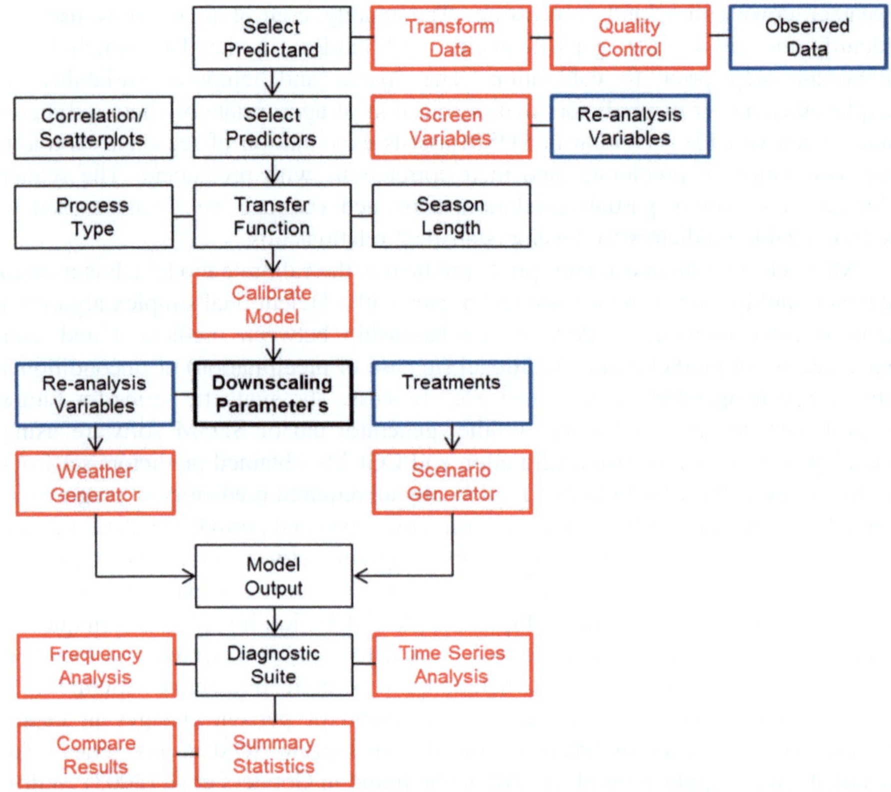

Fig. 1 Work flow in SDSM-DC (reproduced from Wilby et al. 2014)

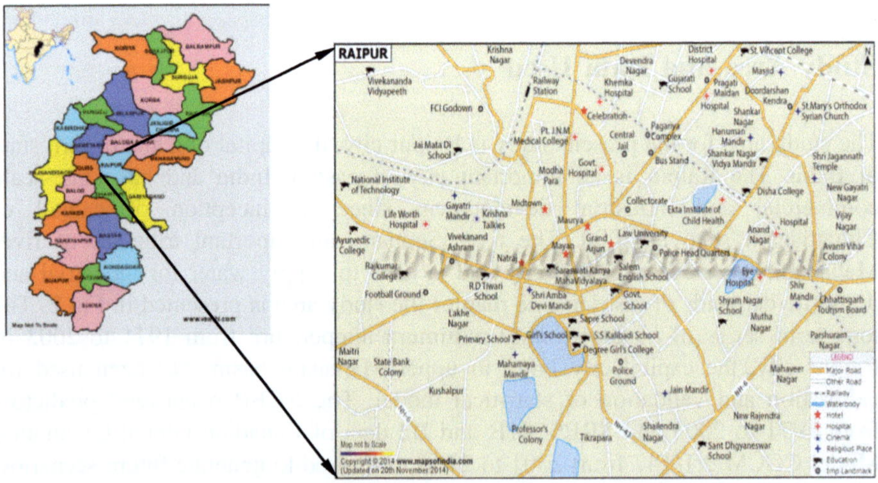

Fig. 2 Location of Raipur in Chhattisgarh state of India

Methodology

The methodology for application of SDSM for generation of minimum temperature series for future climatic scenarios consists of verification of predictand and predictors series, analysis of predictand, predictor relationship, and selection of appropriate predictors which can explain the temporal and spatial variability of predictand with reasonable degree of agreement, calibration and validation of model, generation of future time series using GCM predictors series, computation of statistics and comparison of statistics of present and future scenarios. In the present study, correlation coefficient, partial correlation coefficient, and *P*-value-based method along with scatter diagram were used. The correlation coefficients between predictand (precipitation) and predictors (26 NCEP rescaled parameters) were computed using unconditional approach for annual, monthly, and monsoon season (Mahmood and Babel 2013, 2014). The correlation coefficients were then arranged in descending order and top ten predictors were selected for further analysis. The predictors ranked first in this process can be termed as super predictor (SP) and using this super predictor, absolute correlation coefficient, absolute partial correlation, and the *P*-value were computed for remaining nine predictors with predictand. In order to avoid multicollinearity, all predictors having *P*-value more than 0.05 and other predictors having high individual correlation with super predictor (more than 0.70 for this study) were removed from consideration. The percentage reduction (PR) was then computed for remaining predictors using following equation (Pallant 2007).

$$\mathrm{PR} = \frac{\mathrm{Pr} - R}{R}, \tag{1}$$

where Pr and *R* are the partial and absolute correlation coefficient, respectively. At the end, a predictor having lowest PR value was considered the second super predictor. Similar approach was applied to get third, fourth, and other predictors. In general, one to three predictors are sufficient to model climatic variability (Xu 1999; Chu et al. 2010). After selecting the appropriate predictors, empirical relations between predictand and selected predictors were developed considering appropriate transformation, process (conditional for precipitation and unconditional for other climatic parameters), *k*-fold cross validation, and model types (monthly, seasonal or annual model). The whole series of predictor and selected predictands of base period is divided in two parts considering *k*-fold cross validation technique available in SDSM-DC. In this technique, the whole series can be divided into k equal size subsamples, where one sample is used for calibration while remaining for testing or validation (Bedia et al. 2013; Casanueva et al. 2014). If results of calibration and validation were found appropriate, the weather generator in SDSM can be used to generate future predictor series using predictors obtained from different GCM scenarios.

Analysis of Results

In the present study, SDSM 5.2 software has been employed to generate minimum temperature series for current and future climate forcing. For the present study, 26 NCEP rescaled predictors for the period of 1971–2003 and predictor as minimum temperature series of Raipur (Chhattisgarh) for the same period were analyzed. The scatter diagram, correlation coefficient, and partial correlation based percentage reduction were used to identify an appropriate combination of predictors which can forecast predictand with acceptable degree of error. The scatter diagram of few predictors was presented in Fig. 3. The specific humidity at 850 hpa (nceps850gl) displayed the highest correlation coefficient as 0.649 and was considered as the first super predictor. The PR values of remaining nine predictors having next highest correlation were computed and 500 hpa geopotential height (ncepp500gl) and surface airflow strength (ncep_fgl) were shortlisted as second and third super predictor for calibration. The threefold cross-validation was used which divided the whole series of data from 1971 to 2003 into two parts where first two-third parts were considered for calibration while remaining one-third part for validation.

The coefficient of determination (*R*2) and standard error for different months during calibration and validation obtained from analysis are presented in Table 1. From the analysis, it has been observed that the standard error varies from 1.01 to 3.45 in calibration and 1.20 to 3.60 in validation. The Nash–Sutcliff efficiency of minimum temperature for calibration and validation was computed as 71.55 and 73.89%, respectively, which may be considered as reasonably acceptable match. The comparison of observed versus calibrated and validated mean monthly minimum temp of Raipur has been presented in Fig. 4. The finally selected parameters in calibration were further used to synthetically generate time series of minimum temperature under CGCM supplied data of A1B and A2 forcing conditions.

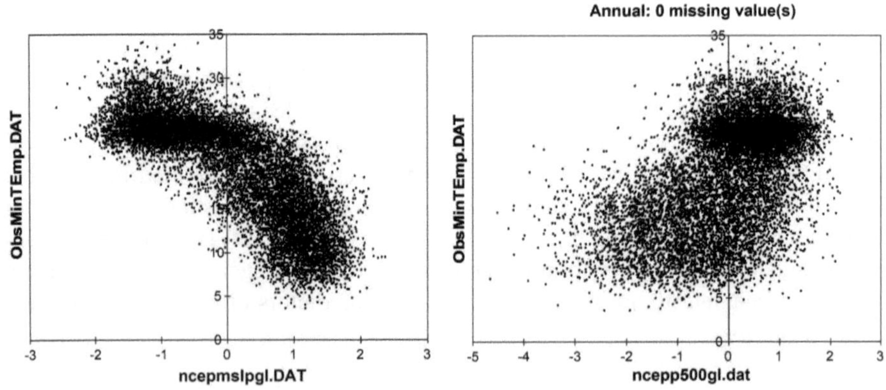

Fig. 3 Scatter diagram for few predictors used in calibration

Table 1 Coefficient of determination and standard error during calibration and validation

Month	Calibration		Validation	
	R2	Std. error	*R2*	Std. error
January	0.0241	3.1167	0.0025	3.1807
February	0.0042	2.9721	0.0014	3.0024
March	0.0183	2.7412	0.0021	2.7810
April	0.0682	2.6013	0.0140	2.7221
May	0.0212	2.5334	0.0020	2.6341
June	0.1180	2.3174	0.0648	2.4051
July	0.0031	1.2724	0.0749	1.3794
August	0.0199	1.1167	0.0302	1.2006
September	0.0146	1.1001	0.0487	1.1999
October	0.1072	2.5883	0.0162	2.8488
November	0.0667	3.4525	0.0190	3.5963
December	0.0462	2.7824	0.0020	2.9153
Mean	0.0426	2.3829	0.0232	2.4888

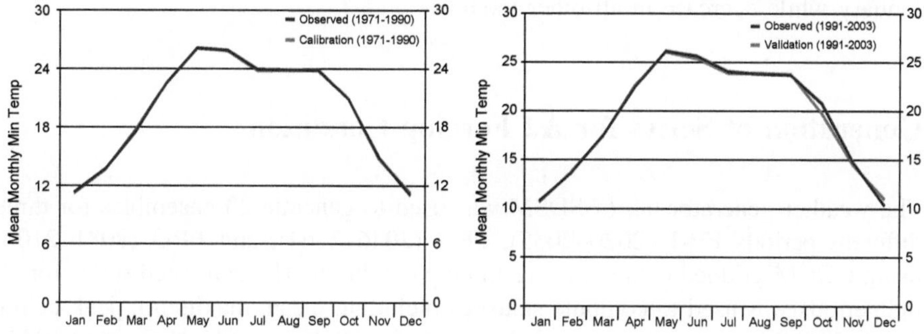

Fig. 4 Observed and calibrated/validated mean monthly minimum temperature in SDSM

CGCM A1B Forcing Condition

The finally selected combination of variables with calibrated parameters was used to synthetically generate 20 series for three future periods FP-1 (2020–2035), FP-2 (2046–2064) and FP-3 (2081–2100) using gridded predictors obtained from Canadian general circulation model (CGCM) under climatic forcing condition of A1B. The statistics including mean monthly minimum temperature, peak below threshold (PBT: number of days/year below 10 °C minimum temperature), variance, inter-quantile range, etc., were computed. The mean monthly minimum temperature and PBT of base period (1971–2003) and all three periods FP-1, FP-2, and FP-3 can be seen in Fig. 5. From the analysis, it has been observed that mean monthly minimum temperature may increase by 1.1–11.2% during summer months

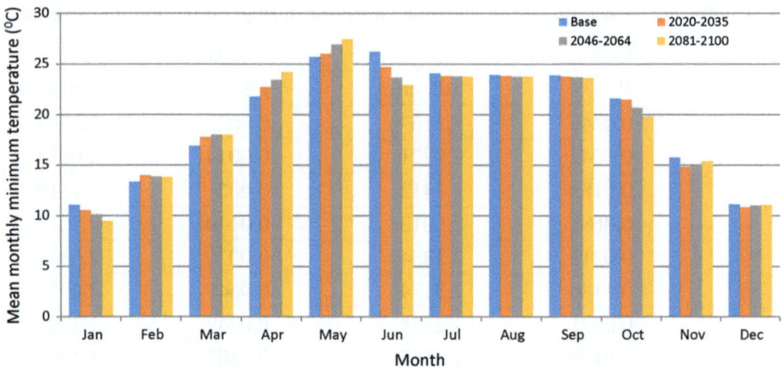

Fig. 5 Comparison of mean monthly minimum temperature for different periods of generated data with observed data under A1B climatic scenario

(February to May) in all three future predictive periods while decrease by 0.5–14.2% in remaining months (June to January) under A1B climate forcing condition. The average number of days/year below 10 °C may increase in November and January while decrease in all other months.

Generation of Series for A2 Forcing Condition

The weather generator tab of SDSM was used to generate 20 ensembles for three different periods FP-1 (2020–2035), FP-2 (2046–2064), and FP-3 (2081–2100) using CGCM gridded data under A2 forcing condition. The generated series for all the periods was used to compute statistics including mean, maximum, peak below threshold (10 °C), variance, etc., and compared with the same for the period 1973–2003. The mean monthly minimum temperature series for different periods with observed data has been presented in Fig. 6. From the analysis, it has been found that the mean monthly minimum temperature may increase by 2.88–24.61% in most of the months except June to October where there may be slight decrease of minimum temperature. The increased minimum temperature during summer and winter months may increase user demands and water requirements of crops in rabi season. The number of cold days below 10 °C may increase significantly in November and January while decrease slightly in December and February in all three future periods.

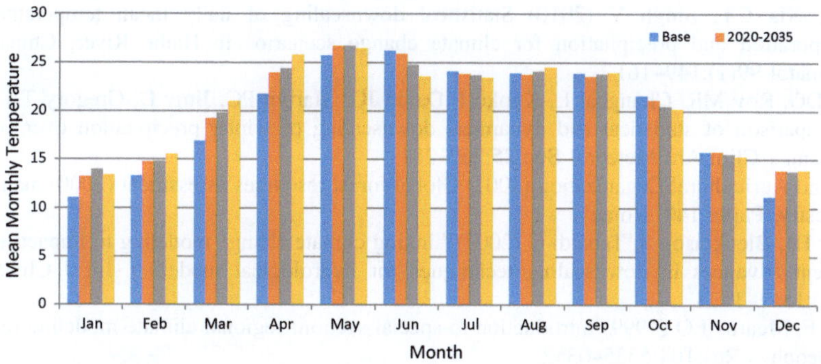

Fig. 6 Comparison of mean monthly minimum temperature for different periods of generated data with observed data under A2 climatic scenario

Conclusion

The changing climate of the world has adverse effect on different facets of life and there is need to develop adaptation strategy for water resource management, agriculture, health, and many more areas of life. The changes in minimum temperature and extreme events are now clearly visible in different parts of earth. The statistical downscaling model (SDSM) has been used to generate several future predictive series for three different periods 2020–2035 (FP-1), 2046–2064 (FP–2), and 2081–2100 (FP-3). The different goodness of fit criterions including scatter diagram, correlation coefficient, and percentage reduction confirmed specific humidity at 850 hpa (nceps850gl), 500 hpa geopotential height (ncepp500gl) and surface airflow strength (ncep_fgl) were found the most appropriate parameters to generate future scenarios. The multiple series for each three predictive periods for A1B and A2 climate forcing conditions were generated and compared statistically with base series (1971–2003). It may be concluded that mean monthly temperature may increase significantly during summer months in both A1B and A2 climate scenarios. The winter months may observe decrease of minimum temperature in A1B condition while slight increase under A2 climate condition.

References

Anandhi A, Srinivas VV, Nanjundiah RS, Nagesh Kumar D (2008) Downscaling precipitation to river basin in India for IPCC SRES scenarios using support vector machine. Int J Climatol 28:401–420

Bedia J, Herrera S, San-Martín D, Koutsias N, Gutiérrez JM (2013) Robust projections of fire weather index in the Mediterranean using statistical downscaling. Clim Change 120:229–247

Casanueva A, Frías MD, Herrera S, San-Martín D, Kaminovic K, Gutiérrez JM (2014) Statistical downscaling of climate impact indices: testing the direct approach. Clim Change 127:547–560

Chu J, Xia CY, Singh V (2010) Statistical downscaling of daily mean temperature, pan evaporation and precipitation for climate change scenarios in Haihe River, China. Appl Climatol 99(1):149–161

Ethan DG, Roy MR, Changhai L, Kyoko I, David JG, Martyn PC, Jimy D, Gregory TA (2011) Comparison of statistical and dynamical downscaling of winter precipitation over complex terrain. J Clim Am Meteorol Soc 25:262–281

Food and Agricultural Organization (2001) Global forest resources assessment (2000) main report forestry. Paper 140, Rome

Fowler HJ, Blenkinsop S, Tebaldi C (2007) Linking climate change modeling to impacts studies: recent advances in downscaling techniques for hydrological modeling. Int J Climatol 27 (12):1547–1578

Giorgi F, Mearns LO (1999) Introduction to special section: regional climate modeling revisited. J Geophys Res 104:6335–6352

Gleick PH (1987) The development and testing of a waterbalance model for climate impact assessment: modeling the Sacramento Basin. Water Resour Res 23(6):1049–1061

Goodess C, Osborn T, Hulme M (2003) The identification and evaluation of suitable scenario development methods for the estimation of future probabilities of extreme weather events. Tyndall Centre for Climate Change Research, Technical report 4

Hewitson BC, Crane RG (2006) Consensus between GCM climate change projections with empirical downscaling: precipitation downscaling over South Africa. Int J Climatol 26: 1315–1337

Huntington TG (2006) Evidence for intensification of the global water cycle: review and synthesis. J Hydrol 319(104):83–95

Intergovernmental Panel on Climate Change (IPCC) (2003) Good practice guidance for land use, land-use 27 change and forestry. In: Penman J, Gytarsky M, Hiraishi T, Kruger D, Pipatti R, Buendia L, Miwa K, Ngara T, Tanabe K, Wagner F (eds) IPCC/IGES, Hayama, Japan

Intergovernmental Panel on Climate Change (IPCC) (2007) Climate change 2007: the physical science basis. Contribution of working group I to the fourth assessment report of the Intergovernmental Panel on Climate Change. In: Solomon S, Qin D, Manning M, Chen Z, Marquis M, Averyt KB, Tignor M, Miller HL (eds). Cambridge University Press, Cambridge, 996 pp

Lopes PG (2008) Assessment of statistical downscaling methods-application and comparison of two statistical methods to a single site in Lisbon. Master thesis, University of Lisbon

Mahmood R, Babel M (2013) Evaluation of SDSM developed by annual and monthly sub-models for downscaling temperature and precipitation in the Jhelum basin, Pakistan and India. Theor Appl Climatol 113(1–2):27–44

Mahmood R, Babel M (2014) Future changes in extreme temperature events using the statistical downscaling model (SDSM) in the trans-boundary region of the Jhelum river basin. Weather Clim Extremes 5–6:56–66

McCarthy JJ, Canziani OF, Leary NA, Dokken DJ, White KS (eds) (2001) Climate change 2001: impacts, adaptation and vulnerability, Inter-Governmental Panel on Climate Change (IPCC), Work group II input to the third assessment report. Cambridge University Press, Cambridge

Menzel A, von Vopelius J, Estrella N, Schleip C, Dose V (2006) Farmers annual activities are not tracking speed of climate change. Clim Res 32:201–207

Milly PCD, Wetherald RT, Dunne KA, Delworth TL (2002) Increasing risk of great floods in a changing climate. Nature 415(514):517

Milly PCD, Dunne KA, Vecchia AV (2005) Global pattern of trends in streamflow and water availability in a changing climate. Nature 438(347):350

Pallant J (2007) SPSS survival manual: a step by step guide to data analysis using the SPSS for windows. Open University Press Buckinghum. http://www.mheducation.co.uk/openup/chapters/0335208908.pdf. Accessed 25 Dec 2015

Ravindran PN, Nirmal Babu K, Sasikumar B, Krishnamurthy KS (2000) Botany and crop improvement of black pepper (in) Black pepper. In: Ravindran PN (ed). Harwood Academic Publishers, pp 23–142

Salathe EP Jr (2003) Comparision of various precipitation downscaling methods for the simulatsion of streamflow in a rainshadow river basin. Int J Climatol 23:887–901

Sivakumar T, Thennarasu A, Rajkumar JSI (2012) Effect of season on the incidence of infectious diseases of bovine in Tamilnadu. Elixir Meteorol 47:8874–8875

Tisseuil C, Vrac M, Lek S, Wade AJ (2010) Statistical downscaling of river flows. J Hydrol (Amst) 385:279–291

Tripathi S, Srinivas VV. Nanjundiah RS (2006) Downscaling of precipitation for climate change scenarios: a support vector machine approach. J Hydrol. doi:10.1016/j.jhydrol.2006.04.030

Wilby RL, Dettinger MD (2000) Streamflow changes in the Sierra Nevada, CA simulated using a statistically downscaled General Circulation Model scenario of climate change. In: McLaren SJ, Kniveton DR (eds) Linking climate change to land surface change. Kluwer Academic Publishers, Netherlands, pp 99–121

Wilby RL, Wigley TML (1997) Downscaling general circulation model output: a review of methods and limitations. Prog Phys Geogr 21:530–548

Wilby RL, Dawson CW, Barrow EM (2001) SDSM—a decision support tool for the assessment of regional climate change impacts. Environ Model Soft 17:145–157

Wilby RL, Tomlinson OJ, Dawson CW (2003) Multi-site simulation of precipitation by conditional resampling. Clim Res 23:183–194

Wilby RL, Dawson CW, Murphy C, O' Connor P, Hawkins E (2014) The statistical downscaling model—decision centric (SDSM-DC): conceptual basis and applications. Clim Res 61: 251–268

Xu CY (1999) From GCMs to river flow: a review of downscaling methods and hydrologic modeling approaches. Prog Phys Geogr 23(1):57–77

Statistical Downscaling of Daily Temperatures and Precipitation Data from Coupled Model Inter-comparison Project 5 (CMIP5)-RCPs Experiment: In Weyib River Basin, Southeastern Ethiopia

Abdulkerim Bedewi Serur and Arup Kumar Sarma

Abstract Impact of climate change on the temperature and precipitation characteristics of Weyib River basin in Ethiopia has been investigated using CanESM2 model for the RCP2.6, RCP4.5, and RCP8.5 scenarios. The statistical downscaling model calibrated and validated using the observed daily data of 12 meteorological stations was used to generate the future scenario. The change in mean annual maximum temperature from the base period has indicated an increment of 0.16, 0.14, and 0.15 °C for RCP2.6, 0.12, 0.19, and 0.21 °C for RCP4.5, and 0.12, 0.22 and 0.32 °C for RCP8.5 for the 2020s, 2050s, and 2080s, respectively. Mean annual minimum temperature has shown an increment of 0.30, 0.43, and 0.39 °C for RCP2.6, 0.31, 0.48, and 0.57 °C for RCP4.5, and 0.34, 0.66 and 1.04 °C for RCP8.5 for the 2020s, 2050s and 2080s, respectively. For the percentage change in mean annual precipitation from the base period, the increment has been 8.68, 12.93, and 11.34% for RCP2.6, 9.54, 14.36, and 16.94% for RCP4.5, and 14.70, 19.14, and 28.69% for RCP8.5 for the 2020s, 2050s, and 2080s, respectively. There was a significantly (at 5% significant level) increasing trend of both temperatures and precipitation in all the three RCPs for future until the year 2100. The increment of rainfall in the study area was comparatively higher in the dry season 20.68% in the 2020s, 33.65% in 2050s, and 53.74% in 2080s for RCP8.5 which might have positive impact on pastoral region of the study area and it might affect the highland areas negatively since this season is expressly main crop harvesting period.

Keywords CMIP5-ESM-RCP experiment · SDSM · Climate change scenarios Weyib river basin · Ethiopia

---

A.B. Serur (✉) · A.K. Sarma
Civil Engineering Department, Indian Institute of Technology Guwahati, Guwahati 781039, Assam, India
e-mail: abdulkerimb1@gmail.com

A.K. Sarma
e-mail: aks@iitg.ernet.in

V.P. Singh et al. (eds.), *Climate Change Impacts*, Water Science and Technology Library 82, https://doi.org/10.1007/978-981-10-5714-4_5

Introduction

It is observed that there is a substantial scale gap between what Global Circulation Models (GCMs) supply the predictors and what the hydrological models require climatic inputs to simulate hydrological processes. This scale discrepancy sources to a sizeable trouble for the valuation of climate change effect through hydrological models. Hence, significant awareness should be drawn to the development of downscaling methodologies so as to obtain local scale climate variables (mainly precipitation and temperatures) from coarse resolution ESMs. There are two prime methodologies accessible for the downscaling of large-scale resolution ESMs output to the finer (local) scale resolution (Wilby and Dawson 2007) namely dynamical (a higher resolution regional climate model is enforced to use ESMs output) and statistical (forms empirical relationships between large-scale atmospheric ESM variables, the predictors and local (finer or catchment) scale climate variables, the predictands) downscaling methodologies.

The projected mean annual temperature in Ethiopia are found to be in the ranges 0.9–1.1, 1.7–2.1, and 2.7–3.4 °C by 2030, 2050, and 2080 time slices, respectively (Ethiopian National Meteorological Agency 2007). The projected mean annual maximum and minimum temperature shows rising trend in southeastern part of Ethiopia (Shawul et al. 2016). The increment of large-scale mean surface temperature by the end of the twenty-first century are found to be in the ranges 2.6–4.8 °C (RCP8.5), 1.1–2.6 °C (RCP4.5), and 0.3–1.7 °C (RCP2.6) (IPCC 2013). In general, temperatures revealed an increasing trend (Kruger and Shongwe 2004; New et al. 2006; Unganai 1996) which tends to glacier to melt, sea level to rise, and alteration in circulation pattern which influence precipitation, water availability, and extremes of floods and droughts; just to name a few. There has been observed substantial variability in rainfall (rise about 20% and also declined by about 20%) in the globe (Bates et al. 2008). Larger spatial variation of precipitation (from a drop of about 25–50% to rise to 25–50%) has been reported in East Africa (Faramarzi et al. 2013). Increase in rainfall has been observed (Shongwe et al. 2009) in the tropics. Rainfall variability is more in the African continent and resulted in variation in water availability. For instance, a decline in water availability (streamflow) (Beck and Bernauer 2011) by 2050s has been reported. Nevertheless, the rise of water availability (Graham et al. 2011) has also been stated. As we have seen in various literatures that there is an argument on the amount of decreasing or increasing of water availability on the globe.

The present unpredictable climate is a striking plenteous threat to Ethiopia by mainly disturbing water resources and agricultural sectors. Recent flooding incidences, as well as the widespread drought in Ethiopia, could remain placed as visible indications for these influences (Ethiopian National Meteorological Agency 2007). The Weyib River basin is used for several purposes (i.e., various water resources schemes involved to the flow of the river) but rises in temperature and change in rainfall magnitude and pattern affect the basin negatively. Therefore, in this study, statistical downscaling methodology using multiple linear regression

(MLR) based statistical downscaling model (SDSM) has been used to downscale daily temperatures (maximum and minimum) and precipitation data for 12 arbitrary meteorological stations found inside study area of Weyib River basin using CMIP5-CanESM2 for the RCP8.5, RCP4.5, and RCP2.6 scenarios. These future downscaled temperatures and precipitation data can be used as an input for hydrological models to simulate future, various surfaces, and subsurface hydrological processes. The analysis of future maximum and minimum temperature and precipitation was carried out on an annual, seasonal, and monthly basis for three (the 2020s; represents 2011–2040 time series data, 2050s; represents 2041–2070 time series data and 2080s; represents 2071–2100 time series data) time slices in future periods.

Materials and Methods

Study Area (Weyib River Basin)

The Weyib River basin (Fig. 1) has an area of 4215.93 km^2 and is situated between 6.50 and 7.50°N latitude and 39.50–41.00°E longitude. The altitude variation ranges around 4389 m (a.m.s.l.) at the highest point to 898 m at the confluence point. Mean annual maximum and minimum temperature of the study area are 22.30 and 7.60 °C, respectively. The average rainfall in the study area ranges 749.34–1368.90 mm (mean of 1037.40 mm) per annum. The Eutric Vertisol and Dystric Cambisol are the two main soil types, and agriculture is a leading land use type. Roughly, 70.54% of the basin area covered with 0–15% land slope. Mean annual total water availability (in the simulation period 1984–2004) has to be 553.46 mm.

Earth System Model (ESM) and RCP Scenarios

In this study, a bias-corrected CMIP5-ESMs climate model for the RCP8.5 (very high emission scenario), RCP4.5 (an intermediate emission scenario), and RCP2.6 (very low emission scenario) scenarios has been used. The historical and future predictor variables have been downloaded through official website (http://ccds-dscc.ec.gc.ca/?page=pred-canesm2) of Canadian Centre for Climate Modelling and Analysis (http://ccds-dscc.ec.gc.ca/?page=pred-canesm2CCCMA). These predictors are assigned in zip file format and have five files inside (CanESM2_historical_1961-2005, CanESM2_rcp26_2006-2100, CanESM2_rcp45_2006-2100, CanESM2_rcp85_2006-2100 and NCEP-NCAR_1961-2005) the predictors prepared, this technique was used as an input in SDSM.

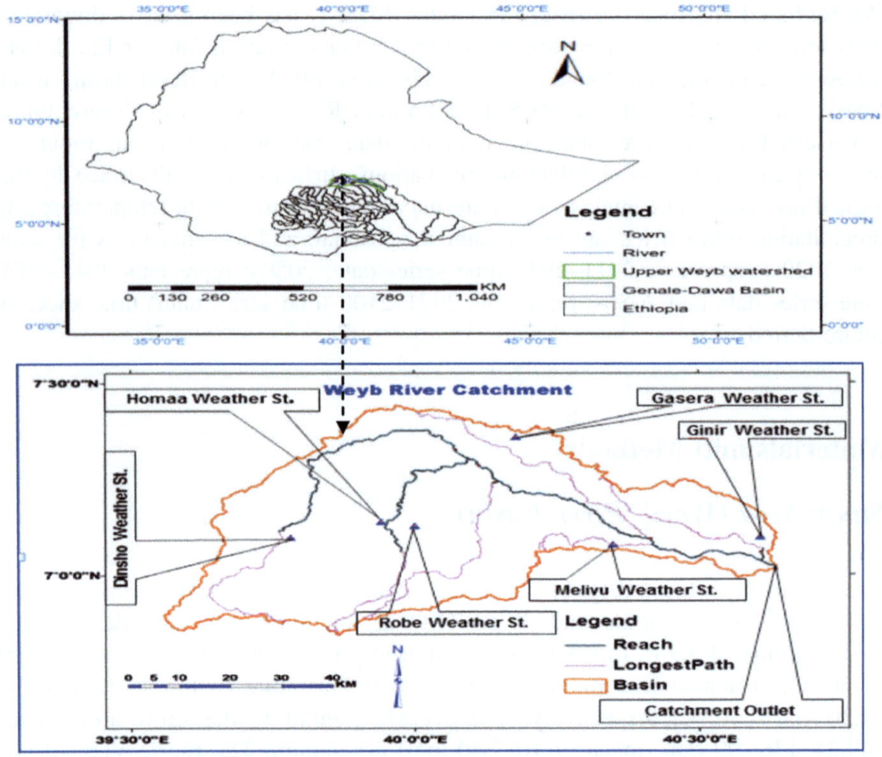

Fig. 1 Study area: location map, reach, basin, and selected weather stations of the Weyib River basin

Types of Data Used

Daily weather data and ESM data were utilized for this study. The daily weather data (daily precipitation, Tmax, Tmin, mean wind speed, relative humidity and sunshine hours) for 12 weather stations (detailed in Table 1) were collected from National Meteorological Service Agency of Ethiopia (NMSA). The source of ESM data is the official website of Canadian Centre for Climate Modelling and Analysis mentioned above in 'earth system model (ESM) and RCP Scenarios' section. The mean monthly rainfall (mm) and temperature (°C) characteristics of Weyib River basin are depicted in Fig. 2.

Table 1 Details of the 12 meteorological stations and their data records

S. No.	Station name	Data recording periods	Latitude (°N)	Longitude (°E)	Altitude (m)	Total annual mean precipitation (mm)	Mean annual max. temp (°C)	Mean annual min. temp (°C)
1	Robe	1984–2011	7.133	40.000	2464	804.14	21.79	8.00
2	Goba	1998–2007	7.017	40.000	2613	980.03	20.13	6.52
3	Dinsho	1981–2007	7.100	39.783	3072	1368.90	17.43	3.43
4	Agarfa	1988–1997	7.267	39.817	2465	762.36	22.38	8.13
5	Sinnana	1982–2008	7.067	40.217	2364	894.35	21.39	8.02
6	Adaba	1980–2010	7.000	39.383	2415	823.65	23.87	5.10
7	Homa	1988–2010	7.133	39.933	2505	846.58	21.75	7.33
8	Ali	1988–2005	7.017	40.350	2460	1264.17	20.47	6.68
9	Gassera	1980–2010	7.367	40.183	2337	1181.12	20.67	6.95
10	Goro	1981–2005	7.000	40.467	1806	887.14	26.77	7.89
11	Ginnir	1980–2012	7.133	40.700	1929	1030.04	23.82	13.20
12	Hunte	1980–2011	7.100	39.417	2413	749.34	23.71	6.73

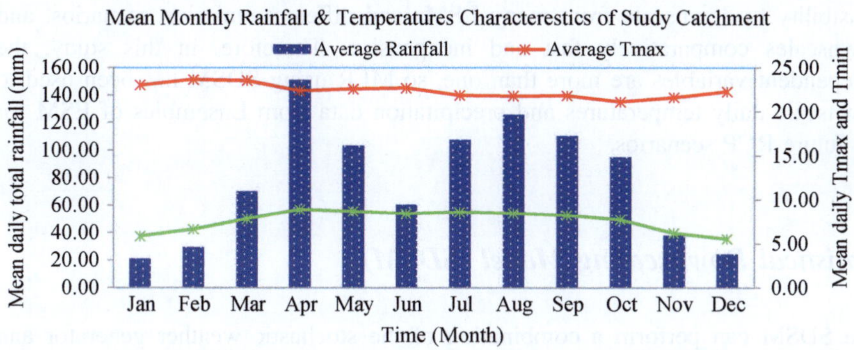

Fig. 2 Mean (12 stations) monthly rainfall and temperature characteristics of Weyib River basin

Downscaling Methods

There are two prime methodologies accessible for the downscaling of large-scale spatial ESMs output to the local scale spatial resolution (Wilby and Dawson 2007) namely dynamical (a higher resolution regional climate model is enforced to use ESMs output) and statistical downscaling methodologies. The statistical downscaling method establishes empirical relationships between large-scale ESM outputs, the predictors, and local scale climatic variables (for instance, Tmax, Tmin, and precipitation), the predictands with the help of some transfer function as shown in the following relation.

$$Y = f(X), \quad (1)$$

where Y is local Tmax, Tmin, and precipitation that were being downscaled, X stands for a set of large-scale potential predictor variables (for instance, mean sea level pressure, geopotential heights and specific humidity at the surface and 850 hpa), and f represents a stochastic function that relates the predictands and predictors.

The "f" function is determined empirically from historical observations by training and validating the model. Thus, the achievement of the statistical downscaling method was based on the relationship used and choice of potential predictor variables, whose performance can be verified through estimation of various statistical indices (for instance, R^2, RMSE, and NSE). It is roughly divided into three classes (weather typing, weather generators, and regression-based downscaling). Regression analysis is very powerful for forecasting (Ghosh and Mujumdar 2008); it is divided into two categories namely simple regression and multiple regressions. The statistical downscaling method has its own merits and demerits. Key demerits of statistical downscaling contain the assumption that observed relations between large-scale predictors and local predictands will continue in a changing climate. Similarly, some merits of statistical downscaling contain: it is easy to apply, has the possibility to downscale from many ESMs and different emission scenarios, and downscales comparatively fast and inexpensive. Therefore, in this study, the independent variables are more than one, so MLR using SDSM has been used to downscale daily temperatures and precipitation data from Ensembles of ESM for the future RCP scenarios.

Statistical Downscaling Model (SDSM)

The SDSM can perform a combination of the stochastic weather generator and regression-based on the family of the transfer function (Liu et al. 2011). It performs the spatial downscaling through daily predictor–predictand relationships using MLR and creates predictands that represent the local weather condition. Regression-based in the family of transfer function method is the well-known technique of downscaling (Ghosh and Mujumdar 2008) that depends on the direct, measurable link between predictand and predictors through some form of regression. SDSM version 4.2.9 a decision support tool (Wilby and Dawson 2007) has been used for this study to downscale daily future Tmax, Tmin, and precipitation. This model was downloaded from the website http://co-public.lboro.ac.uk/cocwd/SDSM/. There are seven major steps to be followed in developing the best performing MLR equation for the downscaling processes in this version of SDSM. Detailed discussions of the steps are given in (Wilby et al. 2002; Wilby and Dawson 2007).

The choice of appropriate downscaling potential predictor variables has been done with the help of the screen variables option of the SDSM using correlation analysis, partial correlation analysis, and scatter plot. Screening the potential predictors has been done through choosing seven or eight predictors at a time and their explained variance has been analyzed, thereby choosing those predictors which have greater explained variance and drop the rest. Then partial correlation analysis has been done for nominated predictors to see the level of association with each other; these statistics identify the extent of the descriptive power of the predictor to describe the predictand. Therefore, the predictors used for downscaling should be reliably generated by ESMs, freely accessible from ESM outputs archive and strongly linked with the local climate variables of concern (Tmax, Tmin, and precipitation in this case).

The calibration process in SDSM constructs downscaling model based on MLR equations, with given daily station wise weather data (predictand) and potential predictors. The model calibration operation has been run for station wise precipitation, Tmax and Tmin along with a set of possible predictor variables, and computes the parameters of MLR equations through an optimization algorithm (Dual simplex has used in this case). SDSM's weather generator enables to produce ensembles of synthetic current daily weather data based on inputs of the measured time series data and the MLR parameters generated during the calibration step. Finally, station wise Tmax, Tmin, and precipitation scenario have been generated until the year 2100 using CMIP5-CanESM2 model outputs (potential predictor variables for the future period) for the RCP8.5, RCP4.5, and RCP2.6 emission scenarios. Twenty ensembles of synthetic daily Tmax, Tmin and precipitation time series data were generated for the period from 2006 to 2100 for all stations and the mean of these 20 ensembles was used as final daily Tmax, Tmin, and precipitation data for the stated period. Moreover, then the future Tmax, Tmin, and precipitation scenarios have been established by dividing the later date series into three (the 2020s, 2050s, and 2080s) time slices for the 12 averaged arbitrary spatial weather stations.

Precipitation and Temperatures Scenario Statistics

Percentage and absolute change have been used to calculate three-time slices of 30 years precipitation and temperatures, respectively.

A percentage change has been used for precipitation;

$$\Delta 2020\text{s} = \frac{(\text{V2020s} - \text{Vbase}) \times 100}{\text{Vbase}} \tag{2}$$

$$\Delta 2050\text{s} = \frac{(\text{V2050s} - \text{Vbase}) \times 100}{\text{Vbase}} \tag{3}$$

$$\Delta 2080\text{s} = \frac{(\text{V2080s} - \text{Vbase}) \times 100}{\text{Vbase}}. \quad (4)$$

Absolute difference has been used for temperatures,

$$\Delta 2020\text{s} = \text{V2020s} - \text{Vbase} \quad (5)$$

$$\Delta 2050\text{s} = \text{V2050s} - \text{Vbase} \quad (6)$$

$$\Delta 2080\text{s} = \text{V2080s} - \text{Vbase}, \quad (7)$$

where, Vbase is the mean of 20 ensembles of Tmax, Tmin and precipitation for the base period for each ESM-RCP and each station. V2020s, V2050s, and V2080s are the average of 20 ensembles of Tmax, Tmin, and precipitation for the period of 2011–2040, 2041–2070, and 2071–2100, respectively, for each ESM-RCP experiment and each station.

Mann–Kendall Trend Test

A nonparametric rank-based procedure has frequently been used to evaluate if there is a rise or decline trend in the time series of meteorological and hydrological data (Hamed 2008; Karpouzos et al. 2010). The Mann–Kendall test was applied in this study to see the existing trends (rise or decline) of Tmax, Tmin, and precipitation for the RCP8.5, RCP4.5, and RCP2.6 scenarios in future periods.

SDSM Performance Evaluation

In order to evaluate the SDSM performance relative to the observed Tmax, Tmin, and precipitation data, the following three statistical model performance evaluation measures, in addition to graphical technique, were used during the calibration and validation periods.

Coefficient of Determination (R^2)

It was given by (Krause and Boyle 2005) as shown in Eq. 8

$$R^2 = \frac{(\sum[\text{Xi} - \text{Xav}][\text{Yi} - \text{Yav}])^2}{\sum(\text{Xi} - \text{Xav})^2 \sum(\text{Yi} - \text{Yav})^2}, \quad (8)$$

where, Xi is measured value, Xav is average measured value, Yi is simulated value, and Yav is average simulated value.

Nash–Sutcliffe Coefficient (*E*)

It was given by (Nash and Sutcliff 1970) as shown in Eq. 9

$$E = 1 - \frac{\sum_{i=1}^{n} \left(X_{\text{obs},i} - X_{\text{model}}\right)^2}{\sum_{i=1}^{n} \left(X_{\text{obs},i} - \overline{X_{\text{obs}}}\right)^2}, \tag{9}$$

where X_{obs} is observed values and X_{model} is modeled values at time/place *i*.

Root Mean Square Error (RMSE)

It was given by (Singh et al. 2004) as shown in Eq. 10

$$\text{RMSE} = \sqrt{\frac{\sum_{i=1}^{n} \left(X_{\text{obs},i} - X_{\text{model},i}\right)^2}{n}}, \tag{10}$$

where X_{obs} is observed values, X_{model} is modeled values at time/place *i,* and *n* is number of observation.

Results and Discussion

Selected Potential Predictor Variables

The lists of selected common potential predictor variables in the entire basin that gave better correlation results at $p < 0.05$ for CanESM2 are listed in Table 2.

Results revealed that different atmospheric variables affect different local variables. For instance, precipitation is more sensitive to mean sea level pressure, specific humidity (at surface and 850 hPa), zonal velocity (at 500 and 850 hPa), and geopotential heights (at 500 hPa). Mean sea level pressure, geopotential heights (at 500 and 850 hPa), average temperature (at 2 m height), specific humidity (at near surface and 850 hPa), and wind direction (at 850 hPa) affect both the maximum and minimum temperature under the CanESM2-historical model.

Table 2 List of selected potential predictor variables that provided better correlation results at $p < 0.05$ from CanESM2-historical model for study area of Weyib River basin

Predictand	Predictor full name	Notations	Parti.cor. (*r*-value)	*p*-value
Precipitation	Mean sea level pressure	ceshmslpgl.dat	0.050	0.020
	Specific humidity at 850 hPa	ceshs850gl.dat	0.090	0.000
	Surface specific humidity	ceshshumgl.dat	0.077	0.002
	500 hPa zonal velocity	ceshp5_ugl.dat	−0.094	0.000
	850 hPa zonal velocity	ceshp8_ugl.dat	−0.119	0.000
	500 hPa geopotential height	ceshp500gl.dat	0.056	0.016
Maximum temperature	Mean sea level pressure	ceshmslpgl.dat	0.148	0.000
	500 hPa geopotential height	ceshp500gl.dat	0.134	0.000
	Specific humidity at 850 hPa	ceshs850gl.dat	−0.268	0.000
	Mean temperature at 2 m	ceshtempgl.dat	−0.235	0.000
	850 hPa wind direction	ceshp8thgl.dat	0.110	0.000
	850 hPa geopotential height	ceshp850gl.dat	−0.200	0.000
	Surface specific humidity	ceshshumgl.dat	−0.274	0.000
Minimum temperature	Mean sea level pressure	ceshmslpgl.dat	−0.308	0.000
	500 hPa geopotential height	ceshp500gl.dat	0.222	0.000
	Surface specific humidity	ceshshumgl.dat	0.146	0.000
	Mean temperature at 2 m	ceshtempgl.dat	0.323	0.000
	Specific humidity at 850 hPa	ceshs850gl.dat	−0.116	0.000
	850 hPa geopotential height	ceshp850gl.dat	−0.103	0.000
	850 hPa wind direction	ceshp8thgl.dat	0.078	0.000
	850 hPa zonal velocity	ceshp8_ugl.dat	−0.116	0.000

Note The partial correlation coefficient (r) shows the explanatory power that is specific to each predictor. All are significant at $p \leq 0.05$. *hpa* is a unit of pressure, 1 hPa = 1 mbar = 100 Pa = 0.1 kPa

Calibration and Validation of SDSM for Both Temperatures and Precipitation

For downscaling of maximum temperature and minimum temperature, and precipitation MLR, using SDSM, was used to calibrate and validate the model. The entire length of the observed data was available from 1981 to 2005. This data was divided into two parts for calibration and validation. Data from 1981 to 1993 was used for the calibration whereas data from 1994 to 2005 was used for the validation of the model.

Calibration and validation results for 12 averaged spatial stations maximum temperature downscaling model are shown in (Fig. 3a, b). The coefficients of determination (R^2), RMSE and NSE values were 0.95, 0.44, and 0.79, respectively, for calibration period whereas R^2, RMSE and NSE values of 0.94, 0.46, and 0.82, respectively, for validation period. For minimum temperature (Fig. 3c, d), the R^2, RMSE, and NSE values were 0.92, 0.58, and 0.86, respectively, for calibration period whereas R^2, RMSE, and NSE values were 0.93, 0.58, and 0.88 for validation period. Calibration and validation results for 12 averaged spatial stations precipitation downscaling model are shown in (Fig. 3e, f). The R^2, RMSE, and NSE values were 0.83, 0.98, and 0.78, respectively, for calibration period whereas R^2, RMSE, and NSE values of 0.86, 0.79, and 0.84, respectively for validation period.

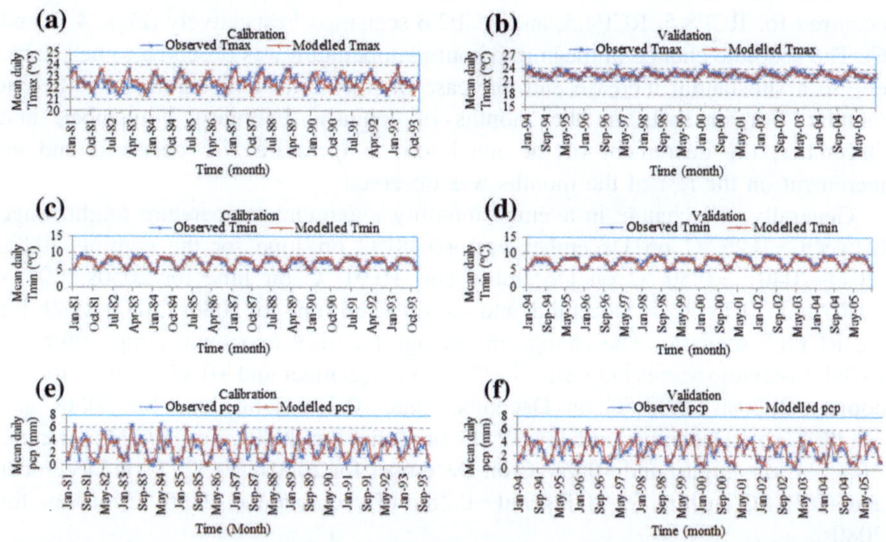

Fig. 3 **a** Calibration result of SDSM for maximum temperature from average of 3 ESMs (1981–1993) (*upper left*), **b** same as (**a**) but for validation period (1994–2005) (*upper right*), **c** calibration result of SDSM for minimum temperature (*middle left*), **d** same as (**c**) but for validation period (*middle right*), **e** calibration result of SDSM for precipitation (*bottom left*), **f** same as (**e**) but for validation period (*bottom right*)

From these results, we can argue that the model is well performed for the maximum and minimum temperature, and precipitation downscaling for both calibration and validation period.

Scenarios Developed for Future Temperatures and Precipitation

Maximum Temperature Scenarios

The projected mean seasonal maximum temperature shows a decreasing trend in the dry season in all the three future time slices for RCP2.6, RCP4.5, and RCP8.5 scenarios except for RCP2.6 scenario which is an increasing trend in the dry season of the 2020s. However, it has shown an increasing trend for both an intermediate and wet seasons in all the three future time slices for RCP2.6, RCP4.5, and RCP8.5 scenarios except for RCP2.6 scenario which is a decreasing trend in a wet season of the 2020s. The absolute changes of maximum temperature from the base period for three scenarios in future time slice for each season are presented in Table 3.

The projected mean monthly maximum temperature has a larger magnitude of increment on the month of June 2080s which was 0.58, 0.80, and 1.37 °C for RCP2.6, RCP4.5, and RCP8.5 scenarios, respectively. On the other hand, the larger decrement on December 2080s 0.75, 0.38 °C, and on December 2050s 0.32 °C occurred for RCP8.5, RCP4.5, and RCP2.6 scenarios, respectively (Figs. 4, 5 and 6). The absolute change in mean maximum temperature was observed to be sizable due to a substantial increase and decrease of maximum temperature on different months. For instance, in the months of January, February, November, and December, the decrement of the maximum temperature was observed and an increment on the rest of the months was observed.

Generally, the change in average monthly maximum temperature might range between −0.25 °C on December and +0.48 °C on June for the coming 2020s (2011–2040); −0.50 °C on December and +0.91 °C on June for 2050s (2041–2070) and −0.75 °C on December and +1.37 °C on June for 2080s (2071–2100) for the RCP8.5 scenario. The change in average monthly maximum temperature for RCP4.5 scenario varies between −0.25 °C on December and +0.47 °C June for the coming 2020s; −0.33 °C on December and +0.70 °C on June for 2050s and −0.38 °C on December and +1.37 °C on June for 2080s. For RCP2.6, ranges −0.12 °C on August and +0.42 °C on December for 2020s; −0.32 °C on December and +0.58 °C on June for 2050s and −0.28 on December and +0.55 °C on June for 2080s.

For each time slice, the change in mean annual maximum temperature has indicated a slight increment from the base period, 0.16, 0.14, and 0.15 °C for 2020s, 2050s, and 2080s, respectively, for RCP2.6 scenario, 0.12, 0.19, and 0.21 °C for 2020s, 2050s, and 2080s, respectively, for RCP4.5 scenario, and 0.12, 0.22, and 0.33 °C for 2020s, 2050s, and 2080s, respectively, for RCP8.5 scenario.

Table 3 Absolute change in mean seasonal maximum temperature (°C) at different time horizons from the base period

Scenario	Seasons in the 2020s			Seasons in the 2050s			Seasons in the 2050s		
	Dry	Intermediate	Wet	Dry	Intermediate	Wet	Dry	Intermediate	Wet
CanESM2-RCP2.6	+0.275	+0.194	−0.002	−0.017	+0.228	+0.215	−0.002	+0.224	+0.208
CanESM2-RCP4.5	−0.012	+0.192	+0.167	−0.001	+0.301	+0.257	−0.001	+0.338	+0.283
CanESM2-RCP8.5	−0.007	+0.202	+0.168	−0.036	+0.370	+0.326	−0.051	+0.529	+0.495

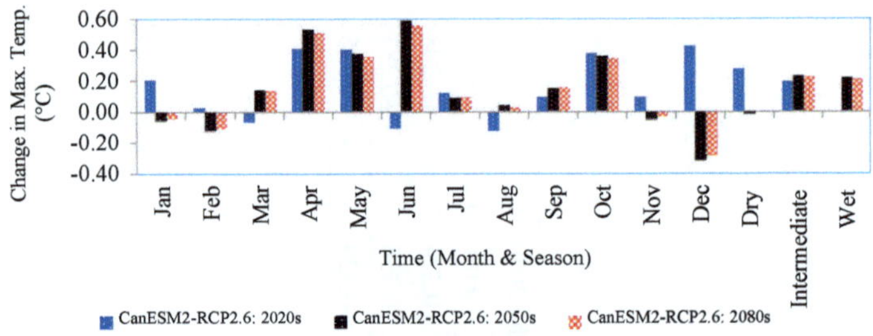

Fig. 4 Change in average monthly and seasonal maximum temperature in the future from the base period for CanESM2-RCP2.6 scenario

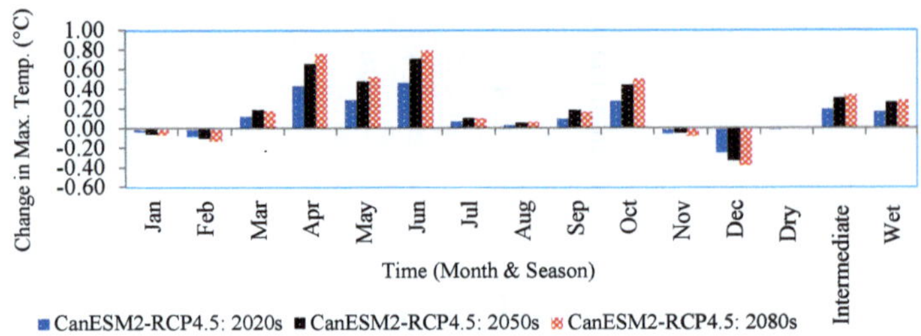

Fig. 5 Change in average monthly and seasonal maximum temperature in the future from the base period for CanESM2-RCP4.5 scenario

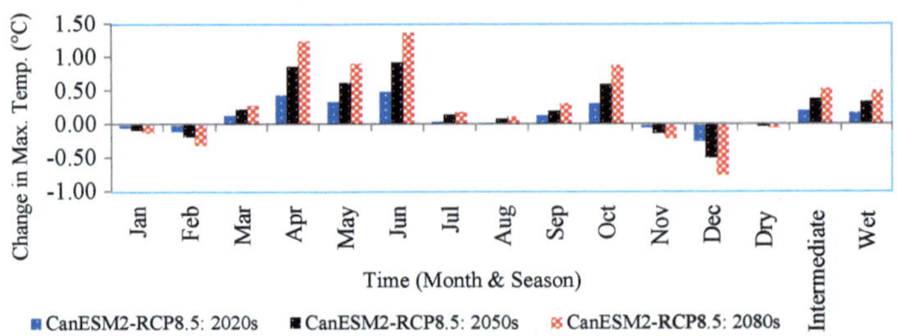

Fig. 6 Change in average monthly and seasonal maximum temperature in the future from the base period for CanESM2-RCP8.5 scenario

The variability of maximum temperature is higher for RCP8.5 than RCP4.5 and RCP2.6, and the linear trend line for three scenarios has indicated a significantly (at 5% significant level) increasing trend of average annual maximum temperature until the end of the century (Table 6 and Fig. 7). Comparatively, RCP8.5 (very high emission scenario) prevails higher change in maximum temperature trend at the end of the century than the RCP4.5 (an intermediate emission scenario) and RCP2.6 (very low emission scenario).

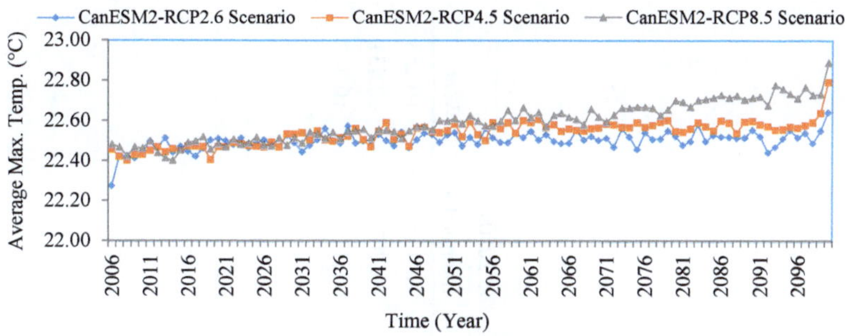

Fig. 7 Future pattern of average annual maximum temperature

Minimum Temperature Scenarios

The projected mean seasonal minimum temperature shows an increasing trend in all the seasons (Dry, intermediate, and wet) in all the three future time horizons for all the RCP (RCP2.6, RCP4.5, and RCP8.5) scenarios. The changes of minimum temperature from the base period for the three scenarios in future time slice for each season are presented in Table 4. The projected mean monthly minimum temperature has a larger magnitude of increment on the month of October 2080s which was 2.13, 1.24, and 0.95 °C for RCP8.5, RCP4.5, and RCP2.6 scenarios, respectively, at the end of the century. On the other hand, the larger decrement on February 2080s 0.51, 0.48, and 0.36 °C occurred for RCP8.5, RCP4.5, and RCP2.6 scenarios, respectively, at the end of the century (Figs. 8, 9 and 10). The absolute change from the base period in mean minimum temperature was observed to be significant due to a substantial increase and decrease of minimum temperature on different months. For instance, in the months of February, September, and December, the decrement of the minimum temperature was observed and an increment on the rest of the months was observed. In both extreme conditions (rise or decline), the change in minimum temperature was higher in the 2080s.

Commonly, the change in average monthly minimum temperature might range between −0.30 °C on February and +0.87 °C on October for the coming 2020s; −0.41 °C on February and +1.38 °C on October for 2050s and −0.51 °C on February and +2.13 °C on October for 2080s for the RCP8.5 scenario. For RCP4.5 scenario, it varies between −0.32 °C on February and +0.69 °C October for the coming 2020s; −0.42 °C on February and +1.12 °C on October for 2050s and −0.48 °C on February and +1.24 °C on October for 2080s. For RCP2.6 scenario, ranges become −0.30 °C on February and +0.74 °C on October for the 2020s; −0.31 °C on February and +0.94 °C on October for 2050s and −0.36 °C on February and +0.95 °C on October for 2080s.

For each time slice, the change in mean annual minimum temperature has indicated a slight increment from the base period, 0.30, 0.43, and 0.39 °C for 2020s, 2050s, and 2080s, respectively, for RCP2.6 scenario, 0.31, 0.48, and 0.57 °C for 2020s, 2050s, and 2080s, respectively, for RCP4.5 scenario and 0.34,

Table 4 Absolute change in mean seasonal minimum temperature (°C) at different time horizons from the base period

Scenario	Seasons in the 2020s			Seasons in the 2050s			Seasons in the 2050s		
	Dry	Intermediate	Wet	Dry	Intermediate	Wet	Dry	Intermediate	Wet
CanESM2-RCP2.6	+0.307	+0.371	+0.213	+0.431	+0.544	+0.303	+0.418	+0.467	+0.283
CanESM2-RCP4.5	+0.299	+0.390	+0.240	+0.488	+0.598	0.354	+0.567	+0.702	+0.431
CanESM2-RCP8.5	+0.366	+0.425	+0.238	+0.682	+0.809	+0.473	+1.090	+1.332	+0.702

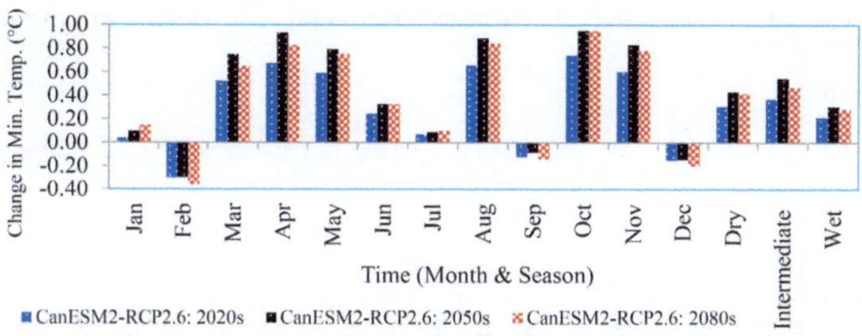

Fig. 8 Change in average monthly and seasonal minimum temperature in the future from the base period for CanESM2-RCP2.6 scenario

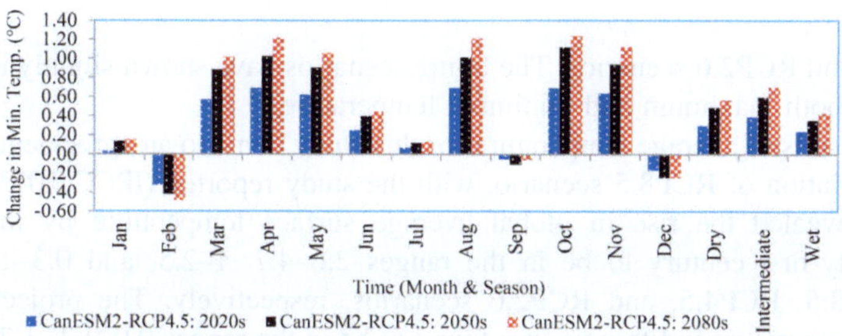

Fig. 9 Change in average monthly and seasonal minimum temperature in the future from the base period for CanESM2-RCP4.5 scenario

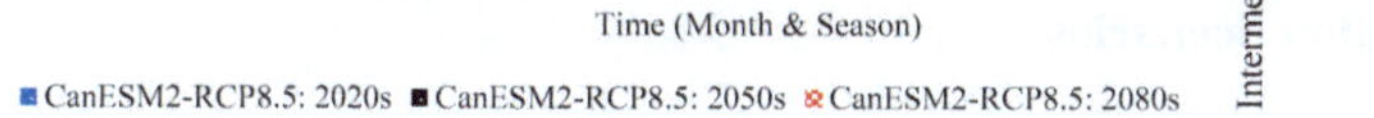

Fig. 10 Change in average monthly and seasonal minimum temperature in the future from the base period for CanESM2-RCP8.5 scenario

0.66, and 1.04 °C for 2020s, 2050s, and 2080s, respectively, for RCP8.5 scenario. The variability of minimum temperature is higher for RCP8.5 than RCP4.5 and RCP2.6, and the linear trend line for three scenarios has indicated a significantly (at 5% significant level) increasing trend of average annual minimum temperature until the end of the century (Table 6 and Fig. 11). Comparatively, RCP8.5 prevails higher change in minimum temperature trend at the end of the century than the

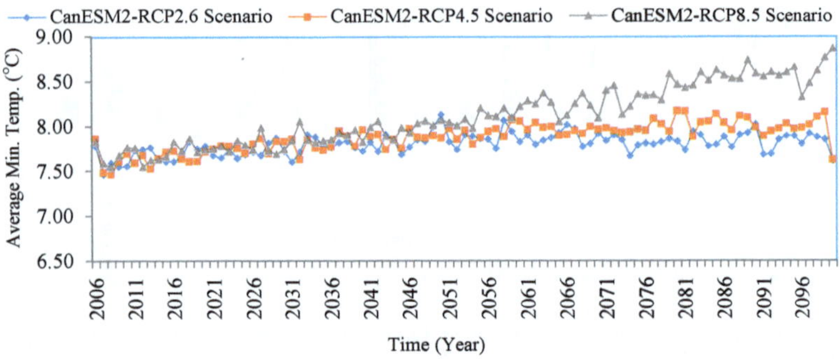

Fig. 11 Future pattern of average annual minimum temperature

RCP4.5 and RCP2.6 scenarios. The future scenarios have shown slightly increasing trend on both maximum and minimum temperature.

The results of average temperature for this study come to an agreement, with the slight variation of RCP8.5 scenario, with the study reported (IPCC 2013) and the results revealed the rise in global average surface temperature by the end of the twenty-first century to be in the ranges 2.6–4.7, 1–2.5, and 0.3–1.6 °C for the RCP8.5, RCP4.5, and RCP2.6 scenarios, respectively. The projected mean annual temperature in Ethiopia was found to be in the ranges 0.9–1.1, 1.7–2.1, and 2.7–3.4 °C by 2030, 2050, and 2080 time slices, respectively (Ethiopian National Meteorological Agency 2007). The projected mean annual maximum and minimum temperature show rising trend in southeastern part of Ethiopia (Shawul et al. 2016). The mean annual temperature of this study comes to an agreement with all the literature given above regarding direction (pattern), but with slight variation regarding magnitude (amount). This slight variation of mean annual temperature increment might arise due to the types of GCM/ESM and emission scenario used, a method of downscaling, and spatial variation of temperature.

Precipitation Scenarios

The projected mean seasonal precipitation scenarios have indicated an increase of precipitation in all the seasons (Dry, intermediate, and wet) in all the three future time slices for RCP2.6, RCP4.5, and RCP8.5 scenarios except for an intermediate period of the 2020s (a decreasing trend) for RCP2.6 and RCP4.5 scenarios. The percentage changes of precipitation from the base period for the three scenarios in future time horizon for each season are presented in Table 5.

The projected mean monthly precipitation has a larger magnitude of increment on the month of October 2080s 81.02, 54.66, and 42.77% for RCP8.5, RCP4.5, and RCP2.6 scenarios, respectively, at the end of the century. Conversely, the larger decrement on February 2020s 8.20 '(RCP8.5)', 10.25% '(RCP4.5)', and 10.90% '(RCP2.6)' was observed (Figs. 12, 13 and 14). The percentage change from the

Table 5 Percentage difference in mean seasonal precipitation at different time horizons from the base period

Scenario	Seasons in the 2020s			Seasons in the 2050s			Seasons in the 2050s		
	Dry	Intermediate	Wet	Dry	Intermediate	Wet	Dry	Intermediate	Wet
CanESM2-RCP2.6	+15.19	−1.35	+8.83	+24.11	+2.57	+11.44	+21.14	+1.02	+9.81
CanESM2-RCP4.5	+18.20	−0.32	+9.16	+26.63	+1.96	+13.50	+30.71	+4.55	+15.65
CanESM2-RCP8.5	+20.68	+1.37	+9.10	+33.65	+6.23	+18.85	+53.74	+18.90	+27.06

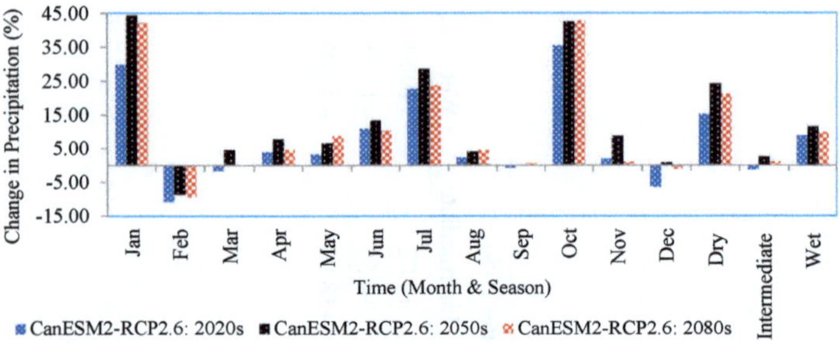

Fig. 12 Percentage change in average monthly and seasonal precipitation in the future from the base period for CanESM2-RCP2.6 scenario

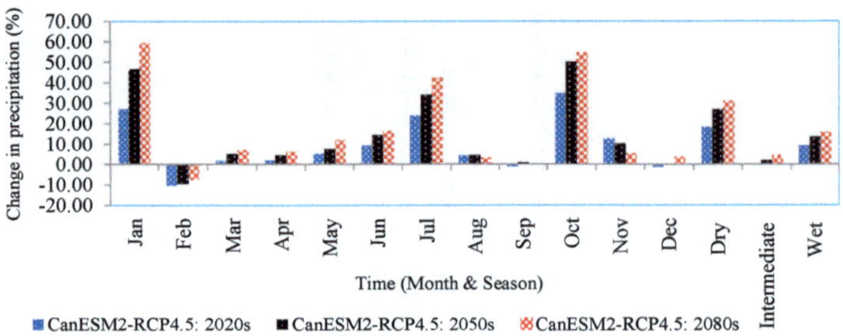

Fig. 13 Percentage change in average monthly and seasonal precipitation in the future from the base period for CanESM2-RCP4.5 scenario

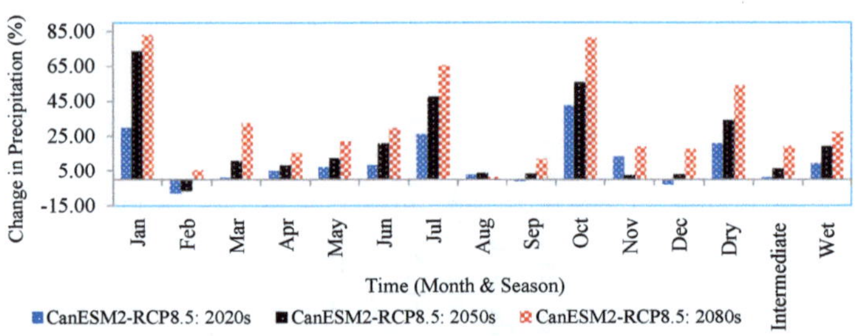

Fig. 14 Percentage change in average monthly and seasonal precipitation in the future from the base period for CanESM2-RCP8.5 scenario

base period in mean monthly precipitation was observed to be significant due to a substantial rise and decline of precipitation on different months. For instance, in the months of February, September, and December, the decrement of precipitation was observed and an increment in the rest of the months was shown.

Characteristically, the percentage change in average monthly precipitation might range between −8.20% on February and +42.59% on October for the coming

2020s; −6.50% on February and +55.61% on October for 2050s and +1.58% on August and +81.02% on October for 2080s for the RCP8.5 scenario. For RCP4.5 scenario, the range tends to be between −10.35% on February and +34.53% on October for the coming 2020s; −9.36% on February and +49.90 on October for 2050s and −7.42% on February and +54.66% on October for 2080s. For RCP2.6 scenario, percentage change in average monthly precipitation ranges between −10.80% on February and +35.41% on October for 2020s; −8.59% on February and +42.55% on October for 2050s and −9.37% on February and +42.77% on October for 2080s.

For each time slice, the percentage change in mean annual precipitation has indicated a considerable increment from the base period, 8.68, 12.93, and 11.34% for 2020s, 2050s, and 2080s, respectively, for RCP2.6 scenario, 9.54, 14.36, and 16.94% for 2020s, 2050s, and 2080s, respectively for RCP4.5 scenario and 14.70, 19.14, and 28.69% for 2020s, 2050s, and 2080s, respectively, for RCP8.5 scenario. The variability of precipitation is higher for RCP8.5 than RCP4.5 and RCP2.6, and the linear trend line for three scenarios has indicated a significantly (at 5% significant level) increasing trend of average annual total precipitation until the end of the century (Table 6 and Fig. 15). Comparatively, RCP8.5 prevails higher change in precipitation trend at the end of the century than the RCP4.5 and RCP2.6 scenarios.

Figure 15 indicated the pattern of future total mean annual precipitation with a range of 1124.00–1540.16 mm in the year 2028 and 2099, respectively, for RCP8.5 scenario, 1065.43–1359.59 mm in the year 2018 and 2064, respectively, for RCP4.5 scenario and 1100.74–1300.24 mm in the year 2096 and 2050, respectively, for RCP2.6 scenario. It has shown the substantial variability of total mean annual precipitation from year to year throughout simulation period. There has been observed substantial variability in rainfall (rises about 20% and also declined by about 20%) in the globe (Bates et al. 2008). Larger spatial variation of precipitation (from a reduction of 25–50% to an increase of 25–50%) has been reported in East Africa (Faramarzi et al. 2013). Increase in rainfall has been observed

Table 6 Mann–Kendall trend test for future average annual both temperatures and precipitation under three RCP scenarios

scenarios	Kendall's tau	*p*-value	Alpha	Sen's slope	Trend
Tmax for rcp2.6	0.492	<0.0001	0.05	0.117	Significantly increasing
Tmax for rcp4.5	0.521	<0.0001	0.05	0.117	Significantly increasing
Tmax for rcp8.5	0.634	<0.0001	0.05	0.211	Significantly increasing
Tmin for rcp2.6	0.196	<0.0001	0.05	0.106	Significantly increasing
Tmin for rcp4.5	0.256	<0.0001	0.05	0.109	Significantly increasing
Tmin for rcp8.5	0.435	<0.0001	0.05	0.118	Significantly increasing
pcp for rcp2.6	0.186	0.0003	0.05	0.378	Significantly increasing
pcp for rcp4.5	0.193	0.0005	0.05	0.546	Significantly increasing
pcp for rcp8.5	0.201	0.0009	0.05	0.607	Significantly increasing

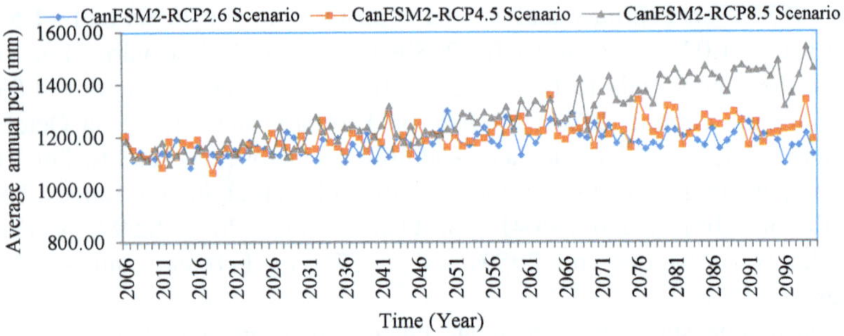

Fig. 15 Future pattern of average annual total precipitation

(Shongwe et al. 2009) in the tropics, which is also the case in this study. Generally, increment of rainfall in this study is comparatively higher in the dry season 20.68% in the 2020s, 33.65% in 2050s, and 53.74% in 2080s for RCP8.5 scenario which might have positive impact on pastoral region of the study area and it might affect the highland areas negatively since this season is expressly main crop harvesting period.

Mann–Kendall Trend Test of Future Temperatures and Precipitation

Based on the standardized test statistic, it is possible to infer that Mann–Kendall test has revealed a statistically significant trend in the study area for both future precipitation and temperatures at the 5% significant level. The maximum and minimum temperature and precipitation for RCP2.6, 4.5 and 8.5 scenarios have revealed a significantly (at 5% significant level) increasing trend for future until the year 2100 as shown in Table 6.

Summary and Conclusion

This study tried to downscale daily temperatures and precipitation data from the CMIP5-CanESM2 output for the RCP8.5, RCP4.5, and RCP2.6 emission scenarios for future periods until year 21000 in the study area of Weyib River basin. The SDSM was used to generate future possible local Tmax, Tmin, and precipitation in the study area and it has a good ability to replicate the baseline Tmax, Tmin, and precipitation for the baseline period.

For each time slice, the change in mean annual maximum temperature has indicated a slight increment from the base period, 0.16, 0.14, and 0.15 °C for 2020s, 2050s, and 2080s, respectively, for RCP2.6 scenario, 0.12, 0.19, and

0.21 °C for 2020s, 2050s, and 2080s, respectively, for RCP4.5 scenario, and 0.12, 0.22, and 0.33 °C for 2020s, 2050s, and 2080s, respectively, for RCP8.5 scenario. For mean annual minimum temperature, the increment from the base period has been found to be 0.30, 0.43, and 0.39 °C for 2020s, 2050s, and 2080s, respectively, for RCP2.6 scenario 0.31, 0.48, and 0.57 °C for 2020s, 2050s, and 2080s, respectively, for RCP4.5 scenario and 0.34, 0.66, and 1.04 °C for 2020s, 2050s, and 2080s, respectively, for RCP8.5 scenario. The percentage change in mean annual precipitation has indicated a considerable increment from the base period 8.68, 12.93, and 11.34% for 2020s, 2050s, and 2080s, respectively, for RCP2.6 scenario, 9.54, 14.36, and 16.94% for 2020s, 2050s, and 2080s, respectively, for RCP4.5 scenario and 14.70, 19.14, and 28.69% for 2020s, 2050s, and 2080s respectively for RCP8.5 scenario.

The variability of both temperatures (maximum and minimum) and precipitation is higher for RCP8.5 than RCP4.5 and RCP2.6, and the linear trend line for all the three scenarios has indicated a significantly (at 5% significant level) increasing trend of both temperatures and precipitation for future until the year 2100. Comparatively, RCP8.5 prevails higher change in both temperatures and precipitation trend at the end of the century than the RCP4.5 and RCP2.6 scenarios. The increment of rainfall in the study area is comparatively higher in the dry season 20.68% in the 2020s, 33.65% in 2050s, and 53.74% in 2080s for RCP8.5 which might have positive impact on pastoral region of the study area and it might affect the highland areas negatively since this season is specifically main crop harvesting period.

Acknowledgements We express our heartfelt gratitude to National Meteorological Service Agency of Ethiopia (NMSAE) for providing us the meteorological data to be considered for this study.

References

Bates B, Kundzewicz Z, Wu S (2008) Climate change and water. Technical paper of the Intergovernmental Panel on Climate Change, IPCC Secre (Geneva), 210. http://www.citeulike.org/group/14742/article/8861411

Beck L, Bernauer T (2011) How will combined changes in water demand and climate affect water availability in the Zambezi river basin? Glob Environ Change 21(3):1061–1072

Ethiopian National Meteorological Agency (2007) Climate Change National Adaptation Programme of Action (Napa) of Ethiopia (June). pp 1–73

Faramarzi M, Abbaspour KC, Ashraf Vaghefi S, Farzaneh MR, Zehnder AJB, Srinivasan R, Yang H (2013) Modeling impacts of climate change on freshwater availability in Africa. J Hydrol 480:85–101. http://dx.doi.org/10.1016/j.jhydrol.2012.12.016

Ghosh S, Mujumdar PP (2008) Statistical downscaling of GCM simulations to streamflow using relevance vector machine. Adv Water Resour 31(1):132–146

Graham LP, Andersson L, Horan M, Kunz R, Lumsden T, Schulze R, Warburton M, Wilk J, Yang W (2011) Using multiple climate projections for assessing hydrological response to climate change in the Thukela River Basin, South Africa. Phys Chem Earth 36(14–15):727–735. http://dx.doi.org/10.1016/j.pce.2011.07.084

Hamed KH (2008) Trend detection in hydrologic data: the Mann-Kendall trend test under the scaling hypothesis. J Hydrol 349(3):350–363

IPCC (2013) Climate change 2013: the physical science basis. Contribution of working group I to the fifth assessment report of the Intergovernmental Panel on Climate Change. Intergovernmental Panel on Climate Change, working group i contribution to the IPCC fifth assessment report (AR5).Cambridge Univ Press, New York, p 1535

Karpouzos D, Kavalieratou S, Babajimopoulos C (2010) Trend analysis of precipitation data in Pieria Region (Greece). Eur Water 30(May):31–40

Krause P, Boyle DP (2005) Advances in geosciences comparison of different efficiency criteria for hydrological model assessment. Adv Geosci 5(89):89–97. http://www.adv-geosci.net/5/89/2005/

Kruger AC, Shongwe S (2004) Temperature trends in South Africa: 1960–2003. Int J Climatol 24 (15):1929–1945

Liu L, Liu Z, Ren X, Fischer T, Xu Y (2011) Hydrological impacts of climate change in the Yellow River Basin for the 21st century using hydrological model and statistical downscaling model. Quater Int 244(2):211–220. http://dx.doi.org/10.1016/j.quaint.2010.12.001

Nash JE, Sutcliff JV (1970) River flow forecasting through conceptual models, part I—a discussion of principles. J Hydrol 10:282–290

New M, Hewitson B, Stephenson DB, Tsiga A, Kruger A, Manhique A, Gomez B, Coelho CAS, Masisi DN, Kululanga E, Mbambalala E, Adesina F, Saleh H, Kanyanga J, Adosi J, Bulane L, Fortunata L, Mdoka ML, Lajoie R (2006) Evidence of trends in daily climate extremes over southern and west Africa. J Geophys Res Atmos 111(14):1–11

Shawul AA, Alamirew T, Melesse AM, Chakma S (2016) Climate change impact on the hydrology of Weyb River watershed, Bale mountainous area, Ethiopia. In: Landscape dynamics, soils and hydrological processes in varied climates. Springer, Berlin, pp 587–613

Shongwe ME, Van Oldenborgh GJ, Van Den Hurk BJJM, De Boer B, Coelho CAS, Van Aalst MK (2009) Projected changes in mean and extreme precipitation in Africa under global warming. Part I: Southern Africa. J Clim 22(13):3819–3837

Singh J, Knapp HV, Arnold JG, Demissie M (2004) Hydrologic modeling of the Iroquois river watershed using HSPF and SWAT. J Am Water Resour Assoc 41(2):343–360

Unganai LS (1996) Historic and future climatic change in Zimbabwe. Climate Res 6(2):137–145

Wilby RL, Dawson CW (2007) SDSM 4.2-A decision support tool for the assessment of regional climate change impacts, Version 4.2 user manual. Lancaster University: Lancaster/Environment Agency of England and Wales (August), pp 1–94

Wilby RL, Dawson CW, Barrow EM (2002) SDSM-a decision support tool for the assessment of regional climate change impacts. Environ Model Softw 17(2):145–157. http://linkinghub.elsevier.com/retrieve/pii/S1364815201000603

Global Climate Pattern Behind Hydrological Extremes in Central India

Kironmala Chanda and Rajib Maity

Abstract The concurrent influence of large-scale, coupled oceanic–atmospheric circulation patterns was established to have an effect on hydrologic variability across the world. El Niño–Southern Oscillation (ENSO) and Indian Ocean Dipole (IOD) are, in particular, important for Indian hydroclimatology. However, it is now established that rather than just a few well-known teleconnection patterns, a Global Climate Pattern (GCP) comprising of a global field of several climate anomalies are responsible for above-normal and below-normal precipitation events over entire India. The existence of a GCP for hydrological extremes in an even smaller spatial scale is illustrated in this study. The central part of India, consisting of the contiguous homogeneous meteorological subdivisions—West Madhya Pradesh, East Madhya Pradesh, Vidarbha, and Chattisgarh (hereinafter 'central India'), is selected as the study area. Hydrological extremes (this study focus on precipitation) in the study area are identified in terms of the Standardized Precipitation Anomaly Index (SPAI), which is suitable for quantifying extreme events in a monsoon-dominated climatology. After investigation of the global anomaly fields of five climate variables, a set of 19 specific zones of climate anomalies from across the world are found to constitute the GCP for the hydrological extremes in the study region. The identified GCP is further utilized in a Support Vector Machine (SVM) model to investigate the potential of the GCP in foreseeing dry and wet extremes over the study area.

K. Chanda
Indian Institute of Technology (Indian School of Mines), Dhanbad 826004, Jharkhand, India
e-mail: kironmala.iitkgp@gmail.com

R. Maity (✉)
Department of Civil Engineering, Indian Institute of Technology Kharagpur, Kharagpur 721302, West Bengal, India
e-mail: rajib@civil.iitkgp.ernet.in; rajibmaity@gmail.com

V.P. Singh et al. (eds.), *Climate Change Impacts*, Water Science and Technology Library 82, https://doi.org/10.1007/978-981-10-5714-4_6

Introduction

The association of large-scale atmospheric–oceanic circulation patterns and hydrologic variables across the world has been established through several studies. Recent studies have confirmed that asymmetry in the response of rainfall anomalies in different parts of the world result from opposite phases of low variability oceanic circulation patterns (King et al. 2013; Qiu et al. 2014). For example, observed changes in the frequency and intensity of precipitation extremes in Europe are now largely explained by the persistence in atmospheric circulation patterns over the North Atlantic (Willems 2013). Changes in large-scale circulation patterns are found to be responsible for the observed long-term warming and drying in central Europe (Philipp et al. 2007). For the past two decades, the role of specific oceanic–atmospheric circulation phenomenon such as, El Niño–Southern Oscillation (ENSO), Indian Ocean Dipole (IOD), Equatorial Indian Ocean Oscillation (EQUINOO), Pacific Decadal Oscillation (PDO), Atlantic Multi-decadal Oscillation (AMO), North Atlantic Oscillation (NAO) in triggering and enhancing droughts and floods on a continental scale have been the focus of research (Chiew and McMahon 2002; Terray et al. 2003; Gadgil et al. 2004; Goswami et al. 2006; Maity and Nagesh Kumar 2006, 2008; Feng and Hu 2008; Li et al. 2008; Mo and Schemm 2008; Ting et al. 2011; Singhrattna et al. 2012; Oubeidillah et al. 2012; Jiang et al. 2013; Rogers 2013; Wang et al. 2013). Some of the recent advances establish the influence of ENSO and AMO in relation to the variability of China's summer precipitation (Gu et al. 2009; Ye 2014) as well as the frequency of its extreme precipitation events (Fu et al. 2013). The effect of NAO is found to affect winter precipitation, river flow, and temperature in the Mediterranean region (Brandimarte et al. 2011).

Some of the most frequently researched phenomena are ENSO and IOD, primarily because they influence rainfall anomalies in a large number of countries across the world. For instance, the unusual warming of the central and eastern tropical Pacific Ocean during the El Niño events is known to be responsible for below-normal precipitation in Indonesia and the surrounding Pacific islands and above-normal precipitation in the western coast of South America. The role of ENSO in biennial relationship of rainfall variability between Central and equatorial South America was also recently identified (Wu and Zhang 2010). In the Indian context, the occurrence of an El Niño event in the Pacific generally indicates that rainfall deficiencies are in the offing—dry and drought conditions may be expected due to poor Indian Summer Monsoon Rainfall (ISMR). However, the relationship between circulation patterns and the occurrence of anomalous continental scale hydrologic behavior is often very complicated. This is due to the possible involvement of a number of factors, some more rare than others, which lead to the optimum conditions for the development of an extreme hydrologic event. Thus, in a recent study, a direct relationship (as opposed to an inverse relationship) of ENSO with the rainfall and streamflow series in Mahanadi river basin of south India was observed (Panda et al. 2013). Again, some studies postulate that the ENSO-ISMR relationship has weakened over the years (Viswambharan and Mohanakumar 2014). Apart from ENSO, the other most significant circulation pattern affecting ISMR is the IOD (Saji et al. 1999; Webster et al. 1999). A positive IOD event, which is accompanied by high SST over the western Indian Ocean, is known to affect ISMR positively with abundant rainfall in the Indian subcontinent and dry and drought conditions in Australia and Indonesia.

Existing literature indicates that most of the previous studies had investigated the role of specific large-scale Oceanic–Atmospheric Circulation Patterns (OACPs) in causing extreme hydrologic events such as droughts and floods. However, it is now established that apart from the well-known teleconnection patterns such as ENSO, IOD, etc., the concurrent effect of global anomaly fields of several climate variables influences hydrologic events in a regional scale (Chanda and Maity 2016). Considering the entire Indian landmass (also referred as all-India) as the test bed, it was demonstrated that a distinct Global Climate Pattern (GCP) consisting of 15 climate anomaly zones is responsible for the occurrence of dry and wet events. The potential of the GCP in predicting dry and wet events on an all-India scale is also established. In fact, the GCP is found to be more useful as precursor of hydrologic extremes in India compared to the most commonly used hydroclimatic teleconnection patterns in the Indian context. The objective of this study is to explore the existence of a distinct GCP for hydrologic events on a smaller spatial scale. Central India is selected as the target area to explore the association of regional dry/wet events with global anomaly fields of five climate variables—sea surface temperature, surface pressure, air temperature, wind speed, and total precipitable water. The specific GCP for central India, once identified, is utilized as an input to a prediction model for categorizing hydrologic events into dry, normal, and wet. Based on experience from the previous study, a temporal scale of three months is adopted for this study as the climate anomaly zones are found to be sufficiently well-defined at this scale for identification of GCP.

Study Area and Data

The India Meteorological Department (IMD) divides India into 36 homogenous meteorological subdivisions. Out of these, four contiguous subdivisions, namely, West Madhya Pradesh, East Madhya Pradesh, Vidarbha, and Chattisgarh are considered together as the study area (Fig. 1). The study area is referred as 'central India'. The monthly precipitation data of the aforementioned four subdivisions are obtained from IMD for the period 1959–2010. The datasets are downloaded from the website of Indian Institute of Tropical Meteorology (IITM) (ftp://www.tropmet.res.in/pub/data/rain/iitm–regionrf.txt). The method of development of the dataset along with the information of the raingauge distribution may be found in Parthasarathy et al. (1995), Rajeevan et al. (2006). The climate variables used for this study are the global fields of Sea Surface Temperature (SST), Surface Pressure (SP), Air Temperature (AT), Wind Speed (WS), and Total Precipitable Water

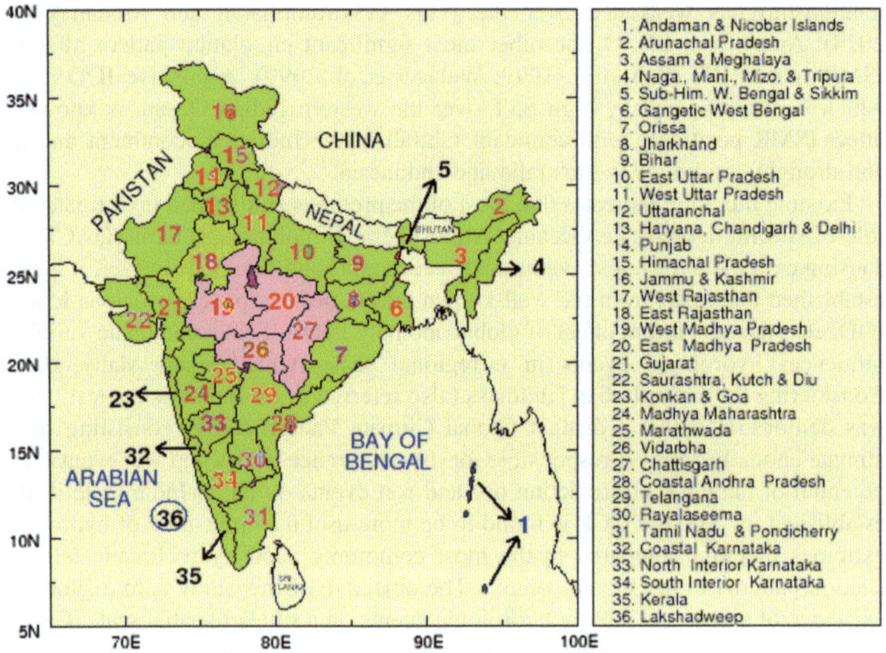

Fig. 1 Study area consisting of the four contiguous homogeneous meteorological Sub divisions of India—West Madhya Pradesh, East Madhya Pradesh, Vidarbha, and Chattisgarh (modified from map provided by Indian Institute of Tropical Meteorology, Pune, web address: www.tropmet.res.in

(TPW). Monthly global gridded datasets of these variables are obtained from National Oceanic and Atmospheric Administration (NOAA) (http://www.esrl.noaa.gov/psd/data/gridded/data.ncep.reanalysis.surface.html) for the period 1958–2010. The spatial resolution of AT, SP, WS and TPW data is 2.5° lat × 2.5° lon and that of SST data is of 2° lat × 2° lon.

Methodology

Quantification of Dry and Wet Events Through Standardized Precipitation Anomaly Index (SPAI)

Since the study area (i.e., central India) encompasses of the four aforementioned meteorological subdivisions, the monthly rainfall of these four meteorological subdivisions is averaged to get the monthly time series rainfall over the study area. In order to identify the GCP for dry and wet extremes, the time series of Standardized Precipitation Anomaly Index (SPAI) is computed from the obtained precipitation time series. Details of the anomaly-based SPAI can be found in Chanda and Maity (2015). The SPAI is established to be a generalized index that is suitable for the characterization of meteorological droughts in monsoon-dominated climatology such as India (Chanda and Maity 2015). A temporal scale of three months is used for SPAI computation and the period 1961–1990 is used to represent the long-term climatology, based on which the rainfall anomalies are calculated. Following the guideline of the U.S. Drought Monitor regarding the threshold value of Standardized Precipitation Index (SPI) indicating drought, the criteria adopted for this study are: SPAI–3 values less than −0.8 are designated as *dry* events, those greater than 0.8 are designated as *wet* events and those in between are designated as *normal* events.

Identification of the Global Climate Pattern (GCP) for Central India

Since a temporal scale of 3 months is used for identifying the dry and wet events, the global climate anomaly fields are also considered at a temporal scale of three months. For each dry event (i.e., SPAI-3 < −0.8), the climate anomaly field at the preceding 3-month period is considered. For instance, for a dry event comprising the months April–May–June, the global climate anomaly field is obtained from the period January–February–March. For each of the climate variables, the global anomaly field corresponding to all observed dry events during the period 1959–2000 is obtained and they are averaged event-wise to get the mean global gridded climate anomaly field for dry events (Chanda and Maity 2016). A similar procedure is followed to get the mean global gridded climate anomaly field for wet events for each climate variable. For a given climate variable, the grid-wise difference of anomalies between dry and wet events is computed and maps showing the anomaly differences are plotted. On inspection of these maps, it is found that contrasting (above-normal/below-normal) features of climate anomalies are revealed during dry and wet events at the target location. A particular zone on the globe consisting of opposite anomalies of a climate variable corresponding to dry and wet events at the study area is considered as one of the variables constituting the GCP. All such variables, together forming the GCP, are used as input to a prediction model for categorizing dry and wet events.

Utilization of the Identified GCP for Prediction of Dry and Wet Events in Central India

The potential of the GCP in prediction of dry and wet events in India has been recently established (Chanda and Maity 2016). In this study, a smaller target area, i.e., central India has been selected and the GCP responsible for dry and wet extremes in this region is investigated. The potential of the identified GCP for prediction of hydrologic extremes in central India is assessed through the following steps.

Reduction of Dimensionality of GCP

Since the curse of dimensionality of inputs could affect the prediction process, it is wise to reduce the large number of variables constituting the GCP. However, the information contained in the identified inputs must not be lost in the process. Hence, principal component analysis (PCA) (Jolliffe 1986) is used to orthogonally transform the dataset from a number of observed correlated variables to a number of uncorrelated components which explain the variance of the target variable in a gradually decreasing order. The number of principal components considered should be such that it should be large enough to substantially explain the variability. However, it should not be too large so as to hamper the SVM training owing to high dimensionality.

Model for Classification of Dry and Wet Events

Once a number of principal components of the GCP are identified as inputs, the next step is to devise a prediction model that can classify the events into different categories. The three categories that are considered in this study are *dry* (SPAI < -0.8), *normal* ($-0.8 \leq$ SPAI ≤ 0.8) and *wet* (SPAI > 0.8), respectively. It is true that sometimes more number of categories, indicating different levels of severities of dry and wet events, are of interest. However, any finer categorization is avoided here since the observed number of events in each category would then become too less to train the prediction model as well as to evaluate the prediction performance. Support Vector Machines (SVM) are one of the machine learning techniques that classify data points using a hypothesis space of linear functions in a high-dimensional feature space. It maps the input space to a higher dimensional feature space and selects a hyperplane to attain maximum separation between the different classes. SVMs have been successfully used in hydrological applications (Bray and Han 2004; Qin et al. 2005; She and Basketfield 2005; Tripathi et al. 2006; Lin et al. 2006; Anirudh and Umesh 2007; Kişi and Çimen 2009; Chen et al. 2010; Maity et al. 2010; Samsudin et al. 2011; Bhagwat and Maity 2012; Zakaria and Shabri 2012; Raghavendra and Deka 2014). SVM-based models may be suitably used in classifying dry, normal and wet events based on the selected components (Chanda and Maity 2016).

Following Chanda and Maity (2016), two SVM models (named SVM-I and SVM-II) are used simultaneously to process the inputs, i.e., the selected principal components. SVM-I performs classification into two categories—*dry* (SPAI < −0.8) and *not dry* (SPAI ≥ −0.8), while SVM-II performs classification into two categories—*not wet* (SPAI ≤ 0.8) and *wet* (SPAI > 0.8). After training the two SVM models during the model development period (1959–2000), they are used for classification for both model development (1959–2000) and testing period (2001–2010). At any given time step, the output from the two SVM models is logically joined to obtain the final output. When the output of SVM-I is dry and that of SVM-II is not wet, then the event is categorized as dry. When the output of SVM-I is not dry and that of SVM-II is wet, then the event is categorized as wet. When the outputs of SVM-I and SVM-II are not dry and not wet, respectively, the event is categorized as normal. If the outputs of the two SVM models are contradictory, i.e., SVM-I classifies the event as dry and SVM-II classifies it as wet, then the model fails to categorize the event. However, these events are also categorized into normal to prevent loss of data during evaluation of prediction performance, which is a little deviation from Chanda and Maity (2016). The categorization procedure is illustrated with the help of a flowchart (Fig. 2). It may be noted that three-way classification (say, groups A, B, and C) through SVMs is also possible. However, in such cases, two steps need to be followed. First, the classification has to be performed between 'Group A' versus 'Group B and Group C'. In the next step, the candidates falling in the second category may be classified further into 'Group B' and 'Group C'. For such classification, training the SVM becomes computationally too intensive. Thus, in this study, two separate SVM models for bi-category classification are used simultaneously and their outputs are logically joined to obtain a three-way classification.

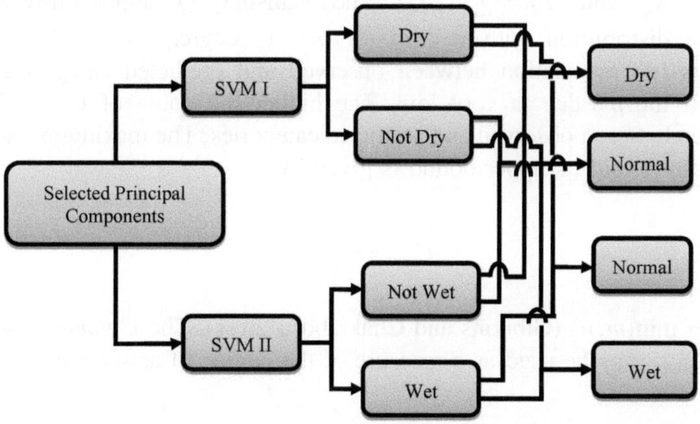

Fig. 2 Flowchart showing logical combination of the outputs of two SVMs to obtain final categorization of hydrological events

Evaluation of the Prediction Performance Using GCP as Input

After obtaining the final output obtained from the logical combination of the two SVM models, the prediction performance may be evaluated by constructing a contingency table for both the development and the testing period. The potential of classification of dry, normal, and wet events in the target area using the GCP as input may be assessed by inspecting the number of events (in the three-way contingency table) which are categorized correctly. Quantitatively, the model performance may be assessed in terms of Contingency Coefficient (*C*) (Pearson 1904), which is used to measure the degree of association in a contingency table for *N* samples (Gibbons and Chakraborti 2011). It is expressed as

$$C = \sqrt{\frac{Q}{Q+N}}, \tag{1}$$

where Q is a statistic that tests the null hypothesis that there is no association between the observed and predicted categories. Q is expressed as

$$Q = \sum_{i=1}^{m}\sum_{j=1}^{n}\frac{\left(NX_{ij} - X_{i.}Y_{.j}\right)^2}{NX_{i.}Y_{.j}}, \tag{2}$$

where X_{ij} is the number of cases falling in *i*th observed and *j*th predicted category, m and n are the number of observed and predicted categories respectively, and $X_{i.} = \sum_{j=1}^{n} X_{ij}$ and $Y_{.j} = \sum_{i=1}^{m} Y_{ij}$. The statistics Q approximately follows chi-square distribution with $v = (m-1)(n-1)$ degrees of freedom. The null hypothesis (no association between observed and predicted categories) may be rejected if the *p*-value is very low. The higher the value of C, the better the association between observed and predicted categories. The maximum value of C is theoretically 1, but its upper bound is given by

$$C_{\max} = \sqrt{\frac{t-1}{t}}, \tag{3}$$

where $t = \min(m, n)$ (Gibbons and Chakraborti 2011). The C value as well as the ratio $C/C_{\max}$ may be used as a measure of the degree of association (Maity et al. 2013).

Results and Discussions

Identification of the Global Climate Pattern for Central India

The SPAI values computed from the trimonthly rainfall series of central India are used to categorize each time step in the development (1959–2000) and testing period (2001–2010) as dry, normal and wet event. During the development period, the number of dry, normal and wet events are found to be 90, 308, and 106 respectively. During the testing period, the same is found to be 27, 72, and 21 respectively.

As mentioned earlier, for each of the climate variables (SST, SP, AT, WS, TPW), the global anomaly field corresponding to all observed dry events during the period 1959–2000 is obtained and they are averaged event-wise to get the mean global gridded climate anomaly field for dry events. Following a similar procedure, the mean global gridded climate anomaly field for wet events is also obtained for each variable. The grid-wise anomaly difference maps are subsequently investigated for identifying the zones with contrasting anomaly features during dry and wet events. For many grid locations, the difference in anomalies is found to be statistically significant at 99% confidence level. From the large contiguous zones of statistically significant anomaly differences, the core areas are selected as constituent variables forming the GCP. In all, the GCP for central India is characterized by 19 globally distributed zones from five climate variables. The spatial location and extent of the zones are discussed in comparison to those identified in case of 'all-India' analysis (Chanda and Maity 2016) in the following subsections.

Global Fields of Sea Surface Temperature (SST)

Figure 3a reproduces the SST anomaly difference map for 'all-India' analysis from Chanda and Maity (2016). For dry and wet extremes in central India, which is the target area for this study, the global SST anomaly difference map is shown in Fig. 3b.

The large positive anomaly zone (5°N–5°S and 100°W–140°W) in the equatorial Pacific Ocean is evident in the case of all-India analysis as well as 'central India' analysis. However, for central India, this zone is not as strong as in the case of the SST patterns for all-India. The lessening of this zone in extent as well as magnitude may be indicative of the fact that the effect of El Niño on dry and wet extremes in central India are weaker than the same on dry and wet extremes occurring on all-India scale. In both Fig. 3a, b, the positive anomaly differences in northern Pacific Ocean (40°N–48°N and 150°W–165°W) are even stronger than those in the equatorial region. The two negative anomaly zones in the Pacific (20°S–26°S and 160°E–170°E; 40°S–50°S, 114°W–124°W) are found to be potent in both Fig. 3a, b. However, the negative anomaly region (26°N–34°N, 136°E–144°E) along the

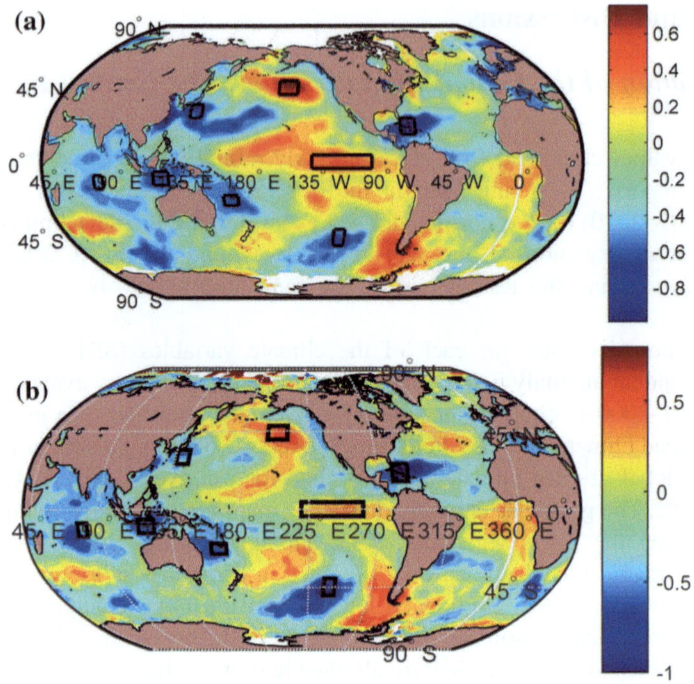

Fig. 3 Differences in mean SST anomalies during dry events (SPAI-3 < −0.8) and wet events (SPAI-3 > 0.8) over **a** India (Chanda and Maity 2016) and **b** central India

coast of Japan is found to be relatively less well defined in case of central India (Fig. 3b). In general, it is observed that warm anomaly pockets in the eastern part of Pacific Ocean and cold anomaly pockets in the western part of Pacific Ocean are associated with dry events in central India as well as 'all-India'.

As in case of the SST over Pacific Ocean, the negative anomaly regions in sub-equatorial Indian Ocean (8°S–16°S and 74°E–80°E) and to the west of Australia (6°S–14°S and 114°E–124°E) are similar in extent and magnitude in Fig. 3a, b. The strong negative anomaly region (16°N–26°N and 70°W–80°W) between the North and South Americas is also equally well defined in both the figures.

Thus, it may be concluded that the global SST zones responsible for dry and wet extremes in central India are identical to those of all-India. Hence, 8 SST zones are selected as constituents of the GCP for central India.

Global Fields of Surface Pressure (SP)

Figure 4a reproduces the SP anomaly difference map for 'all-India' analysis from Chanda and Maity (2016). For dry and wet extremes in central India, the global SP anomaly difference map is shown in Fig. 4b. The positive anomaly zone in northern Pacific (55°N–65°N and 145°W–160°W) as well as the negative anomaly zone in tropical Pacific (15°N–30°N, 145°W–160°W) is found to be very well defined and strong for both 'all-India' and central India. In addition to these two regions in Pacific Ocean, a very prominent positive anomaly zone in the western part of equatorial Pacific (5°S–5°N, 170°E–210°E) is observed in case of central India. The whole of Arabian Sea, Indian Ocean, and Bay of Bengal exhibit positive anomalies in both Fig. 4a, b. The signature of this region is represented through the zone 10°N–20°N, 55°E–65°E. The mild negative anomaly region in the Atlantic is also similar in extent and magnitude in both the figures. Thus, a total of 5 SP zones is selected as constituents of the GCP for central India.

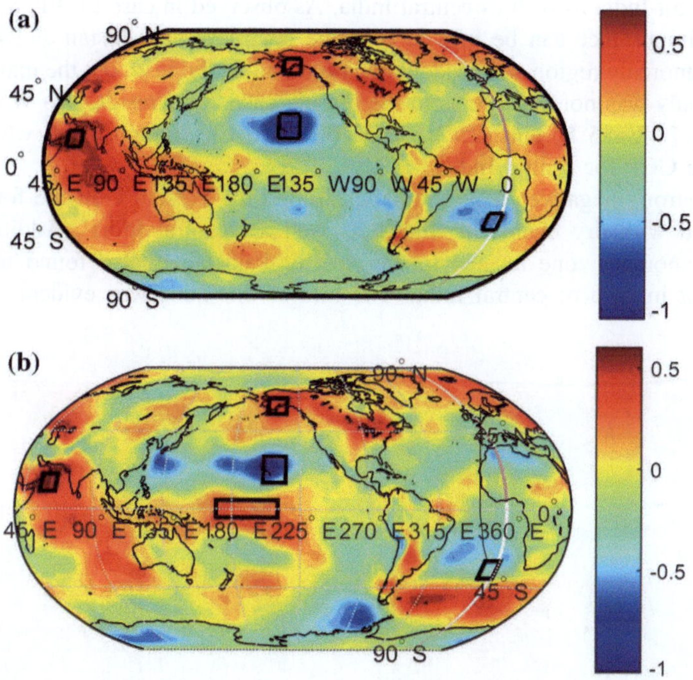

Fig. 4 Differences in mean SP anomalies during dry and wet events in **a** India (Chanda and Maity 2016) and **b** central India

Global Fields of Air Temperature (AT), Wind Speed (WS), and Total Precipitable Water (TPW)

Since the variables AT, WS, and TPW generally influence convective activity on a smaller spatial scale, the anomaly fields of these variables are investigated around the Indian subcontinent region. The patterns of anomaly differences of AT during dry and wet events in case of 'all-India' and in case of 'central India' are shown in Fig. 5a, b, respectively. Similar comparative figures for WS and TPW are shown in Figs. 6a, b and 7a, b, respectively. In case of AT, it is observed that a positive anomaly region at the nook of the Bay of Bengal is associated with dry events for all-India as well as for central India. For central India, a negative anomaly zone below the landmass of Pakistan and Iran is also found to be very much prominent in extent and magnitude, much more than that observed in case of the all-India study. Hence, both the AT zones—(20°N–30°N and 85°E–95°E) and (20°N–25°N and 60°E–65°E) are considered in the pool of GCP for central India.

Positive WS anomalies in the Indian Ocean are found to be associated with dry events in all-India as well as central India. As observed in case of AT, here also, a zone of importance can be located over the landmass of Pakistan and Iran. This positive anomaly region was evident in case of 'all-India' also, but the magnitude of the anomaly was not as large as in the present case. As a result, two WS zones—(0°–5°N, 70°E–85°E) and (25°N–30°N, 60°E–70°E) are considered while developing the GCP for central India.

Very strong negative TPW anomalies around the Persian Gulf are found to be associated with dry events in 'India' as well as 'central India'. Additionally, a positive anomaly zone located to the west of Pakistan is also found to be very prominent in case of central India. The hint of this zone was evident in case of

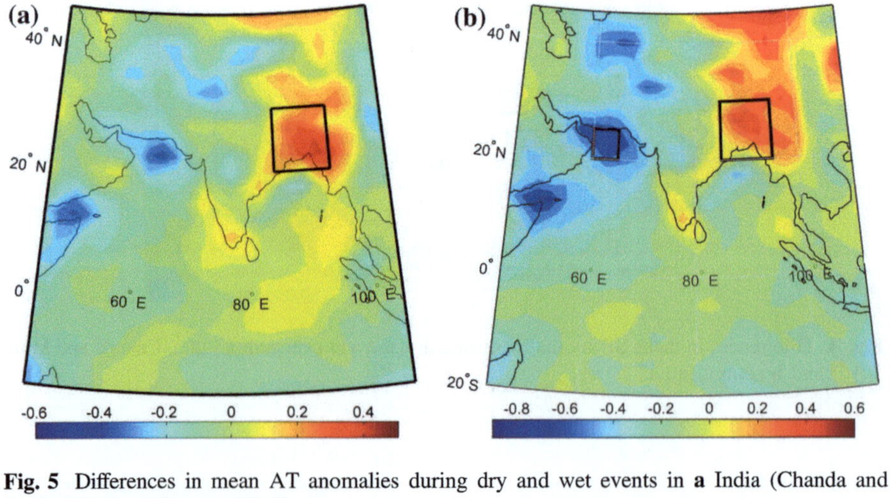

Fig. 5 Differences in mean AT anomalies during dry and wet events in **a** India (Chanda and Maity 2016) and **b** central India

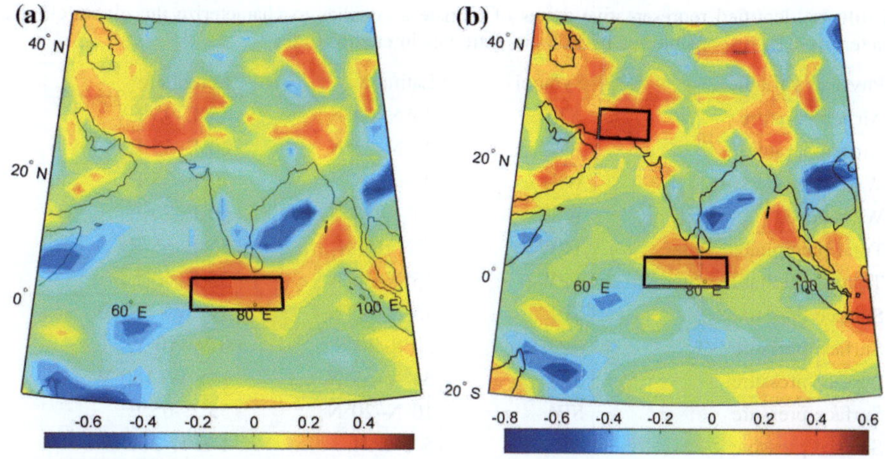

Fig. 6 Differences in mean WS anomalies during dry and wet events in **a** India (Chanda and Maity 2016) and **b** central India

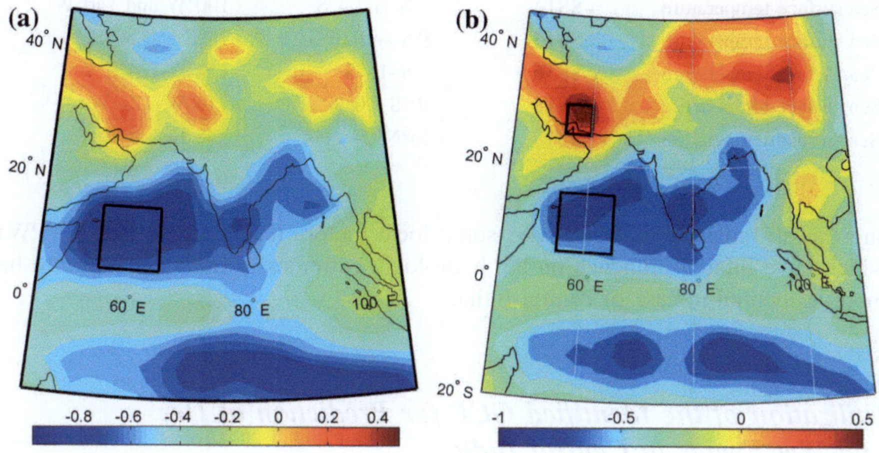

Fig. 7 Differences in mean TPW anomalies during dry and wet events in **a** India (Chanda and Maity 2016) and **b** central India

all-India study also, but was not as well defined. Thus, TPW zones—(5°N–15°N and 55°E–65°E) and (25°N–30°N and 55°E–60°E) are considered for central India.

Thus, in all, a total of 19 variables, each being denoted by a specific climate anomaly from a distinct part of the globe, together constitutes the GCP for dry and wet events in central India. The extent of these zones is specifically mentioned in Table 1. It is observed that many of the variables are equally important factors affecting hydrologic extremes in the 'central India' region as well as on an 'all-India' scale. However, it is noted that often the extent of the anomaly zones as well as their magnitudes differ in case of the present study concerning central India

Table 1 Identified representative zones of climate anomalies to characterize the global climate pattern (GCP) responsible for hydrologic extremes in central India

Physical variable	Symbol	Latitude	Longitude
Air temperature	AT1	20°N–30°N	85°E–95°E
Air temperature	AT2	20°N–25°N	60°E–65°E
Wind speed	WS1	0°N–5°N	70°E–85°E
Wind speed	WS2	25°N–30°N	60°E–70°E
Total precipitable water	TPW1	5°N–15°N	55°E–65°E
Total precipitable water	TPW2	25°N–30°N	55°E–60°E
Surface pressure	SP1	15°N–30°N	145°W–160°W
Surface pressure	SP2	30°S–40°S	0°W–10°W
Surface pressure	SP3	55°N–65°N	145°W–160°W
Surface pressure	SP4	10°N–20°N	55°E–65°E
Surface pressure	SP5	5°S–5°N	170°E–210°E
Sea surface temperature	SST1	40°N–48°N	150°W and 164°W
Sea surface temperature	SST2	16°N–26°N	70°W–80°W
Sea surface temperature	SST3	20°S–26°S	160°E–170°E
Sea surface temperature	SST4	4°N and 4°S	100°W and 140°W
Sea surface temperature	SST5	8°S and 16°S	74°E and 80°E
Sea surface temperature	SST6	6°S–14°S	114°E–124°E
Sea surface temperature	SST7	40°S–50°S	116°W–124°W
Sea surface temperature	SST8	26°N–34°N	136°E–144°E

and the all-India study. Moreover, some local factors (such AT, WS and TPW) which have limited influence on the hydrologic extremes of India are found to be more influential in case of central India.

Utilization of the Identified GCP for Prediction of Dry and Wet Events in Central India

As mentioned earlier, the number of principal components selected should substantially explain the variability and also alleviate the curse of dimensionality. It is found that the first seven principal components together explain about 72.5% of the variability. Hence, the seven selected principal components of the 19-dimensional GCP for the model development period are used as inputs to train the two SVM models. The models are then used for classification of the events at each time step (both development and testing period) into dry, normal, or wet category. The prediction performance is subsequently assessed by inspecting the contingency table (Table 2). During the development period, 59 out of the 90 observed dry events are correctly predicted while 12 and 19 observed droughts are wrongly predicted as normal and wet respectively. Of the 106 observed wet events, 60 are

Table 2 Contingency table for multiclass classification using identified GCP for central India

Performance assessment criteria/statistics	Model performance									
	Development period (1959–2000)					Testing period (2001–2010)				
Contingency tables			Predicted category					Predicted category		
			Dry	Normal	Wet			Dry	Normal	Wet
	Observed category	Dry	59	12	19	Observed category	Dry	11	8	8
		Normal	105	77	126		Normal	9	14	49
		Wet	30	16	60		Wet	2	2	17
Q	43.29					18.27				
v	4					4				
p value	8.9×10^{-9}					0.0011				
C	0.281					0.364				
C_{max}	0.707					0.707				
C/C_{max}	0.398					0.515				

correctly predicted while 30 and 16 observed wet events are wrongly predicted as dry and normal, respectively. During the testing period, the number of observed dry events predicted correctly is 11 (out of a total of 27) and the number of observed wet events predicted correctly is 17 (out of a total of 21). The value of the Contingency Coefficient C is obtained as 0.281 and 0.364 during the development and testing period respectively. The low p-values and reasonably good C/C_{max} ratios indicate a good performance considering that prediction of regional hydrological extremes is immensely complicated due to large uncertainty in the climatic system. Thus, it is observed that GCP identified for hydrological extremes in central India may serve as effective precursors of dry and wet events.

Conclusion

This study reinforces the fact that hydrological extremes at the regional scale are caused by the concurrent effect of several climate anomaly fields across the globe rather than only well-known atmospheric–oceanic circulation patterns. A total of 19 globally distributed anomaly zones of different climate variables are found to constitute the Global Climate Pattern (GCP) responsible for hydrological extremes in central India. For the large-scale variables such as sea surface temperature and pressure, the zones of importance for central India are more or less similar to that of all India. However, for variables like air temperature, wind speed and total precipitable water, the number of influential zones is found to be more in number and relatively better defined for central India than that in case of all-India.

The identified GCP for central India is found to have potential use as precursor of hydrologic extremes in the target area. The SVM-based modeling approach used in this study exhibits reasonably good potential of GCP in foreseeing the above- and below-normal precipitation events. It may be possible to obtain more reliable prediction using GCP by adopting further sophisticated modeling approach. Moreover, as a future extension of this work, the illustrated methodology may be applied for other homogeneous meteorological subdivisions of India having considerably different precipitation regime.

Acknowledgements This study is partially supported by the Ministry of Earth Science (MoES), Government of India, through sponsored Project No. MoES/PAMC/H&C/30/2013-PC-II.

References

Anirudh V, Umes CK (2007) Classification of rainy days using SVM. In: Proceedings of Symposium at HYDRO, Norfolk, Virginia, USA

Bhagwat PP, Maity R (2012) Multistep-ahead river flow prediction using LS-SVR at daily scale. J Water Resour Prot 4:528–539. doi:10.4236/jwarp.2012.47062

Brandimarte L, Di Baldassarre G, Bruni G, D'Odorico P, Montanari A (2011) Relation between the north-atlantic oscillation and hydroclimatic conditions in mediterranean areas. Water Resour Manag 25(5):1269–1279

Bray M, Han D (2004) Identification of support vector machines for runoff modelling. J Hydroinform 6:265–280

Chanda K, Maity R (2015) Meteorological drought quantification with standardized precipitation anomaly index (SPAI) for the regions with strongly seasonal and periodic precipitation. J Hydrol Eng ASCE 06015007-1–06015007-7. doi:10.1061/(ASCE)HE.1943-5584.0001236

Chanda K, Maity R (2016) Uncovering global climate fields causing local precipitation extremes. Hydrol Sci J. doi:10.1080/02626667.2015.1006232 (Taylor and Francis)

Chen H, Guo J, Wei X (2010) Downscaling GCMs using the smooth support vector machine method to predict daily precipitation in the Hanjiang basin. Adv Atmos Sci 27(2):274–284. doi:10.1007/s00376-009-8071-1

Chiew FHS, McMahon TA (2002) Global ENSO-streamflow teleconnection, streamflow forecasting and interannual variability. Hydrol Sci J 47(3):505–522

Feng S, Hu Q (2008) How the North Atlantic multidecadal oscillation may have influenced the Indian summer monsoon during the past two millennia. Geophys Res Lett 35:L01707. doi:10.1029/2007GL032484

Fu G, Yu J, Yu X, Ouyang R, Zhang Y, Wang P, Liu W, Min L (2013) Temporal variation of extreme rainfall events in China, 1961–2009. J Hydrol. doi:10.1016/j.jhydrol.2013.02.021

Gadgil S, Vinayachandran PN, Francis PA, Gadgil S (2004) Extremes of the Indian summer monsoon rainfall, ENSO and equatorial Indian Ocean oscillation. Geophys Res Lett 31: L12213. doi:10.1029/2004GL019733

Gibbons JD, Chakraborti S (2011) Nonparametric statistical inference, 5th edn. Chapman and Hall, Boca Raton

Goswami BN, Madhusoodanan MS, Neema CP, Sengupta D (2006) A physical mechanism for North Atlantic SST influence on the Indian summer monsoon. Geophys Res Lett 33:L02706. doi:10.1029/2005GL024803

Gu W, Li C, Li W, Zhou W, Chan JCL (2009) Interdecadal unstationary relationship between NAO and east China's summer precipitation patterns. Geophys Res Lett 36:L13702. doi:10.1029/2009GL038843

Jiang P, Gautam M, Zhu J, Yu Z (2013) How well do the GCMs/RCMs capture the multi-scale temporal variability of precipitation in the Southwestern United States? J Hydrol 479:75–85. doi:10.1016/j.jhydrol.2012.11.041

Jolliffe IT (1986) Principal component analysis. Springer, New York

King AD, Alexander LV, Donat MG (2013) Asymmetry in the response of eastern Australia extreme rainfall to low-frequency Pacific variability. Geophys Res Lett 40:2271–2277. doi:10.1002/grl.50427

Kişi O, Çimen M (2009) Evapotranspiration modelling using support vector machines. Hydrol Sci J 54(5):918–928

Li S, Perlwitz J, Quan X, Hoerling MP (2008) Modelling the influence of North Atlantic multidecadal warmth on the Indian summer rainfall. Geophys Res Lett 35:L05804. doi:10.1029/2007GL032901

Lin J-Y, Cheng C-T, Chau K-W (2006) Using support vector machines for long-term discharge prediction. Hydrol Sci J 51(4):599–612

Maity R, Nagesh Kumar D (2006) Hydroclimatic association of the monthly summer monsoon rainfall over India with large-scale atmospheric circulations from tropical Pacific Ocean and the Indian Ocean region. Atmos Sci Lett 7:101–107. doi:10.1002/asl.141(RMetS)

Maity R, Nagesh Kumar D (2008) Basin-scale stream-flow forecasting using the information of large-scale atmospheric circulation phenomena. Hydrol Process 22(5):643–650. doi:10.1002/hyp.6630

Maity R, Bhagwat PP, Bhatnagar A (2010) Potential of support vector regression for prediction of monthly streamflow using endogenous property. Hydrol Process 24:917–923. doi:10.1002/hyp.7535

Maity R, Ramadas M, Govindaraju RS (2013) Identification of hydrologic drought triggers from hydro-climatic predictor variables. Water Resour Res Am Geophys Union 49(7):4476–4492. doi:10.1002/wrcr.20346

Mo KC, Schemm JE (2008) Relationships between ENSO and drought over the southeastern United States. Geophys Res Lett 35:L15701. doi:10.1029/2008GL034656

Oubeidillah AA, Tootle G, Anderson S-R (2012) Atlantic Ocean sea-surface temperatures and regional streamflow variability in the Adour-Garonne basin, France. Hydrol Sci J 57(3):496–506

Panda DK, Kumar K, Ghosh S, Mohanty RK (2013) Streamflow trends in the Mahanadi River basin (India): linkages to tropical climate variability. J Hydrol. doi:10.1016/j.jhydrol.2013.04.054

Parthasarathy B, Munot AA, Kothaale DR (1995) All-India monthly and seasonal rainfall series: 1871–1993. Theor Appl Climatol 49:217–224

Pearson K (1904) On the theory of contingency and its relation to association and normal correlation. Draper's Comp Res Mem Biometric Ser I, Dulau and Co., London, U.K

Philipp A, Della-Marta PM, Jacobeit J, Fereday DR, Jones PD, Moberg A, Wanner H (2007) Long-term variability of daily North Atlantic-European pressure patterns since 1850 classified by simulated annealing clustering. J Clim 20:4065–4095

Qin Z, Yu Q, Li J, Wu Z, Hu B (2005) Application of least squares vector machines in modelling water vapor and carbon dioxide fluxes over a cropland. J Zhejiang Univ Sci B 6(6):491–495. doi:10.1631/jzus.2005.B0491

Qiu Y, Cai W, Guo X, Ng B (2014) The asymmetric influence of the positive and negative IOD events on China's rainfall. Sci Rep 4:4943. doi:10.1038/srep04943

Raghavendra NS, Deka PC (2014) Support vector machine applications in the field of hydrology: a review. Appl Soft Comput 19:372–386. doi:10.1016/j.asoc.2014.02.002

Rajeevan M, Bhate J, Kale JD, Lal B (2006) High resolution daily gridded rainfall data for the Indian region: analysis of break and active monsoon spells. Curr Sci 91(3):296–306

Rogers JC (2013) The 20th century cooling trend over the southeastern United States. Clim Dyn 40(1–2):341–352

Saji NH, Goswami BN, Vinayachandran PN, Yamagata T (1999) A dipole mode in the tropical Indian Ocean. Nature 401:360–363

Samsudin R, Saad P, Shabri A (2011) River flow time series using least squares support vector machines. Hydrol Earth Syst Sci 15:1835–1852. doi:10.5194/hess-15-1835-2011

She N, Basketfield D (2005) Long range forecast of streamflow using support vector machine. In: Walton R (ed) Proceedings of the world water and environment resources congress ASCE, 15–19 May 2005, Anchorage: Alaska, USA. doi:10.1061/40792(173)481

Singhrattna N, Babel MS, Perret SR (2012) Hydroclimate variability and long-lead forecasting of rainfall over Thailand by large-scale atmospheric variables. Hydrol Sci J 57(1):26–41

Terray P, Delecluse P, Labattu S, Terray L (2003) Sea surface temperature associations with the late Indian summer monsoon. Clim Dyn. doi:10.1007/s00382-003-0354-0

Ting M, Kushnir Y, Seager R, Li C (2011) Robust features of Atlantic multi-decadal variability and its climate impacts. Geophys Res Lett 38:L17705. doi:10.1029/2011GL048712

Tripathi SH, Srinivas VV, Nanjundiah RS (2006) Downscaling of precipitation for climate change scenarios: a support vector machine approach. J Hydrol 330:621–640

Viswambharan N, Mohanakumar K (2014) Modulation of Indian summer monsoon through northern and southern hemispheric extra-tropical oscillations. Clim Dyn. doi:10.1007/s00382-014-2049-0 (Springer, Article in press)

Wang B, Liu J, Kim H, Yim S, Xiang B (2013) Northern hemisphere summer monsoon intensified by mega-El Niño/southern oscillation and Atlantic multidecadal oscillation. Proc Natl Acad Sci USA. doi:10.1073/pnas.1219405110

Webster PJ, Moore A, Loschnigg J, Leban M (1999) Coupled dynamics in the Indian Ocean during 1997–1998. Nature 401:356–360

Willems P (2013) Multidecadal oscillatory behaviour of rainfall extremes in Europe. Clim Change 120(4):931–944. doi:10.1007/s10584-013-0837-x

Wu R, Zhang L (2010) Biennial relationship of rainfall variability between Central America and equatorial South America. Geophys Res Lett 37:L08701. doi:10.1029/2010GL042732
Ye J-S (2014) Trend and variability of China's summer precipitation during 1955–2008. Int J Climatol 34:559–566. doi:10.1002/joc.3705
Zakaria ZA, Shabri A (2012) Streamflow forecasting at ungaged sites using support vector machines. Appl Math Sci 6(60):3003–3014

Changes in ENSO and IOD Effects on the Extreme Rainfall of Hyderabad City, India

V. Agilan and N.V. Umamahesh

Abstract The global physical processes such as El Niño-southern oscillation (ENSO) and Indian Ocean dipole (IOD) have a significant impact on Indian extreme rainfall. Recent studies show that the impact of ENSO cycle on Indian rainfall has changed. Therefore, understanding the changes in the effects of these physical processes on extreme rainfall of an urban area may help us to reduce the damage caused by urban floods. In this study, the changes in ENSO and IOD effects on the Hyderabad city monsoon/non-monsoon extreme rainfall between 1901–1950 and 1951–2004 are analyzed. The findings of this study indicate that the effect of IOD on non-monsoon months' extreme rainfall of the Hyderabad city is significantly reduced while there is no significant change in monsoon season relationship. In addition, a significant increase in ENSO effects on non-monsoon months' very extreme rainfall of the Hyderabad city is observed.

Keywords ENSO cycle · Extreme rainfall · Indian ocean dipole · Hyderabad city

Introduction

The El Niño-southern oscillation (ENSO) cycle is the fluctuations in temperature between the ocean and the atmosphere in the east-central Equatorial Pacific (Zelle et al. 2004; Agilan and Umamahesh 2015b). In ENSO cycle, the cold phase is referred as La Niña and the warm phase is referred as El Niño. These deviations from normal surface temperatures can have large-scale impacts not only on ocean processes, but also on global weather and climate (Zelle et al. 2004). The ENSO cycle is the most important coupled ocean-atmosphere phenomenon to cause global

V. Agilan (✉) · N.V. Umamahesh
Department of Civil Engineering, National Institute of Technology Warangal, Warangal 506004, Telangana, India
e-mail: agilanvensiv@gmail.com

V.P. Singh et al. (eds.), *Climate Change Impacts*, Water Science and Technology Library 82, https://doi.org/10.1007/978-981-10-5714-4_7

climate variability on interannual time scales. The recent studies show the effect of ENSO cycle on extreme precipitation at local and regional scale (Kenyon and Hegerl 2010; Zhang et al. 2010; Agilan and Umamahesh 2015a). Revadekar and Kulkarni (2008) demonstrated the effect of ENSO cycle on extreme rainfall over India. Recently, Agilan and Umamahesh (2015a) presented the effect of ENSO cycle on the Hyderabad city extreme rainfall.

During the past decade, a dipole mode in the tropical Indian Ocean is discovered (Saji et al. 1999) and this Indian Ocean dipole (IOD) is quantified with Dipole Mode Index (DMI) (Saji et al. 1999). The DMI is the Sea Surface Temperature (SST) difference between the tropical western Indian Ocean (50°E–70°E, 10°S–10°N) and the tropical southeastern Indian Ocean (90°E–110°E, 10° S-Equator) (Saji et al. 1999). When the DMI is positive, it leads to drought over the Indonesia region and heavy rains and floods over the East Africa (Ashok et al. 2001). When the sign of the DMI reverses, these anomalous fluctuations also swing to the opposite phase (Ashok et al. 2001; Ashok and Saji 2007). Ashok et al. (2001) analyzed the impact of the IOD on the relationship between the Indian Monsoon Rainfall and ENSO and discovered that the ENSO-induced Indian summer monsoon rainfall anomalous circulation over the Indian region is either countered or supported by the IOD-induced anomalous meridional circulation. Ashok and Saji (2007) studied the impacts of ENSO and IOD events on the sub-regional Indian summer monsoon rainfall (ISMR) and concluded that the magnitude of ENSO-ISMR correlations is greater than the magnitude of IOD-ISMR correlations. Ajayamohan and Rao (2008) analyzed the effect of IOD on the extreme rainfall events over India and showed the first evidence that the extreme rainfall events in central India in recent decades are strongly modulated by IOD.

From the above two paragraphs, it is clear that the ENSO cycle and IOD have a significant influence on Indian region extreme rainfall. Recently, it is found that the inverse relationship between the ENSO cycle and the Indian summer monsoon has broken down in recent decades (Kumar et al. 1999). In addition, the frequency of extreme Indian Ocean dipole events is increasing due to global warming (Cai et al. 2014). Moreover, the frequency of El Niño events (part of ENSO cycle) will increase if the concentration of future greenhouse-gas is more (Timmermann et al. 1999). In the other hand, urban flooding and the damage to infrastructure and society are problems in both developing and developed countries (Agilan and Umamahesh 2015b). Note that the high-intensity (extreme) rainfall has a significant contribution in urban area flooding and the key challenge in urbanized area is to provide good quality detailed forecasts. Hence, understanding and quantifying the changes in global physical processes' impacts on extreme rainfall of an urban area may help us in urban flood forecasting. In this study, the changes in ENSO and IOD effects on the extreme rainfall of Hyderabad city, India are analyzed.

Study Area and Data

The Hyderabad city is the capital of the state of Telangana in India. Location map of the Hyderabad city is shown in Fig. 1. The Hyderabad city lies between the latitudes of 17.25°N and 17.60°N and longitudes of 78.20°E and 78.75°E and situated at a height of about 500 m above the mean sea level. It is classified as a semi-arid region and the Köppen-Geiger classification is BSh (Peel et al. 2007).

High-resolution gridded (1° Longitude × 1° Latitude) daily rainfall data prepared with the help of more than 1800 gauge observation over India is available for India for the period of 1901–2004 (Rajeevan et al. 2008). This data set is prepared by India Meteorological Department (IMD) and the details about the preparation of this data set is available in Rajeevan et al. (2008). This dataset is increasingly being used in studies on Indian rainfall (Kulkarni et al. 2012; Mondal and Mujumdar 2015). For this study, the high-resolution gridded rainfall is procured from IMD and the Hyderabad city grid's data is extracted for the period of 1901–2004.

The ENSO cycle is represented by several ENSO indices such as Multivariate ENSO Index (MEI), southern oscillation index (SOI) and Sea Surface Temperature (SST), etc. Different studies use different ENSO indices, i.e. SOI (Katz et al. 2002), SST (Mondal and Mujumdar 2015). Further, Revadekar and Kulkarni (2008) reported that the intensity and frequency of extreme precipitation in Southern India have a strong correlation with NINO 3.4 SST anomalies 4–6 months in advance. In addition, some earlier studies reported that the MEI is better for monitoring ENSO than the SOI or various SST indices because the MEI integrates more information than other indices, it reflects the nature of the coupled ocean-atmosphere system better than either component, and it is less vulnerable to occasional data

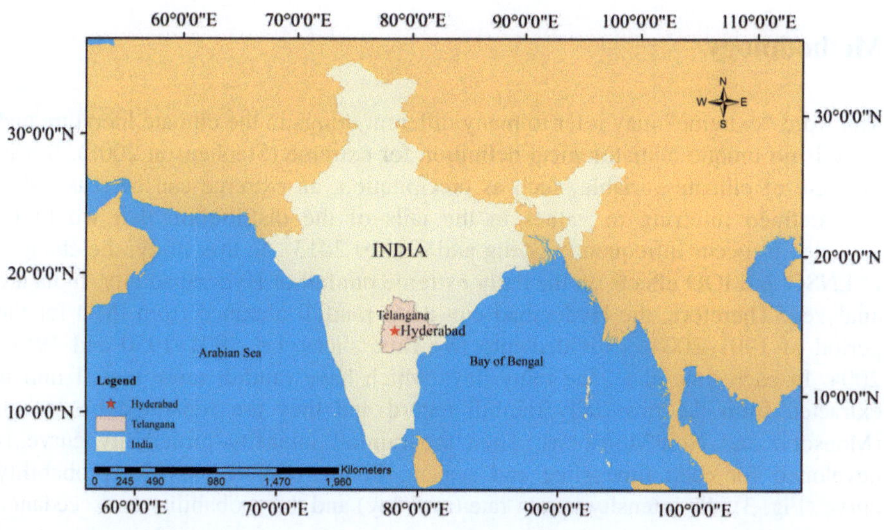

Fig. 1 Location map of Hyderabad city (Agilan and Umamahesh 2015a)

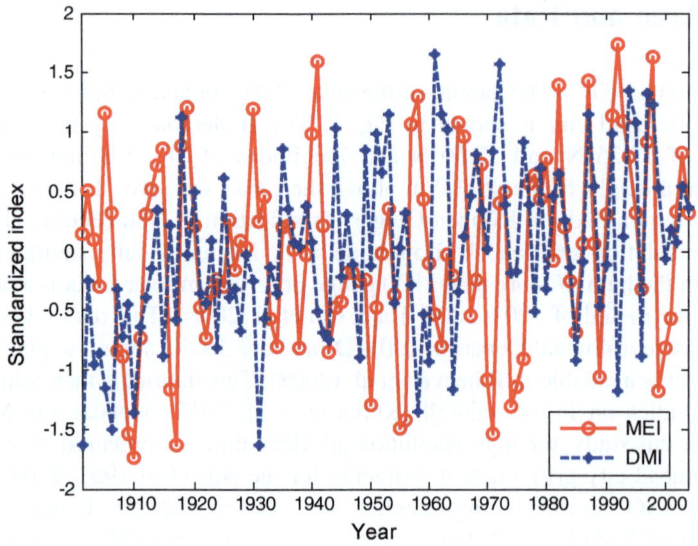

Fig. 2 Annual MEI and DMI

glitches in the monthly update cycles (Wolter and Timlin 1998). Therefore, in this study, MEI, SOI and SST are used as ENSO indicators. The monthly DMI derived from HadISST dataset is downloaded from http://www.jamstec.go.jp/frcgc/research/d1/iod/DATA/dmi.monthly.txt (Accessed on 15-06-2015) to represent IOD. The annual MEI and DMI calculated from the monthly MEI and DMI is plotted in Fig. 2.

Methodology

The word "extreme" may refer to many different things in the climate literature and there is no unique climatological definition for extreme (Stephenson 2008). But in the case of climate variable, such as precipitation, an extreme can be reasonably well defined referring to values in the tails of the distribution that would be expected to occur infrequently (Zeng and Zwires 2013). In this study, the changes in ENSO and IOD effects on the daily extreme rainfall of Hyderabad city, India are analyzed. Therefore, the Hyderabad city daily rainfall obtained from IMD for the period of 1901–2004 is divided into two time slices, i.e. 1901–1950 and 1951–2004. In each time slice, the rainy days which have rainfall more than 1 mm is extracted from the raw daily rainfall record and they are separated by season (Monsoon and Non-Monsoon). Then the rainfall intensity–probability curve is developed for each time slice and season. In the rainfall intensity–probability curve (Fig. 3), the intensity is rain rate (mm/day) and the probability is exceedance probability. The extreme rainfall is defined with different levels of exceedance probability for each time slice and season. In particular, rainfall values which have exceedance probability as 0.1, 0.075, 0.05, 0.025 and 0.01 are calculated from the intensity–probability relationship for the monsoon season. As the number of rainy days in a non-monsoon season of Hyderabad city is less when compared to the monsoon season rainy days, the 0.01 exceedance probability rainfall days will be very less and it will have a statistically insignificant relationship with the physical processes (ENSO and IOD). Thus, only four extreme rainfall values are calculated for the non-monsoon season (i.e. 0.1, 0.075, 0.05, 0.025 exceedance probability rainfall values). Once the threshold values of the extreme rainfall are calculated, the extreme rainfall series (series of rainfall values which have an intensity greater than defined threshold) with different probabilities of exceedance is extracted from the original series for two time slices (1901–1950 and 1951–2004) and two seasons (Monsoon and Non-Monsoon). Further, these extreme rainfall series are used to analyze the changes in ENSO and IOD effects on the extreme rainfall of Hyderabad city, India. In detail, the absolute correlation coefficient (Eq. 1) between these extreme rainfall series and the physical processes (ENSO and IOD) are calculated and the changes in the absolute correlation coefficient value between two periods (1901–1950 and 1951–2004) are presented.

$$r = \left| \frac{\sum_{i=1}^{n} \left(R_i - \overline{R}\right)\left(E_i - \overline{E}\right)}{\sqrt{\sum_{i=1}^{n} \left(R_i - \overline{R}\right)^2 \sum_{i=1}^{n} \left(E_i - \overline{E}\right)^2}} \right| \quad (1)$$

where n is the length of the extreme rainfall series, r is absolute correlation coefficient, R_i is extreme rainfall value, $\bar{R}$ is the mean of extreme rainfall, E_i is the corresponding month index which represents the physical process (ENSO or IOD) and $\bar{E}$ is the index mean value. As mentioned before, the ENSO cycle is represented by several ENSO indices such as MEI, SOI and SST, etc. Different studies use different ENSO indices, i.e. SOI (Katz et al. 2002), SST (Mondal and Mujumdar 2015) and MEI (Singh 2001). Further, some studies considered lag in ENSO index and some of them have not. Therefore, it is not clear that which ENSO index is best for the Hyderabad city and how much lag is to be considered to analyze the relation between ENSO cycle and the Hyderabad city extreme rainfall. Similarly, the lag between IOD and Hyderabad city extreme rainfall is also unknown. Therefore, the best ENSO index with the best lag and the best lag value for DMI are identified for each extreme rainfall series. The changes in ENSO effects on the Hyderabad city extreme rainfall series are analyzed with the best ENSO index and the best lag value and the changes in IOD effects on the Hyderabad city extreme rainfall series is analyzed with the best lag value of DMI.

Results and Discussions

The monsoon months' intensity–probability relationship for 1901–1950 time slice is plotted in Fig. 3 and this relationship is used to estimate the extreme rainfall which has 0.1, 0.075, 0.05, 0.025 and 0.01 exceedance probability. Similarly, the intensity–probability relationships for 1901–1950 time slice non-monsoon months' rainfall, 1951–2004 time slice monsoon and non-monsoon months' rainfall are developed and the extreme rainfall values of different exceedance probability are calculated.

Then the extreme rainfall series with different probabilities of exceedance is extracted from the original series for two time slice (1901–1950 and 1951–2004) and two seasons (Monsoon and Non-Monsoon) using the calculated threshold values. Before calculating the relationship between extreme rainfall series and physical processes (ENSO and IOD), the best ENSO index with best lag value and the best lag value of DMI is identified. The variations in absolute correlation coefficient value between monsoon months' 0.01 exceedance probability extreme rainfall time series of 1901–1950 time slice and physical processes (ENSO and IOD) indices with different lag values are shown in Fig. 4.

From Fig. 4, for monsoon months' 0.01 exceedance probability extreme rainfall time series of 1901–1950 time slice, it is observed that the SOI with 3 months lag is the best ENSO cycle indicator and the best lag value for DMI is 5 months. In a similar way, the best ENSO index with the best lag value and the best lag value of DMI have identified for all extreme rainfall series. From Fig. 4, it is also to be noted that the correlation between extreme rainfall and indices are varying significantly with respect to lag value. Therefore, analyzing the changes in ENSO and IOD effects without identifying the best index and the best lag is perilous.

The absolute correlation coefficient between extreme rainfall of the Hyderabad city and two physical processes, namely, ENSO and IOD is given in Table 1 and

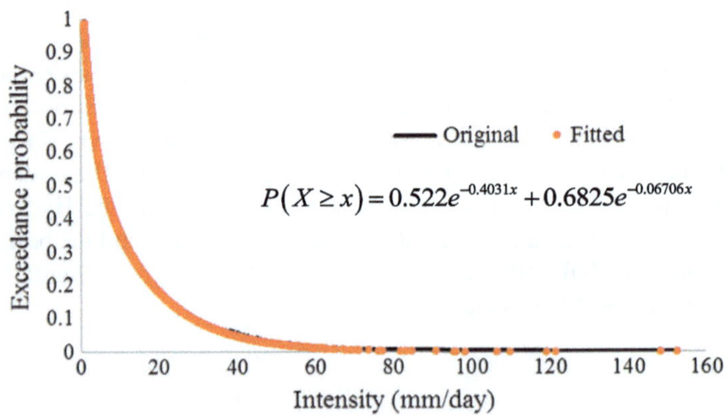

Fig. 3 1901–1950 Monsoon months' intensity–probability relationship

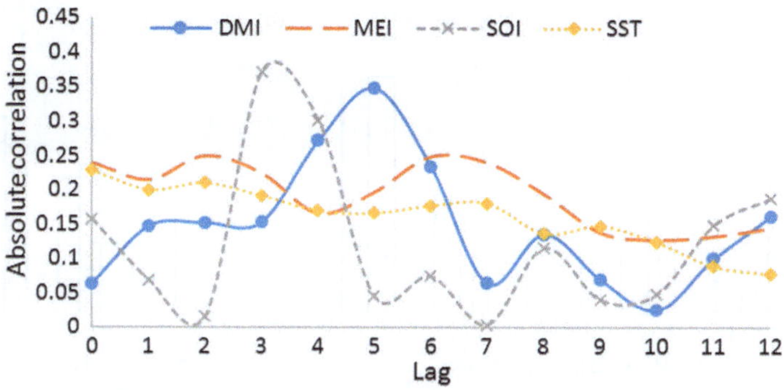

Fig. 4 The variations in absolute correlation coefficient value between monsoon months' 0.01 exceedance probability extreme rainfall time series of 1901–1950 time slice and MEI, SST, SOI and DMI with different lag values

Table 1 The absolute correlation coefficient between extreme rainfall of the Hyderabad city and ENSO and IOD

Rainfall exceedance probability	ENSO		IOD	
	1901–1950	1951–2004	1901–1950	1951–2004
Monsoon				
0.1	0.12	0.10	0.12	0.07*
0.075	0.15	0.15	0.16	0.08*
0.05	0.12*	0.19	0.09	0.13
0.025	0.23	0.19	0.21	0.22
0.01	0.37	0.37	0.35	0.34
Non-monsoon				
0.1	0.17	0.10	0.18	0.11*
0.075	0.23	0.21	0.25	0.12*
0.05	0.27	0.31	0.29	0.22
0.025	0.26*	0.42	0.31	0.19*

*Not significant at 0.1 significance level. Other values are significant

plotted in Fig. 5. From Fig. 5a, it is observed that the relationship between ENSO and monsoon months' extreme rainfall of the Hyderabad city is not changed significantly between two time periods (1901–1950 and 1951–2004). However, the relationship between ENSO and non-monsoon months' very extreme (0.05 and 0.025 exceedance probability) rainfall of the Hyderabad city is increased significantly (maximum of around 60% increase).

Similar to the ENSO cycle relationship, the relationship between IOD and monsoon months' extreme rainfall of the Hyderabad city is not changed significantly between two time periods. But, the relationship between IOD and 1951–2004 time period non-monsoon months' extreme rainfall of the Hyderabad city is

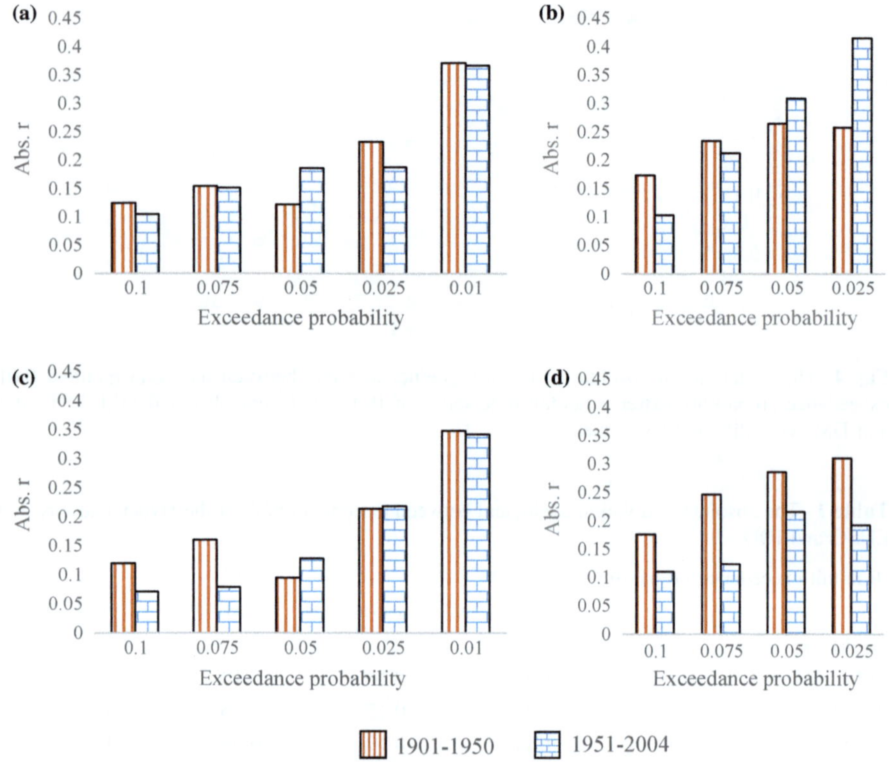

Fig. 5 **a** ENSO-monsoon, **b** ENSO-non-monsoon, **c** IOD-monsoon and **d** IOD-non-monsoon

decreased significantly (maximum of around 50% decrease) when compared to 1901–1950 relationship.

Summary

In this study, with the help of IMD gridded daily rainfall, the changes in ENSO and IOD effects on the Hyderabad city extreme rainfall intensity between two time periods, i.e. 1901–1950 and 1951–2004 are analyzed. The findings of this study indicate that the effect of IOD on non-monsoon months' extreme rainfall of the Hyderabad city is significantly reduced (maximum of 50% decrease) while there is no significant change in monsoon season relationship. In addition, a significant increase (maximum of 60% increase) in ENSO effects on non-monsoon months' very extreme rainfall of the Hyderabad city is observed.

Acknowledgements This work is supported by Information Technology Research Academy (ITRA), Government of India under, ITRA-water grant ITRA/15(68)/water/IUFM/01. We also thank the India Meteorological Department for providing rainfall data.

References

Agilan V, Umamahesh NV (2015a) Effect of el Niño-southern oscillation (ENSO) cycle on extreme rainfall events of Indian urban area. In: Vellore, Proceedings of international conference on sustainable energy and built environment, pp 569–573

Agilan V, Umamahesh NV (2015b) Detection and attribution of non-stationarity in intensity and frequency of daily and 4–h extreme rainfall of Hyderabad, India. J Hydrol 530:677–697

Ajayamohan RS, Rao SA (2008) Indian ocean dipole modulates the number of extreme rainfall events over India in a warming environment. J Meteorol Soc Jpn 86:245–252

Ashok K, Guan Z, Yamagata T (2001) Impact of the Indian Ocean dipole on the relationship between the Indian monsoon rainfall and ENSO. Geophys Res Lett 28(23):4499–4502

Ashok K, Saji NH (2007) On the impacts of ENSO and Indian Ocean dipole events on sub-regional Indian summer monsoon rainfall. Nat Hazards 42:273–285

Cai W et al (2014) Increased frequency of extreme Indian Ocean dipole events due to greenhouse warming. Nature 510:254–258

Katz RW, Parlange MB, Naveau P (2002) Statistics of extremes in hydrology. Adv Water Resour 25(8):1287–1304

Kenyon J, Hegerl GC (2010) Influence of modes of climate variability on global precipitation extremes. J Climate 23(23):6248–6262

Kulkarni MA, Singh A, Mohanty UC (2012) Effect of spatial correlation on regional trends in rain events over India. Theor Appl Climatol 109:497–505

Kumar KK, Rajagopalan B, Cane MA (1999) On the weakening relationship between the Indian monsoon and ENSO. Science 284:2156–2159

Mondal A, Mujumdar PP (2015) Modeling non-stationarity in intensity, duration and frequency of extreme rainfall over India. J Hydrol 521:217–231

Peel MC, Finlayson BL, McMahon TA (2007) Updated world map of the Koppen-Geiger climate classification. Hydrol Earth Syst Sci 11:1633–1644

Rajeevan M, Bhate J, Jaswal AK (2008) Analysis of variability and trends of extreme rainfall events over India using 104 years of gridded daily rainfall data. Geophys Res Lett 35: L18707:1–6

Revadekar JV, Kulkarni A (2008) The El Nino-Southern oscillation and winter precipitation extremes over India. Int J Climatol 28:1445–1452

Saji NH, Goswami BN, Vinayachandran PN, Yamagata T (1999) A dipolemode in the tropical Indian ocean. Nature 401:360–363

Singh OP (2001) Multivariate ENSO index and Indian monsoon rainfall: relationships on monthly and subdivisional scales. Meteorol Atmos Phys 78:1–9

Stephenson DB (2008) Definition, diagnosis, and origin of extreme weather and climate events. In: Climate extremes and society. Cambridge University Press, Cambridge

Timmermann A et al (1999) Increased El NinÄo frequency in a climate model forced by future greenhouse warming. Nature 398:694–697

Wolter K, Timlin MS (1998) Measuring the strength of ENSO events: How does 1997/98 rank? Weather 53(9):315–324

Zelle H, Appeldoorn G, Burgers G, Oldenborgh GJ (2004) The Relationship between sea surface temperature and thermocline depth in the eastern equatorial Pacific. J Phys Oceanogr 34:643–655
Zeng X, Zwires FW (2013) Statistical indices for the diagnosing and detecting changes in extremes. In: Extremes in a changing climate: detection, analysis and uncertainty, s.l.:Springer, pp 1–14
Zhang X, Wang J, Zwiers FW (2010) The influence of large-scale climate variability on winter maximum daily precipitation over North America. J Climate 23:2902–2915

Part II
Rainfall Analysis

Detecting Changes in Regional Rainfall Series in India Using Binary Segmentation-Based Multiple Change-Point Detection Techniques

Shagufta Akbari and M. Janga Reddy

Abstract In this chapter, the rainfall patterns of five homogeneous regions in India namely Northwest, West Central, Central Northeast, Northeast and Peninsular India for the years 1871–2013 were analyzed using two change-point detection (CPD) techniques namely Binary Segmentation based on Cumulative sum and Likelihood ratio approaches. The CPD methodology involves estimation of threshold values based on Monte Carlo simulation. On applying the CPD techniques for five regions, the results showed that for most of the regions there were no significant change-points in annual/monsoon rainfall series except for monsoon rainfall in Northeast India. Further, the study extended to detect the changes in future rainfall for Northeast region by analyzing the outputs from global climate models. The findings of this study can help in better understanding the spatial variations and changing patterns of rainfall across India, and consequently in proper planning and management of water resources in the region.

Keywords Rainfall · Change-point detection · Binary segmentation
Monte Carlo simulation

Introduction

Hydrologists dealing with stochastic time series usually make the stationarity assumption for operations of hydropower systems, reservoirs and construction of hydraulic structures, etc. This assumption is itself questionable as some of the hydrometeorological time series exhibit an abrupt shift in distributional parameters due to recent changes in landuse patterns, construction of hydraulic structures and

S. Akbari (✉) · M. Janga Reddy
Department of Civil Engineering, Indian Institute of Technology Mumbai,
Mumbai 400076, India
e-mail: 124040018@iitb.ac.in

M. Janga Reddy
e-mail: mjreddy@civil.iitb.ac.in

V.P. Singh et al. (eds.), *Climate Change Impacts*, Water Science
and Technology Library 82, https://doi.org/10.1007/978-981-10-5714-4_8

human induced as well as natural climatic change. Hence, there is a pressing need to analyze such series with using change-point detection techniques such as that presented in the current study. Therefore, non-stationarity analysis of hydrometeorological series (such as rainfall, temperature, discharge, etc.) is gaining high significance as it plays a key role in proper planning and management of river basins. Change-point detection (CPD) analysis of hydrometeorological series involves detecting a probable temporal location for a shift in distributional parameters (mean/variance). It could help to take decisions on continuance or change in water management policy. Several techniques are available in literature for CPD analysis in univariate/multivariate time series (Potter 1981; Buishand 1982; Winjngaard et al. 2003; Alexandersson 1986; Tripathi and Govindaraju 2009; Perreault et al. 2000; Tomozeiu et al. 2000), with the assumption that there is at most one CP (AMOC) in the series. However, the problem of identifying multiple CPs in the hydrometeorological time series is largely unexplored with a few exceptions (Seidou and Ouarda 2007; Djibo et al. 2015). Moreover, the assumption that there is at most one CP (AMOC) in the time series can be unrealistic. Hence, multiple change-point detection (MCPD) techniques should be applied to hydro-climatic series for understanding the dynamics of the times series. For the MCPD, some researchers have extended the single CPD techniques to detect at most one CP and applied them iteratively to detect multiple change-points (Rodionov 2005). Few studies also noted the difficulty in extending the methodology to the MCPD setting because of computational complexity. Vostrikova (1981), first proposed Binary Segmentation (BS) in a stochastic process multivariate setting. This study attempted to revisit the BS algorithm for MCPD techniques, i.e. BS based on the Cumulative sum (Cusum) and Likelihood ratio (LR)-based approaches.

Rainfall is the key driver of most hydrological processes. Non-stationarity in rainfall input may lead to inadequate interpretation for future predictions and would directly influence the water resources of a region. Thus, analysis of the spatial and temporal distribution of rainfall and its changing pattern will lead to proper planning and management of water resources in the region. Hence, the present study focusses on analyzing the rainfall patterns of five homogeneous regions in India namely Northwest; West Central; Central Northeast; Northeast and peninsular India for the years 1871–2013 using two change-point detection techniques (Binary Segmentation based on Cumulative sum/Likelihood ratio approaches).

The rest of the chapter is organized as follows. In next section, first, the methodology of the BS-based MCPD techniques and then the procedure for estimating threshold (or penalty) is explained. Then the proposed methodology was applied to the annual and monsoon rainfall time series for five homogeneous regions in India. In the last section, the overall concluding remarks on the performance of BS-based MCPD techniques are presented.

Methodology

Estimation of the Multiple Change-Points Using Binary Segmentation Approach

Binary segmentation (BS) is arguably the most established algorithm for multiple change-point detection (MCPD) analysis in univariate as well as multivariate series and was applied by various researchers (Vostrikova 1981; Killick et al. 2012). Using BS algorithm any single CPD technique can be extended to multiple CPs by iteratively repeating the single CPD technique on different subsets of the sequence. The advantage of BS algorithm includes low computational complexity, conceptual simplicity, and the fact that it is usually easy to code, even in more complex models unlike many of the MCPD techniques. More details of the BS algorithm can be found in Eckley et al. (2011). The BS-based MCPD techniques are discussed in the following sub-sections.

The methodology of Binary segmentation (BS) techniques, (i.e. BS based on the Cumulative sum (BS-Cusum) and Likelihood ratio (BS-LR) approaches) to univariate series can be explained in the following steps:

1. The test statistics $\Lambda(\cdot)$ was estimated using single CPD technique (Cumulative sum approach/Likelihood ratio approach) for the data.
2. The parameters of the distributions were estimated using maximum-likelihood estimation (MLE) method. Best fitted distribution for data is selected employing Akaike Information Criteria, AIC (Akaike 1974; Bozdogan 2000).
3. The threshold value/penalty C^* was estimated using Monte Carlo simulation as given below:
   Let simulated repetitions of the test statistic $\Lambda(\cdot)$ for the selected distribution be denoted as C_t^*, $t = 1, 2, 3\ldots N$, and estimated threshold value (C^*) at the significance level α is $C^*_{(1-\alpha)(N+1)}$, where C_r^* is the rth order value estimation based on $N = 5000$ simulations. From the past literature, it was noticed that in the BS algorithm, the threshold value/Penalty C^* was assumed to be constant when it was compared with the test statistics of the splitted series, i.e. $\Lambda(x_{s:e}) > C^*$; where $[s, e]$ is starting and ending of splitted series.
4. If threshold value/penalty C^* estimated at α level of significance ($\alpha = 5\%$) is greater than the test statistics $\Lambda(\cdot)$, the null hypothesis H_0 was accepted and then no CP is detected and the technique stops. Otherwise, the univariate series are splitted into two segments consisting of the sequence before and after the identified CP, τ and the test statistics $\Lambda(\cdot)$ are estimated using CP technique (Cusum/Likelihood ratio technique) to each new segment. If either or both segments are rejected, data were split into further segments at the newly identified CPs applying the detection technique to each new segment. This procedure is repeated until no further CP is detected. The flowchart for MCPD using BS approach is given in Fig. 1.

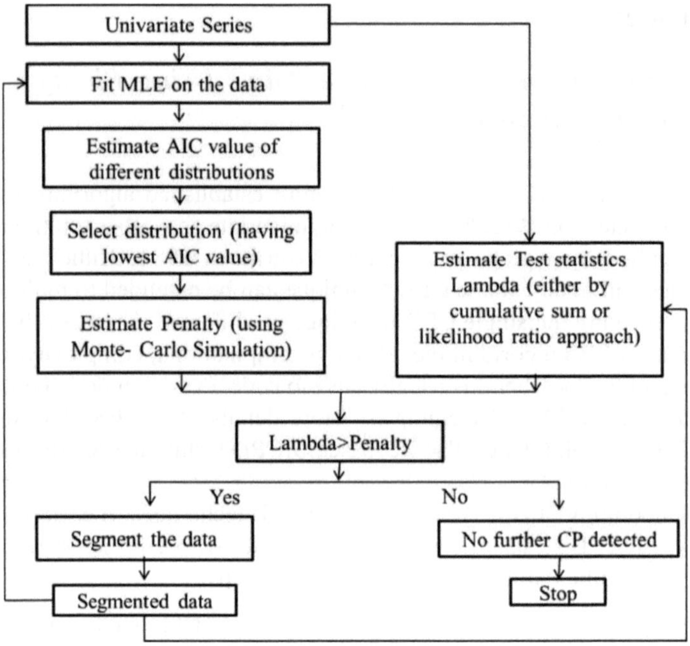

Fig. 1 Methodology for estimating the multiple change-points using binary segmentation approach

Estimation of Test Statistics Lambda

Let $x_1, \ldots x_T$ be the univariate series; $\bar{x}_k = \frac{\sum_{i=1}^{k} x_i}{k}, \bar{x}_{k''} = \frac{\sum_{i=k+1}^{T} x_i}{T-k}$ and $\bar{x}_T = \frac{\sum_{i=1}^{T} x_i}{T}$

The test statistic based on Cumulative sum (Cusum) approach, $\Lambda_{\mathrm{CS}}(\cdot)$ (Page 1954; Scott and Knott 1974; Csorgo and Horvath 1997) is given by

$$\mathrm{CS}_k = \frac{\sum_{1 \le i \le k} \left(x_i - \left(\frac{k \times x_i}{n}\right)\right)}{T} \quad k \in [1, T] \tag{1}$$

$$\Lambda_{\mathrm{CS}}(x_{1:T}) = \max(\mathrm{CS}_k) \tag{2}$$

The test statistic for BS based on Likelihood ratio (LR) approach, $\Lambda_{\mathrm{LR}}(\cdot)$ (Hawkins 1977), is given by

$$\mathrm{LR}_k = \sum_{i=1}^{i=T} (x_i - \bar{x}_T)^2 - \left[\sum_{i=1}^{i=k} (x_i - \bar{x}_k)^2 + \sum_{i=k+1}^{i=T} (x_i - \bar{x}_{k''})^2\right] \quad k \in [1, T] \tag{3}$$

and

$$\Lambda_{\mathrm{LR}}(x_{1:T}) = \max(\mathrm{LR}_k) \tag{4}$$

If $\Lambda(\cdot) > C^*$, then significant CP is detected, where C^* is the threshold value/penalty.

This Cusum and LR-based single CP techniques were extended to multiple CPD using BS algorithm (Eckley et al. 2011), in this study. The following section describes the application of the Binary Segmentation based on Cumulative sum (BS-Cusum) approach and Likelihood ratio (BS-LR) approach to rainfall time series.

Study Area and Data

In this study, the rainfall patterns of five homogeneous regions in India namely Northwest India (NWI); West Central India (WCI); Central Northeast India (CNEI); Northeast India (NEI) and peninsular India (PENSI) were analyzed. Figure 2 shows the location details of the study area.

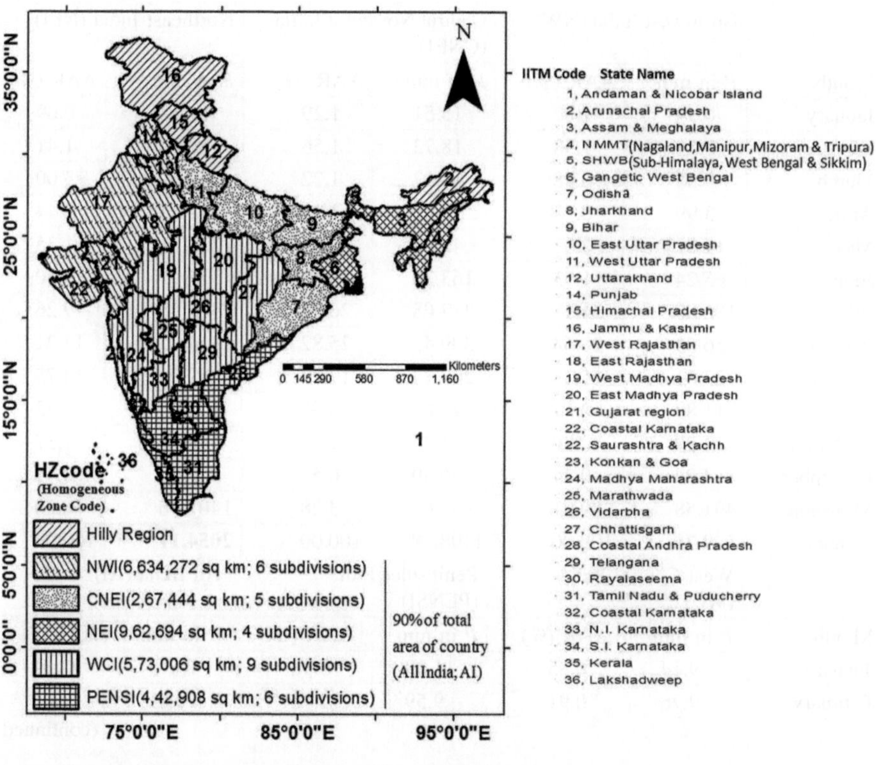

Fig. 2 Location map of study region shows five homogeneous regions in India

Regional monthly rainfall time series for five homogeneous regions for the period 1871–2013 is collected from Indian Institute of Tropical Meteorology (IITM), Pune (http://www.tropmet.res.in), and used for change-point detection analysis. On the basis of climate characteristics, India was divided into five homogeneous regions (Parthasarathy et al. 1993), covering 30 meteorological subdivisions excluding the island subdivisions and hilly areas. The regional rainfall time series for each of the five homogeneous regions were prepared by IITM, assigning the subdivision area as the weight to each of the subdivisions in the region. In past, a number of studies have been carried out with same dataset (Guhathakurta and Rajeevan 2008; Krishnakumar et al. 2009). The annual and monsoon rainfall time series were extracted from the monthly regional rainfall data for five homogeneous regions and taken for the MCPD analysis. The average annual rainfall (AAR) of the homogeneous regions and the percentage (%) contribution of monthly and monsoon rainfalls to AAR of the five homogeneous regions are presented in Table 1. It can be observed that rainfall pattern of Northeast India (NEI) region is the highest rainfall receiving region among the five homogeneous regions.

Table 1 Contribution of monthly rainfalls and seasonal rainfalls to AAR of five homogeneous regions and All India

	Northwest India (NWI)		Central Northeast India (CNEI)		Northeast India (NEI)	
Month	*P* in mm	AAR (%)	*P* in mm	AAR (%)	*P* in mm	AAR (%)
January	6.94	1.27	15.51	1.29	14.12	0.69
February	7.55	1.38	18.73	1.56	28.99	1.41
March	5.35	0.98	14.57	1.22	61.61	3.00
April	3.99	0.73	16.65	1.39	132.34	6.44
May	11.02	2.02	42.64	3.56	233.21	11.35
June	65.24	11.93	163.39	13.63	375.97	18.30
July	186.80	34.17	317.08	26.46	395.58	19.26
August	156.53	28.63	309.45	25.82	353.44	17.21
September	83.01	15.18	208.14	17.37	282.08	13.73
October	11.89	2.17	71.61	5.98	140.13	6.82
November	4.39	0.80	14.14	1.18	140.13	1.37
December	4.04	0.74	6.49	0.54	8.43	0.41
Monsoon	491.58	89.91	998.06	83.28	1407.06	68.50
Annual	**549.7***	100.00	1198.39	100.00	**2054.11****	100.00
	West Central India (WCI)		Peninsular India (PENSI)		All India (AI)	
Month	*P* in mm	AAR (%)	*P* in mm	AAR (%)	*P* in mm	AAR (%)
January	9.14	0.85	11.19	0.96	10.70	0.98
February	9.76	0.91	9.59	0.82	12.81	1.18

(continued)

Table 1 (continued)

	West Central India (WCI)		Peninsular India (PENSI)		All India (AI)	
March	9.31	0.87	13.66	1.17	15.01	1.38
April	12.42	1.16	38.77	3.33	26.60	2.45
May	21.10	1.97	84.81	7.29	52.66	4.84
June	170.25	15.87	165.64	14.24	164.16	15.10
July	305.11	28.44	189.01	16.25	271.99	25.02
August	266.14	24.81	157.24	13.52	241.98	22.26
September	184.52	17.20	148.11	12.73	170.32	15.67
October	59.71	5.57	182.04	15.65	77.83	7.16
November	18.22	1.70	122.08	10.49	31.26	2.88
December	7.14	0.67	41.12	3.54	11.67	1.07
Monsoon	926.03	86.32	660.00	56.74	848.45	78.06
Annual	1072.82	100.00	1163.27	100.00	1086.99	100.00

Note Double and single asterisks (*) represent highest and lowest rainfall receiving regions

Change-point detection analysis becomes more complex for the future than the past because there is not one time series of climate, but rather many future projections from different climate models run with a range of CO_2 emissions scenarios. It is unrealistic to analyze only one climate model for any given emission scenario, but rather multi-model ensembles quantify the range of plausible future climates under different emissions scenarios. Therefore in this study, the future change in monsoon rainfall in the Northeast India Region from the output of seven climate models and multi-model ensemble (simple averaging approach) were studied. For future Rainfall analysis the CMIP3 output data of seven climate models were obtained from the website of SWAT—Texas A&M University, (http://globalweather.tamu.edu/cmip/). The data is available at daily scale for the entire globe (Dile and Srinivasan 2014; Fuka et al. 2014).

The rainfall data was already downscaled to $0.5^0 \times 0.5^0$ grid using the statistical downscaling method with bias correction (Thrasher et al. 2012; Girvetz et al. 2013). Therefore, no further preprocessing needed for the rainfall data. For this study, A1B scenario categorized as 'medium' in greenhouse gas emissions (Errasti et al. 2011), was considered.

The rainfall data available at a $0.5^0 \times 0.5^0$ grid (98 gridpoints) fall in Northeast India region (Fig. 3), was extracted from the region. The summary of the seven climate models namely CCCMA-CGCM3.1; CNRM-CM3; GFDL-CM2.0; GFDL CM2.1; IPSLCM4; MIROC3.2; MRI CGCM2.3.2 selected in this study is given in Table 2.

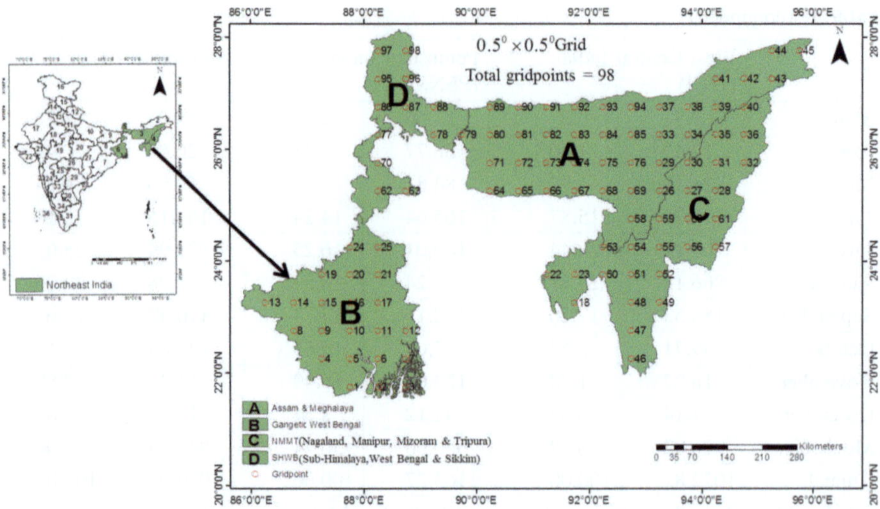

Fig. 3 Locations of rainfall gridpoints from 7 climate models in Northeast India region

Table 2 The summary of seven climate models used in this study

Climate model	Model abbreviation	Organisation
CCCMA-CGCM3.1	CGCM3	Canadian Centre for climate modelling and analysis, Canada
CNRM-CM3	CNRM	Centre National de Recherches Meteorologiques, France
GFDL-CM2.0	GFDL1	Geophysical Fluid Dynamic Laboratory, USA
GFDL CM2.1	GFDL2	Geophysical Fluid Dynamic Laboratory, USA
IPSLCM4	IPSLCM	Institute Pierre Simon Laplace, France
MIROC3.2	MEDRES	Centre for Climate Research, Japan
MRI CGCM2.3.2	MRI	Meteorological Research Institute, Japan

Results and Discussion

The Binary segmentation-based Cusum and Likelihood ratio approaches (i.e. BS-Cusum and BS-LR) are applied for multiple change-point detection (MCPD) in the mean of rainfall time series in five homogeneous regions at annual and seasonal scale. For seasonal analysis, southwest monsoon series was taken (June–September). The temporal variation of annual rainfall and monsoon rainfalls are shown in Fig. 4a–f for Northwest India (NWI); West Central India (WCI); Central Northeast India (CNEI); Northeast India (NEI), peninsular India (PENSI) and All India (AI) regions, respectively.

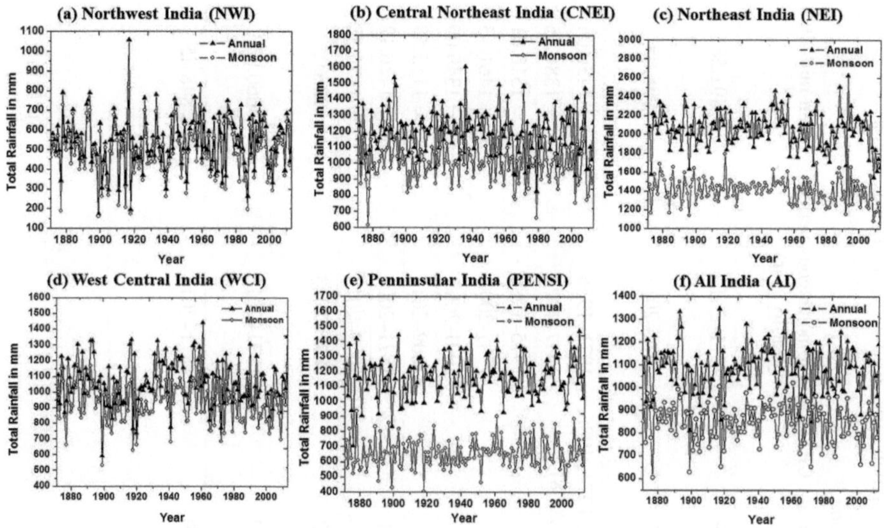

Fig. 4 Temporal variation of annual and monsoon rainfalls in India

The results of BS-Cusum and BS-LR for annual and monsoon rainfall time series are presented in Table 3. From the results of MCPD analysis (Table 3), the following key observations are noted:

Both the BS-Cusum and BS-LR techniques identified no CP for annual rainfall series in all regions.

- Except for monsoon rainfall of Northeast India region, in general the series is stationary for all the homogeneous regions.
- Applying the BS-Cusum technique, the monsoon rainfall in Northeast India region identified CPs in mean and their CP locations (in Year) are 1956, 1974, 1986, 2001 and 2008. Monsoon rainfall is showing an abrupt decrease in mean after 1956 and 1974 (Fig. 5). But, in the year 1986 there was abrupt increase and consequently an abrupt decrease in the year 2008. The last estimated segment for the year 2009–2013 can be viewed as the 'current' regime of stationarity. The quantification of the magnitude of change was estimated to be 49.29; 68.98; −87.15; 38.99 and 210.35 mm for the corresponding six segments (having 5 CPs), respectively. Positive/Negative signs indicate the abrupt decrease/increase in monsoon rainfall, respectively. The current regime (Year 2009 onwards) of monsoon time series has implications in forecasting monsoon rainfall future values for Northeast India regions. However, the BS-LR technique has missed few earlier CPs year 2008 which was actually significant CPs. By visual interpretation also the abrupt changes in mean can be visualized for the Northeast India region (Fig. 5a, b).

Above findings of the study gives an indication of a possible shift of monsoon in the Northeast India region, and the magnitudes of change in mean were quantified.

Table 3 Binary segmentation-based cusum and likelihood ratio approach for annual and monsoon rainfall series for five homogeneous regions

Homogeneous region	MCP Technique	Annual			Monsoon		
		Significant CPs (year)	Homogeneous period (year to year)	Mean rainfall (mm)	Significant CPs (year)	Homogeneous period (year to year)	Mean rainfall (mm)
Northwest India (NWI)	BS-Cusum	No CP No CP	1871–2013	$\mu = 546.75$	No CP	1871–2013	$\mu = 491.58$
	BS-LR						
Central Northeast India (CNEI)	BS-Cusum	No CP	1871–2013	$\mu = 1198.39$	No CP	1871–2013	$\mu = 998.06$
	BS-LR						
Northeast India (NEI)	BS-Cusum	No CP	1871–2013	$\mu = 2054.11$	[1956, 1974, 1986, 2000, 2008]	1871–1956	$\mu_1 = 1436.04$
						1957–1974	$\mu_2 = 1386.75$
						1975–1986	$\mu_3 = 1310.01$
						1987-2000	$\mu_4 = 1461.33$
						2001-2008	$\mu_5 = 1349.09$
						2009–2013	$\mu_6 = 1155.58$
	BS-LR				[2008]	1871–2008	$\mu_1 = 1416.12$
						2009–2013	$\mu_2 = 1155.58$
West Central India (WCI)	BS-Cusum	No CP	1871–2013	$\mu = 1072.82$	No CP	1871–2013	$\mu = 926.03$
	BS-LR						
Peninsular India (PENSI)	BS-Cusum	No CP	1871–2013	$\mu = 1163.27$	No CP	1871–2013	$\mu = 660.003$
	BS-LR						
All India (AI)	BS-Cusum	No CP	1871–2013	$\mu = 1086.99$	No CP	1871–2013	$\mu = 848.45$
	BS-LR						

Note Significant CPs are estimated at 5% significance level

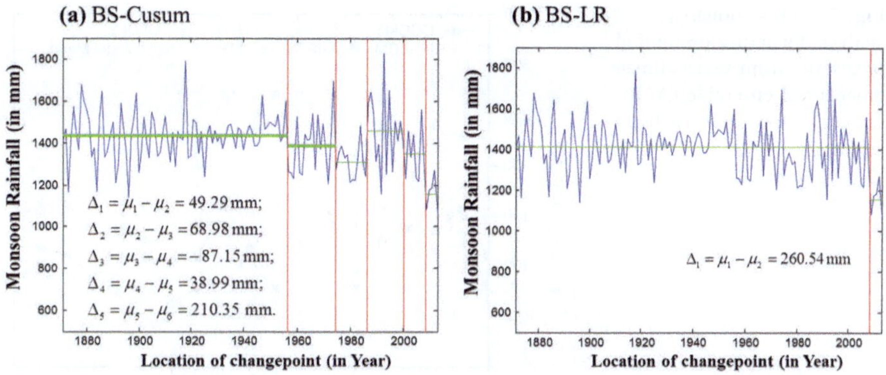

Fig. 5 CP detection analysis for Monsoon rainfalls in Northeast region for a period of 1871–2013

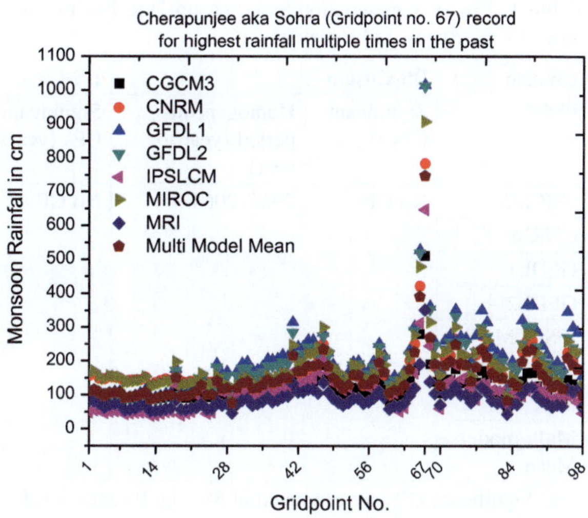

Fig. 6 Mean monsoon rainfall for the period 2046–2064 at different grid points (98 grid points) lying in Northeast India region

Moreover, for understanding the future non-stationary behaviour of monsoon in the Northeast India region, monsoon rainfall data for the period 2046–2064 (19 years) were analyzed. The rainfall data available at a $0.5^0 \times 0.5^0$ grid (98 grid points) fall in Northeast India region (Fig. 3), which were averaged to get weighted average areal monsoon rainfall for the region. Figure 6 shows the mean Monsoon rainfall for the period 2046–2064 at different grid points (98 grid points) lying in Northeast India region.

The change-point analysis using BS-Cusum and BS-LR was applied to the extracted future monsoon rainfall data for Northeast India region. Figure 7 shows the time series of Monsoon Rainfall averaged over Northeast India from seven climate models and Multi-Model Mean for A1B scenario.

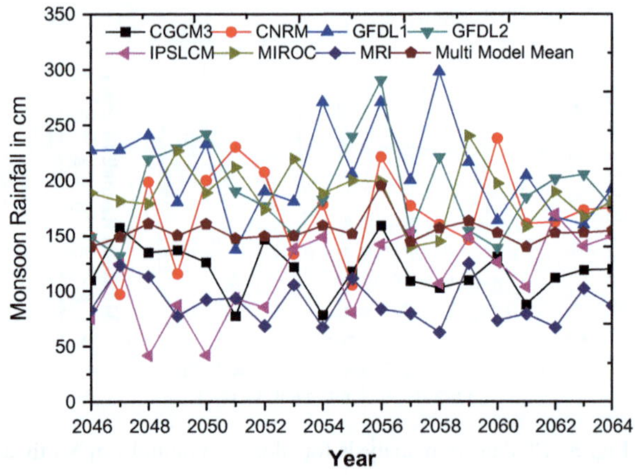

Fig. 7 Non-stationarity analysis for monsoon rainfall time series from seven climate models and ensemble (A1B scenarios) for northeast India region

Table 4 Binary segmentation-based cusum and likelihood ratio approach for monsoon rainfall series for Northeast India regions

Climate model	BS-Cusum		BS-LR		Mean rainfall (cm)
	Significant CPs (year)	Homogeneous period (year to year)	Significant CPs (year)	Homogeneous period (year to year)	
CGCM3	No CP	2046–2064	No CP	2046–2064	118.71
CNRM					169.90
GFDL1					208.78
GFDL2					191.25
IPSLCM					113.46
MIROC					188.05
MRI					89.20
Multi-model Mean					154.19

Note Significant CPs are estimated at 5% significance level

The results of BS-Cusum and BS-LR for monsoon rainfall time series for the seven climate models and multi-model mean are presented in Table 4.

It was seen that the monsoon rainfall time series for future during 2046–2064 is stationary for all the seven climate models and multi-model mean for northeast India region. Results for northeast India region show overall increases in rainfall with climate change which may be due to increased frequency of rainfall events rather than an increase in rainfall intensity. Such results suggest that, regional increases in rainfall are consistent with expected future climate changes. This study can be further extended and analyzed with CMIP5 climate models data with different scenarios and for different periods.

Summary and Conclusions

Hydrologists dealing with stochastic time series usually make the stationarity assumption which is itself questionable as some of the hydrometeorological time series exhibit abrupt change due to changing landuse pattern or climate change. Hence, there is a pressing need to analyze such series using change-point detection techniques. The present work attempted to identify presence and location of a change-point CPs in annual and monsoon rainfall series for five homogeneous regions in India. It is seen that changes in the mean for annual rainfall series are not yet large enough to be detected for all the five regions. Hence the results indicate that stationarity in rainfall series except for monsoon rainfall series of the northeast region. To know the future changes in rainfall patterns for Northeast region the outputs from Global climate models for A1B scenario are analyzed using the two MCPD techniques. It was seen that the monsoon rainfall time series for the future year 2046–2064 is stationary for all the seven climate models and multi-model mean for Northeast India region. It may imply that in Northeast India region, the impacts of climate change and human induced change is not yet warranted for the period 2046–2064. This study suggests that the knowledge of changes in rainfall patterns could be useful for the water resource planners for effective utilization of water resource in the region and to make appropriate decisions for operations of various water resource projects.

References

Akaike H (1974) A new look at the statistical model identification. EEE Trans Autom Contr AC 19 (6):716–723

Alexandersson H (1986) A homogeneity test applied to precipitation data. J Clim 6:661–675

Bozdogan H (2000) Akaike's Information Criterion and recent developments in information complexity. J Math Psychol 44:62–91

Buishand TA (1982) Some methods for testing the homogeneity of rainfall records. J Hydrol 58:11–27

Csorgo M, Horvath L (1997) Limit theorems in change-point analysis. Wiley, Chichester, UK

Dile YT, Srinivasan R (2014) Evaluation of CFSR climate data for hydrologic prediction in data-scarce watersheds: an application in the Blue Nile River Basin. J Am Water Resour Assoc 1–16

Djibo AG, Seidou O, Karambiri H, Sittichok K, Paturel JE, Saley HM (2015) Development and assessment of non-linear and non-stationary seasonal rainfall forecast models for the Sirba watershed, West Africa. J Hydrol Reg Stud 4:134–152

Eckley IA, Fearnhead P, Killick R (2011) Analysis of change-point models. In: Barber D, Cemgil T, Chiappa S (eds) Bayesian time series models. Cambridge University Press, Cambridge

Errasti I, Ezcurra A, Szenz J, Ibarra-Berastegi G (2011) Validation of IPCC AR4 models over the Iberian Peninsula. Theoret Appl Climatol 103(1–2):61–79

Fuka DR, MacAllister CA, Degaetano AT, Easton ZM (2014) Using the climate forecast system reanalysis as weather input data for watershed models. Hydrol Process 28(22):5613–5623

Girvetz EH, Maurer EP, Duffy P, Ruesch A, Thrasher B, Zganjar C (2013) Making climate data relevant to decision making: the important details of spatial and temporal downscaling. The World Bank, pp 1–43

Guhathakurta P, Rajeevan M (2008) Trends in the rainfall pattern over India. Int J Climatol 28 (11):1453–1469

Hawkins DM (1977) Testing a Sequence of Observations for a Shift in Location. J Am Stat Assoc 72(357):180–186

Killick R, Fearnhead P, Eckley I (2012) Optimal detection of change-points with a linear computational cost. J Am Stat Assoc 107:1590–1598

Krishnakumar KN, Rao GSLHVP, Gopakumar CS (2009) Rainfall trends in twentieth century over Kerala, India. Atmos Environ 43:1940–1944

Page ES (1954) Continuous Inspection Schemes. Biometrika 41(1/2):100–115

Parthasarathy B, Rupa Kumar K, Munot A (1993) Homogeneous Indian monsoon rainfall: variability and prediction. Proc Indian Acad Sci 102:121–155

Perreault L, Parent E, Bernier J, Bobee B, Slivitzky M (2000) Retrospective multivariate Bayesian change-point analysis: A simultaneous single change in the mean of several hydrological sequences. Stoch Env Res Risk Assess 14:243–261

Potter KW (1981) Illustration of a new test for detecting a shift in mean in precipitation series. Mon Weather Rev 109:2040–2045

Rodionov SN (2005) A brief overview of the regime shift detection methods. In: Large-scale disturbances (regime shifts) and recovery in aquatic ecosystems: challenges for management toward sustainability, pp 17–24

Scott AJ, Knott M (1974) A cluster analysis method for grouping means in the analysis of variance. Biometrics 507–512

Seidou O, Ouarda TB (2007) Recursion based multiple change-point detection in multiple linear regression and application to river streamflows. Water Resour Res 43(7)

Thrasher B, Maurer EP, McKellar C, Duffy PB (2012) Technical note: bias correcting climate model simulated daily temperature extremes with quantile mapping. Hydrol Earth Syst Sci 16:3309–3314

Tomozeiu R, Busuioc A, Marletto V, Zinoni F, Cacciamani C (2000) Detection of changes in the summer precipitation time series of the region Emilia-Romagna, Italy. Theoret Appl Climatol 67(3–4):193–200

Tripathi S, Govindaraju RS (2009) Change detection in rainfall and temperature patterns over India. In: Proceedings of the third international workshop on knowledge discovery from sensor data. ACM, pp 133–141

Vostrikova LJ (1981) Detecting "disorder" in multidimensional random processes. Soviet Mathematics Doklady 24(1):55–59

Winjngaard JB, Klein Tank AMG, Konnen GP (2003) Homogeneity of 20th century European daily temperature and precipitation series. Int J Climatol 23:679–692

Analyzing Non-stationarity in the Hyderabad City Rainfall Intensity-Duration-Frequency Curves

V. Agilan and N.V. Umamahesh

Abstract The infrastructure design is primarily based on rainfall intensity-duration-frequency (IDF) curves and the current IDF curves are based on the concept of stationary extreme value theory (i.e. occurrence probability of extreme precipitation is not expected to change significantly over time). But, the extreme precipitation events are increasing due to global climate change and questioning the reliability of our current infrastructure design. In this study, the trend in Hyderabad city 1-, 2-, 3-, 6-, 12-, 24- and 48-h duration annual maximum rainfall series are analyzed using the Mann–Kendall (M–K) test, and a significant increasing trend is observed. Further, based on recent theoretical developments in the extreme value theory (EVT), non-stationary rainfall IDF curve for the Hyderabad city is developed by incorporating linear trend in the location parameter of the generalized extreme value (GEV) distribution. The study results indicate that the IDF curves developed under the stationary assumption are underestimating the precipitation extremes.

Keywords Extreme rainfall · IDF curves · Hyderabad city · Non-stationarity

Introduction

The rainfall intensity-duration-frequency (IDF) curves are commonly used in storm water management and other engineering design applications across the world (Endreny and Imbeah 2009) and these curves are developed based on historical rainfall time series data by fitting a theoretical probability distribution to annual maximum rainfall series or partial duration series (Cheng and AghaKouchak 2014). The existing IDF curves are based on the concept of stationary extreme value theory (i.e. occurrence probability of extreme precipitation is not expected to change significantly over time (Jakob 2013)). However, in recent years, the extreme

V. Agilan · N.V. Umamahesh (✉)
Department of Civil Engineering, National Institute of Technology Warangal, Warangal 506004, Telangana, India
e-mail: mahesh@nitw.ac.in

V.P. Singh et al. (eds.), *Climate Change Impacts*, Water Science and Technology Library 82, https://doi.org/10.1007/978-981-10-5714-4_9

precipitation events are escalating due to global climate change (Tramblay et al. 2012; Xu et al. 2015). In specific, in recent years, the impact of different climate processes on the changes in daily extreme precipitation has been analyzed, such as el niño-southern oscillation (ENSO) cycle (Revadekar and Kulkarni 2008; Agilan and Umamahesh 2015a), Global warming (Kunkel et al. 2013; Villafuerte and Matsumoto 2015). Furthermore, reasonable literature reported the possible changes in precipitation due to urbanization (Kishtawal et al. 2009). Especially Burian and Shepherd (2005) hypothesized the possible role of urbanization in diurnal rainfall distribution. In addition, recent studies reported the influence of urbanization in extreme rainfall events as well (Lei et al. 2008; Miao et al. 2011).

Hence, the various physical processes (discussed in the above paragraph) are expected to alter the intensity, duration and frequency of rainfall extremes over time. Consequently, the time series will have a non-stationary component in it and the IDF curves developed based on the stationary extreme value theory may underestimate the extreme event. In other words, the future extreme rainfall events that exceed the capacity of existing drainage systems may occur more frequently if the drainage is designed based on the concept of stationary extreme value theory (Zahmatkesh et al. 2015). In addition, urban flooding and the damage to infrastructure and society are problems in both developing and developed countries. Therefore, the non-stationarity in the rainfall IDF curves of an urban area needs to be analyzed for better urban flood management. In this study, the non-stationarity (trend) in the Hyderabad city extreme rainfall series is analyzed using the Mann–Kendall (M–K) test. Further, based on recent theoretical developments in the extreme value theory (EVT), non-stationary rainfall IDF curve for the Hyderabad city is developed by incorporating linear trend in the location parameter of the generalized extreme value (GEV) distribution and they are compared with stationary IDF curves.

Study Area and Data

The Hyderabad city is the capital of the state of Telangana in India. Location map of the Hyderabad city is shown in Fig. 1. The Hyderabad city lies between the latitudes of 17.25°N and 17.60°N and longitudes of 78.20°E and 78.75°E and situated at a height of about 500 m above the mean sea level. It is classified as a semi-arid region and the Köppen–Geiger classification is BSh (Peel et al. 2007). The major urbanization of Hyderabad city took place after 1990. During 1971–1990, the average rainfall of Hyderabad city was 796 mm per year. But, it has increased to 840 mm per year during 1991–2013. The wettest month of the city is August and the average rainfall in this month is 163 mm.

For this study, the hourly observed rainfall data for Hyderabad city is procured from the India Meteorological Department (IMD) for the period of 1 January

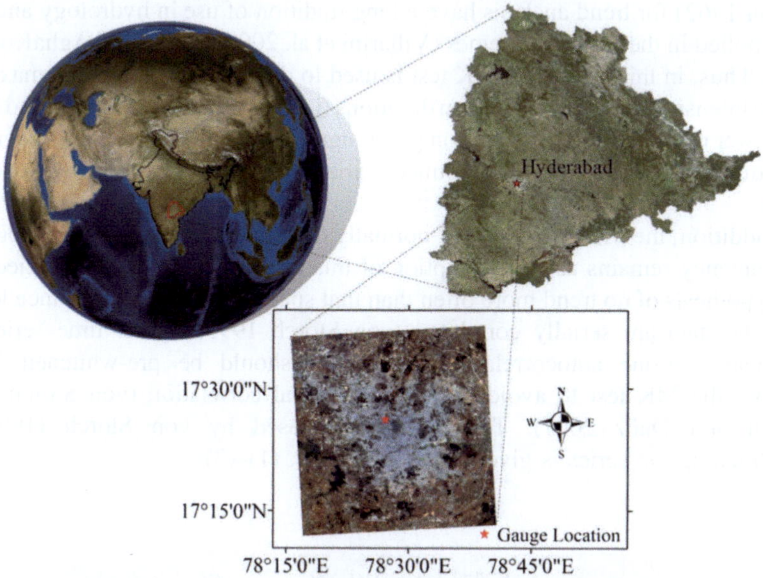

Fig. 1 Location map of Hyderabad city (Agilan and Umamahesh 2015b)

1972–31 December 2013 (42 years). This data is gauge observation and it is observed at the centre of Hyderabad city, i.e. 78.46°E and 17.45°N. The location of this gauge is given in Fig. 1 (star mark).

Methodology

The methodology of this study comprises of following two sections,

1. Analyzing the trend present in the annual maximum rainfall series of 1-, 2-, 3-, 6-, 12-, 24- and 48-h duration using the Mann–Kendall test.
2. Developing non-stationary rainfall IDF relationship for the Hyderabad city.

Trend Analysis

Statistical tests that are used to detect trends in time series are broadly classified into two groups: parametric and nonparametric methods. As the hydrometeorological time series data are generally non-normally distributed and censored, the nonparametric tests are most appropriate for hydrometeorological time series (Bouza-Deaño et al. 2008). The nonparametric Mann–Kendall (M–K) test (Mann 1945;

Kendall 1962) for trend analysis have a long tradition of use in hydrology and have been applied in the case of extremes (Villarini et al. 2009; Cheng and AghaKouchak 2014). Thus, in this study, the MK test is used to detect trends in annual maximum rainfall intensity of the selected storm durations (i.e. 1, 2, 3, 6, 12, 18, 24, 36 and 48 h). The power of the MK test depends on pre-assigned significance level, the magnitude of the trend, sample size and the amount of variations within the time series (Yue et al. 2002).

In addition, the MK test is free of normally distributed data assumption, but data independency remains as an assumption of this test. Thus, the MK test rejects the null hypothesis of no trend more often than that specified by the significance level α when the data are serially correlated (von Storch 1995). If the time series has significant lag-one autocorrelation, the series should be pre-whitened before applying the MK test to avoid the effect of serial correlation (von Storch 1995; Bayazit and Önöz 2007). The method proposed by von Storch (1995) to pre-whitening the series is given as follows (Eqs. (1)–(3)):

$$x_i' = x_i - \rho x_{i-1} \tag{1}$$

$$\rho = \frac{\frac{1}{n-1}\sum_{i=1}^{n-1}(x_i - E(x_i))(x_{i+1} - E(x_i))}{\frac{1}{n}\sum_{i=1}^{n}(x_i - E(x_i))^2} \tag{2}$$

$$E(x_i) = \frac{1}{n}\sum_{i=1}^{n} x_i \tag{3}$$

where x' is pre-whitened data series, x is original series and n is the size of data series. If the original series has significant lag-one autocorrelation, the MK test is applied to pre-whitened data series. Otherwise, the MK is applied to the original data series.

The Non-stationary IDF Curves

Consider an annual maximum series of n independent and identically distributed (iid) random variable x_1, $x_2,\ldots,x_n$. The annual maximum series converges to generalized extreme value (GEV) distribution and the cumulative distribution function is given by Eq. (4) (Coles 2001).

$$F(x;\mu,\sigma,\xi) = \exp\left\{-\left[1+\frac{\xi(x-\mu)}{\sigma}\right]^{-1/\xi}\right\}, \quad \sigma > 0, 1+\frac{\xi(x-\mu)}{\sigma} > 0, \quad \xi \neq 0$$
$$\exp\left\{-\exp\left[-\frac{(x-\mu)}{\sigma}\right]\right\}, \quad \sigma > 0, \quad \xi = 0 \tag{4}$$

where μ, σ and ξ are; location, scale and shape parameters of the GEV distribution, respectively. As the precise estimation of shape parameter is difficult, it is unrealistic to assume it as a smooth function of time (Coles 2001) and modelling temporal changes in scale parameter (σ) requires long-term observations (Cheng et al. 2014). Thus, the scale and shape parameters are kept constant and the non-stationarity is introduced only in the location parameter of GEV. The non-stationary setting for the location parameter of the GEV as a function of the covariate (*f*(*T*)) is given by Eq. (5).

$$\mu(t) = \begin{bmatrix} 1 & f(T) \end{bmatrix} \begin{bmatrix} \mu_0 \\ \mu_1 \end{bmatrix} \tag{5}$$

The slope parameters μ_1 represent the trend in the location parameter due to the covariate *f*(*T*). For computational simplicity, *f*(*T*) must be standardized before non-stationary setting. As an approximation, the linear trend is often used to represent a long-term trend (Cheng et al. 2014; Cheng and AghaKouchak 2014; Yilmaz and Perera 2014). In particular, Sugahara et al. (2009) analyzed the non-stationarity in the extreme daily rainfall in the Sao Paulo city, Brazil by introducing linear trend in scale parameter of the Generalized Pareto distribution. Cheng and AghaKouchak (2014) developed a non-stationary rainfall IDF curves by incorporating linear trend in the location parameter of the GEV distribution. Yilmaz and Perera (2014) investigated the non-stationarity in the IDF curves of Melbourne, Australia by incorporating linear trend in location and shape parameters of the GEV distribution. The covariate for linear trend in the location parameter of the GEV is given by Eq. (6).

$$f(T) = T \tag{6}$$

where $T = 1, 2, 3\ldots n$ (n is the size of the annual maximum rainfall series). In this study, the parameters of the non-stationary GEV distribution are estimated by the method of maximum likelihood, as the method of maximum likelihood can be easily extended to the non-stationary case (Coles 2001; Katz 2013). The non-stationary rainfall intensity is estimated using the model parameters of the non-stationary GEV model. But, unlike the stationary model, the location parameter value will vary over the time. In this study, the low risk (more conservative) approach (Cheng et al. 2014) is used to calculate location parameter value, i.e. the 95 percentiles of the location parameter values and it is given by Eq. (7). The 95 percentiles of the location parameter value give the effective return level (Cheng et al. 2014).

$$\widehat{\mu}_{95} = Q_{95}(\widehat{\mu}_{t1}, \widehat{\mu}_{t2}, \ldots, \widehat{\mu}_{t100},) \tag{7}$$

Estimation of $1/p$ return level is given by Eq. (8) (Coles 2001; Cheng et al. 2014).

$$z_T = \begin{cases} \hat{\mu} + \frac{\hat{\sigma}}{\hat{\xi}}\left[(-\log(1-p))^{-\hat{\xi}} - 1\right], & \hat{\xi} \neq 0 \\ \hat{\mu} + \hat{\sigma}[-\log(-\log(1-p))], & \hat{\xi} = 0 \end{cases} \tag{8}$$

where p (the exceedance probability) is the frequency (i.e. how frequently an extreme rainfall of specified intensity and duration is expected to occur).

Results and Discussion

Before analyzing the trend present in the annual maximum rainfall series of 1-, 2-, 3-, 6-, 12-, 24- and 48-h duration using the MK test, the presence of serial correlation in the rainfall series are analyzed. Significant lag-one autocorrelation is detected in the data series of 1-, 24- and 48-h rainfall durations. The remaining rainfall durations do not show any statistically significant lag-one autocorrelation and they are considered to be independent series. Then the autocorrelated series (i.e. 1-, 24- and 48-h durations rainfall series) are pre-whitened using the method discussed in Sect. 4.1. The MK test is then applied to the pre-whitened 1-, 24- and 48-h duration time series and the original data sets of the remaining rainfall durations (i.e. 2-, 3-, 6- and 12-h). Table 1 shows the results of MK test with different duration annual maximum rainfall series. The 'Tau' value of the MK test is similar to correlation coefficient and its value varies from −1 to 1 (i.e. positive Tau value indicate increasing trend and negative Tau value indicate decreasing trend).

In this study, all duration rainfall series are having positive Tau value and it indicates that the annual maximum rainfall series of all duration (i.e. 1-, 2-, 3-, 6-, 12-, 24- and 48-h) are having increasing trend. The two-tailed p-value given in Table 1 indicates that the increasing trend present in the 12-h duration rainfall series is statistically significant at 0.05 significance level, and the increasing trend present in the 6- and 48-h duration rainfall series is statistically significant at 0.1 significance level. The increasing trend in remaining duration data series is not significant at 0.1 significance level. In addition, the previous studies indicate that the precipitation extremes in several regions of the world have increased (Westra et al. 2013).

Table 1 MK test results

S. No.	Duration (h)	Tau	*p*-value
1	1	0.06	0.60
2	2	0.10	0.35
3	3	0.15	0.17
4	6	0.20	0.06
5	12	0.23	0.03
6	24	0.16	0.14
7	48	0.18	0.09

But, most ground-based stations do not exhibit a strong trend and only limited stations show a statistically significant non-stationary behaviour (Westra et al. 2013; Cheng and AghaKouchak 2014). Moreover, the purpose of analyzing trend using the MK test is to avoid implementing a time varying extreme value analysis on a data that does not show a significant change in extremes over time. The recent studies highlight the need to go beyond subjective criteria for significance analysis, especially in a non-stationary world (Rosner et al. 2014). Thus, the non-stationarity can be applied to all data sets regardless of their trend, avoiding a subjective significance measure (Cheng and AghaKouchak 2014). In this study, the non-stationarity is applied to annual maximum rainfall series of all duration (i.e. 1-, 2-, 3-, 6-, 12-, 24- and 48-h) and compared with stationary models of each duration.

The rainfall IDF relationship for 2, 5 and 10 year return period are developed based on the non-stationary GEV models. The IDF curves of Hyderabad city for 2, 5 and 10 year return periods are shown in Fig. 2a–c, respectively. In this study, it is observed that the IDF curves derived from the stationary models are underestimating the extreme events. If such an IDF curve (developed from the stationary model) is used for an infrastructure design, the project may not be able to withstand the very extreme events.

For example, for an event with a return period of 2 years and duration of 2-h, the difference between the non-stationary (35.74 mm/h) and stationary (26.73 mm/h) extreme rainfall is about 8.98 mm/h. Even for a small watershed, this extra 8.98 mm/h rainfall will lead to a significant increase in peak runoff. In addition, for an event with a return period of 5 years and 2-h duration, the non-stationarity and

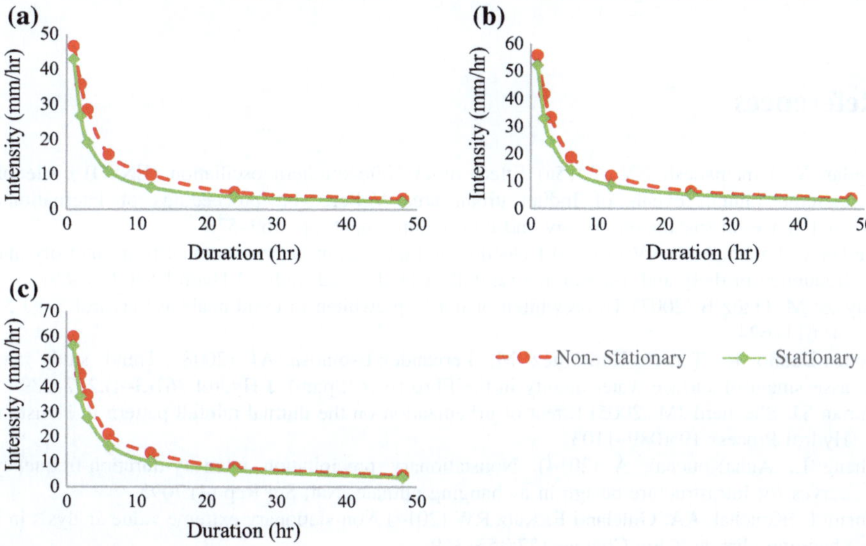

Fig. 2 Non-stationary intensity-duration curves of Hyderabad city for **a** 2-year, **b** 5-year and (10) 10-year return periods

stationarity extreme rainfall are 41.81 and 32.88 mm/h, respectively. The non-stationary extreme rainfall of 2 year return period (35.74 mm/h) is more than the stationary extreme rainfall of 5 year return period (32.88 mm/h). In other words, the return of period of an extreme rainfall of the Hyderabad city is reducing. In other words, the future extreme rainfall events that exceed the capacity of existing drainage systems may occur more frequently if the drainage is designed based on the concept of stationary extreme value theory.

Summary

In this study, with the help of IMD 42 years' hourly rainfall observations, the trend in the Hyderabad city 1-, 2-, 3-, 6-, 12-, 24- and 48-h duration annual maximum rainfall series are analyzed using the Mann–Kendall (M–K) test and a significant increasing trend is observed. Further, based on recent theoretical developments in the extreme value theory (EVT), non-stationary rainfall IDF curve for the Hyderabad city is developed by incorporating linear trend in the location parameter of the generalized extreme value (GEV) distribution. The study results indicate that the IDF curves developed under the stationary assumption are underestimating the precipitation extremes. In other words, the return of period of an extreme rainfall of the Hyderabad city is reducing.

Acknowledgements This work is supported by Information Technology Research Academy (ITRA), Government of India under, ITRA-water grant ITRA/15(68)/water/IUFM/01. We also thank the India Meteorological Department for providing rainfall data.

References

Agilan V, Umamahesh NV (2015a) Effect of el Niño-southern oscillation (ENSO) cycle on extreme rainfall events of Indian urban area Vellore. In: proceedings of international conference on sustainable energy and built environment, pp. 569–573

Agilan V, Umamahesh NV (2015b) Detection and attribution of non-stationarity in intensity and frequency of daily and 4-h extreme rainfall of Hyderabad, India. J Hydrol 530:677–697

Bayazit M, Önöz B (2007) To prewhiten or not to prewhiten in trend analysis? Hydrol Sci J 52 (4):611–624

Bouza-Deaño R, Ternero-Rodríguez M, Fernández-Espinosa AJ (2008) Trend study and assessment of surface water quality in the Ebro river (Spain). J Hydrol 361(3–4):227–239

Burian SJ, Shepherd JM (2005) Effect of urbanization on the diurnal rainfall pattern in Houston. Hydrol Process 19:1089–1103

Cheng L, AghaKouchak A (2014). Nonstationary precipitation intensity-duration-frequency curves for infrastructure design in a changing climate. Nat: Sci Rep (4):7093

Cheng L, Kouchak AA, Gilleland E, Katz RW (2014) Non-stationary extreme value analysis in a changing climate. Clim Change 127:353–369

Coles S (2001) An introduction to statistical modelling of extreme values. Springer, London

Endreny TA, Imbeah N (2009) Generating robust rainfall intensity–duration–frequency estimates with short-record satellite data. J Hydrol 371:182–191

Jakob D (2013) Nonstationarity in extremes and engineering design. In: AghaKouchak A et al (eds) Extremes in a changing climate: detection, analysis and uncertainty. Springer, Dordrecht, pp 363–417

Katz RW (2013) Statistical methods for nonstationary extremes. In: Kouchak AA et al (eds) Extremes in a changing climate: detection, analysis and uncertainty. Springer, Dordrecht, pp 15–37

Kendall M (1962) Rank correlation methods, 3rd edn. Hafner Publishing Company, New York

Kishtawal CM et al (2009) Urbanization signature in the observed heavy rainfall climatology over India. Int J Climatol 30(13)

Kunkel KE et al (2013) Probable maximum precipitation and climate change. Geophys Res Lett 40:1402–1408

Lei M et al (2008) Effect of explicit urban land surface representation on the simulation of the 26 July 2005 heavy rain event over Mumbai. India Atmos Chem Phys 8:5975–5995

Mann HB (1945) Nonparametric tests against trend. J Econ Soci 13(3):245–259

Miao S, Chen F, Fan QL (2011) Impacts of urban processes and urbanization on summer precipitation: a case study of heavy rainfall in Beijing on 1 August 2006. J Appl Meteor Climatol 50(4):806–825

Peel MC, Finlayson BL, McMahon TA (2007) Updated world map of the Koppen-Geiger climate classification. Hydrol Earth Syst Sci 11:1633–1644

Revadekar JV, Kulkarni A (2008) The El Nino-Southern oscillation and winter precipitation extremes over India. Int J Climatol 28:1445–1452

Rosner A, Vogel RM, Kirshen PH (2014) A risk-based approach to flood management decisions in a nonstationary world. Water Resour Res 50:1928–1942

Sugahara S, Rocha RP, Silveira R (2009) Non-stationary frequency analysis of extreme daily rainfall in Sao Paulo, Brazil. Int J Climatol 29:1339–1349

Tramblay Y, Neppel L, Carreau J, Sanchez-Gomez E (2012) Extreme value modelling of daily areal rainfall over Mediterranean catchments in a changing climate. Hydrol Process 26 (25):3934–3944

Villafuerte MQ, Matsumoto J (2015) Significant Influences of global mean temperature and ENSO on extreme rainfall in Southeast Asia. J Climate 28:1905–1919

Villarini G, Serinaldi F, Smith JA, Krajewski WF (2009) On the stationarity of annual flood peaks in the continental United States during the 20th century. Water Resour Res 45:W08417

von Storch H (1995) Misuses of statistical analysis in climate research. In: von Storch H, Navarra A (eds) Analysis of climate variability: applications of statistical techniques. Springer, New York, pp 11–26

Westra S, Alexander LV, Zwiers FW (2013) Global increasing trends in annual maximum daily precipitation. J Clim 26(11):3904–3918

Xu L et al (2015) Precipitation trends and variability from 1950 to 2000 in arid lands of Central Asia. J Arid Land 7(4):514–526

Yilmaz AG, Perera JC (2014) Extreme rainfall nonstationary investigation and intensity–frequency–duration relationship. J Hydrol Eng 19(6):1160–1172

Yue S, Pilon P, Cavadias G (2002) Power of the Mann-Kendall and Spearman's rho tests for detecting monotonic trends in hydrological series. J Hydrol 259(1–4):254–271

Zahmatkesh Z, Karamouz M, Goharian E, Burian SJ (2015) Analysis of the effects of climate change on urban storm water runoff using statistically downscaled precipitation data and a change factor approach. J Hydrol Eng 27(7):0501–4022

Development of Finer Resolution Rainfall Scenario for Kangsabati Catchment and Command

P.M. Dhage, N.S. Raghuwanshi and R. Singh

Abstract Study of the regional scale hydrology of a basin due to impacts of climate change has gained the attention of researchers in the recent past. The general circulation model (GCM), widely used tools for assessing the impacts of climate change, usually predict the hydrological variables of interest at a large-scale, which necessitates the use of downscaling techniques. In this study, Multi-Linear Regression (MLR) and Kernel Regression (KR) two downscaling techniques were compared on the basis of performance statistics, Mean, root mean square error (RMSE), and Nash Sutcliffe Efficiency (NSE). The eleven predictors used from the NCEP/NCAR reanalysis and four GCMs namely, CanESM2, MPI-ESM-LR, CNRM-CM5, and IPSL-CM5A-LR. The best-performed technique among these two techniques was used to develop multi-GCM ensemble daily rainfall scenario at rain gauge stations of Kangsabati study area for high emission scenario (RCP8.5).

Keywords Statistical downscaling · Rainfall scenario · Kangsabati RCP8.5 scenario

Introduction

Studying the impact of climate change on is a challenge facing the climate science community currently. The General Circulation Model (GCM)s are good in simulating the large climate variables accurately but cannot be directly used for hydrologic impact assessment because of their coarse resolution. The objective of downscaling is to predict regional scale hydrologic variables of interest (e.g., rainfall) based on large-scale climatological variables (e.g., specific humidity, mean sea level pressure, temperature, etc.) simulated by a GCM. Climate change study basically involves the downscaling of coarse resolution variables data which are

P.M. Dhage (✉) · N.S. Raghuwanshi · R. Singh
Agricultural and Food Engineering Department, Indian Institute of Technology Kharagpur, Kharagpur 721302, West Bengal, India
e-mail: prandhage@agfe.iitkgp.ernet.in

V.P. Singh et al. (eds.), *Climate Change Impacts*, Water Science and Technology Library 82, https://doi.org/10.1007/978-981-10-5714-4_10

well simulated by GCMs by using two well-known techniques viz. Dynamic downscaling and Statistical Downscaling (Salvi et al. 2013). Every GCM is simulating different climatic predictions because of their physics and varied assumptions from one to another (Benioff et al. 1996). Accordingly, the downscaled (dynamical or statistical) GCMs output tends to develop different predictions at the station scale. Multi-GCM ensemble prediction is well accepted in climate change studies (Yun et al. 2005). Therefore, Four GCMs (CanESM2, MPI-ESM-LR, CNRM-CM5, and IPSL-CM5A-LR) were used from CMIP5 data archive for developing multi-GCM ensemble daily rainfall scenario. The present study is conducted to use the effective statistical downscaling technique by comparing the performance of Multi-Linear Regression (MLR) (linear regression) and Kernel Regression (KR) (nonlinear regression based) techniques for developing daily rainfall scenario at station scale in the Kangsabati reservoir catchment and command in West Bengal (India). Further, the best-performed statistical downscaled technique is applied to predict the rainfall with multi-GCM ensemble outputs for the 2011–2040, 2041–2070 and 2081–2100 time slices under RCP8.5 (high emission) scenario. These predictions are important so that appropriate decisions on adaption and mitigation strategies can be worked under climate change.

Study Area and Data Acquisition

The Kangsabati reservoir, located in the Kangsabati Kumari river basin in West Bengal, India was constructed in 1965, as a multipurpose reservoir for agriculture, water supply and flood control. The annual average rainfall of catchment and command of Kangsabati reservoir is 1302 and 1652 mm. The historical long term daily rainfall (predictand) season at 8 stations for 1971–2005 period is obtained from Indian Meteorological Department (IMD), Pune (Fig. 1). National Centers for Environmental Prediction (NCEP) of National Oceanic and Atmospheric Administration (NOAA) data for downscaling model are downloaded from http://www.esrl.noaa.gov/psd/data/gridded/data.ncep.reanalysis.html link. The four GCMs namely, CanESM2, IPSL-CM5A-LR, MPI-ESM-LR, and CNRM-CM5 predictors data are collected at daily time scale for historical and Representative Concentration Pathways (RCP)8.5 scenario from CMIP5 portal. The change in predicted rainfall for three future time slices 2011–2040, 2041–2070 and 2071–2100 are estimated using an observed reference period of 1976–2005.

Methodology

Eight stations located in the study area were considered for downscaling station rainfall data (Fig. 1). The MLR and KR downscaling techniques were employed for monsoon (June, July, August, and September) season.

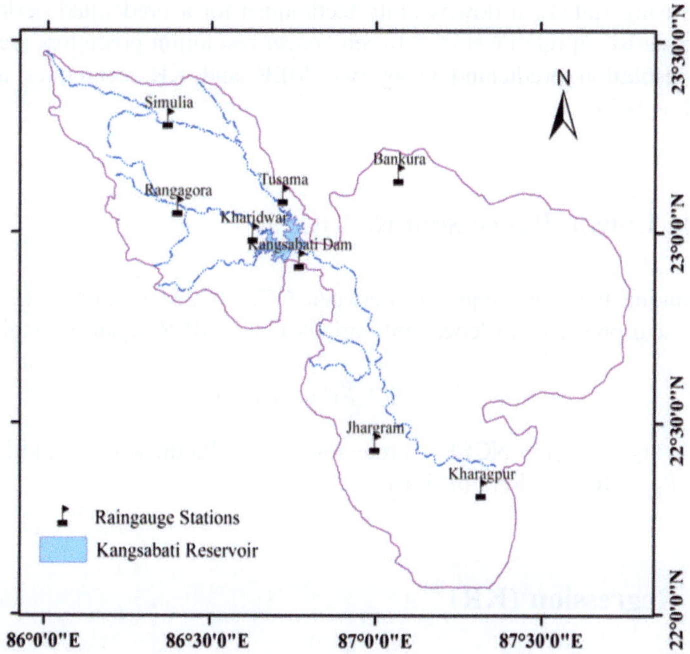

Fig. 1 Location of eight raingauge stations in Kangsabati reservoir catchment and command

Development of Downscaling Model for NCEP Reanalysis Outputs

The NCEP predictors and rainfall time series for the period 1971–1990 and 1991–2005 were used for calibration and validation of MLR and KR downscaling techniques. Ideally, the geophysical processes are associated with fine resolution rainfall. Predictors that directly affect rainfall processes are used in statistical downscaling as input variables. The large-scale atmospheric 11 predictors (precipitation flux, surface zonal and meridional wind, surface based air temperature, mean sea level pressure, geopotential height at 500 hPa, meridional wind at 500 and 850 hPa, specific humidity at 500 and 850 hPa and air temperature at 500 hPa) were selected on the basis of literature and variability of predictors from GCM. These predictors followed the three criteria (Wilby et al. 1999) in statistical downscaling. The selected NCEP and GCM predictors for the calibration and validation period were standardized for monsoon season to remove the units of predictors and scale down to single uniform scale. Standardized data suffers from collinearity. It cannot be directly used for the development of statistical relationship due to instability and unnecessarily complications in statistical downscaling. Therefore, Principal Component analysis (PCA) was used to convert them into a set of uncorrelated variables. The PC based large-scale predictors were calibrated and validated using statistical downscaling techniques for a predefined period. In this study, the statistical relationship between coarse resolution predictors are linked to the fine resolution predictand using two MLR and KR techniques are briefly described as below.

Multiple Linear Regression (MLR)

This technique fits linear equation between NCEP/GCM predictors (independent variables) and predictand (dependent variable). The MLR equation is given as:

$$Y = \beta_0 + \beta_1 x_1 + \beta_{11} x_{11} \tag{1}$$

where $x_1, x_2, \ldots, x_{11}$ = NCEP predictors, Y = Predictand (Rainfall), and $\beta_1, \beta_2, \ldots \beta_{11}$ = Regression coefficients.

Kernel Regression (KR)

The KR is a nonparametric approach to estimate the conditional expectation of random variable. The Gaussian kernel function with a zero mean and unit standard deviation is used here.

$$E(Y|X) = f(X) \tag{2}$$

$$f(x) = (nh)^{-1} \sum_{i=1}^{n} K((x - x_i|h)) \tag{3}$$

where, $K(\cdot)$ = kernel function of predictand, x_i = ith rainfall observation, n = number of observations, and h = smoothing parameter known as bandwidth.

Performance Evaluation of MLR and KR Downscaling Techniques

In order to examine the effectiveness of these two techniques in simulating daily rainfall, the performance evaluation of all techniques for eight raingauge stations was carried out using three statistical measures namely, mean, Root Mean Square Error (RMSE) and Nash–Sutcliffe Efficiency (NSE) (Nash and Sutcliffe 1970). The use of three statistical indicators, in this study, facilitated to explore the relative potential of these indicators in analyzing technique results. The best-performed technique for the study area was used to estimate the rainfall for future years 2011–2100. The mathematical expressions for these measures are given as follows:

$$\text{Mean} = \frac{1}{n}\sum_{i=1}^{n} X_i \tag{4}$$

$$\text{Correlation Coefficent} = \frac{\sum \left(P_i - \overline{P}\right)\left(O_i - \overline{O}\right)}{\sqrt{\sum \left(P_i - \overline{P}\right)^2}\sqrt{\sum \left(O_i - \overline{O}\right)^2}} \tag{5}$$

$$\text{RMSE} = \sqrt{\frac{1}{N}\sum\nolimits_{i=1}^{N} \left(P_i - O_i\right)^2} \tag{6}$$

$$\text{NSE} = 1 - \frac{\sum_{i=1}^{n} \left(P_i - O_i\right)^2}{\sum_{i=1}^{n} \left(O_i - \overline{O}\right)^2} \tag{7}$$

where, X_i = ith observation, n = number of observations, P = Predicted rainfall, O = Observed rainfall, $\overline{O}$ = Mean of observed rainfall.

Establishment of Downscaling Technique for Multi-GCM Ensemble Outputs

The well-performed downscaling technique with NCEP outputs was used for development of the downscaling model with the same atmospheric domain, the same set of selected GCM predictors for the historical observed period (1971–2005). The assumption was appropriated that selected predictors are valid for both NCEP and GCM downscaling models as the raingauge station remained same for both models. The GCM predictors were preprocessed through interpolation, standardization, bias correction and PCA before statistically linked with the predictand. Bilinear interpolation was performed on GCM predictors to match its outputs on NCEP (2.5° × 2.5°) grid. The bias correction was performed for removing the systematic error of GCM with reference of NCEP outputs using equidistant quantile mapping method (Li et al. 2010) for the past 1971–2005 and future 2006–2100 periods. Thereafter, the bias corrected four individual GCMs predictors are prepared through preprocessing steps as same as of NCEP predictors for the historical period. Thereafter, the best-performed technique was used to downscale the outputs of predefined four GCMs (CanESM2, IPSL-CM5A-LR, MLP-ESM-LR, CNRM-CM5) with historical climate (1976–2005) and future climate (2006–2100) data for prediction of future rainfall scenario. The simplest averaging method was used for deriving an ensemble prediction. The outputs of four individual GCMs pertaining to historical and future climate were introduced to MLR downscaling technique to generate multi-GCM ensemble outputs for predicting the rainfall at station scale into future.

Results and Analysis

Performance Comparison of MLR and KR Statistical Downscaling Techniques

The MLR and KR techniques were calibrated with monsoon season and validated for the predefined period using NCEP predictors. The performance of MLR and KR downscaling techniques were evaluated by using mean, RMSE and NSE for the validation period (Table 1). The mean of observed rainfall showed good agreement with the MLR simulated mean for monsoon season as compared to the results obtained from KR technique. In order to have a better idea about the performance of MLR and KR techniques, RMSE and NSE were also calculated. All two techniques showed the higher value of RMSE at all station. But, among these techniques, MLR was obtained lower RMSE than KR technique. Another performance statistic NSE was achieved higher for MLR technique than the other technique, whereas KR showed poor performance for developing daily rainfall, due to higher RMSE, lowest NSE and higher variability of mean rainfall (Table 1). The results of statistical indicators, i.e., mean, RMSE and NSE were better in the case of simple MLR technique under monsoon season. Therefore, it can be concluded from the above analysis that the MLR technique performs relatively better than KR technique at the eight stations. Sachindra et al. (2014) also concluded that MLR technique has the capability to reproduce finer resolution rainfall. Therefore, the MLR technique was successfully implemented in obtaining GCM historical and future predictions with RCP8.5 scenario at 8 stations of West Bengal.

Table 1 Performance statistic for daily rainfall amount (mm) for validation period 1991–2005

Statistic	RMSE		NSE		Mean (mm)		
Technique	MLR	KR	MLR	KR	Observed	MLR	KR
Simulia	6.41	8.63	0.49	0.18	9.35	8.56	8.24
Rangagora	5.28	8.31	0.46	0.19	8.00	8.04	7.16
Kharidwar	4.62	7.54	0.45	0.14	8.91	8.53	7.41
Tusama	5.30	6.91	0.49	0.14	8.87	9.99	8.14
Kangsabati Dam	4.51	9.16	0.44	0.23	9.35	8.85	8.71
Bankura	5.69	7.61	0.46	0.29	11.85	12.07	9.76
Jhargram	9.33	12.42	0.45	0.32	12.83	12.36	11.43
Kharagpur	4.43	6.00	0.45	0.21	9.77	10.39	7.74

Multi-GCM Ensemble Rainfall Prediction for Historical and RCP8.5 Scenarios

From the time series plot (Fig. 2a), it can be easily observed that the mean and standard deviation of observed data showed a good match with simulated mean and standard deviation obtained from MLR downscaling technique. The rainfall variability is captured better way using MLR downscaling technique with 11 NCEP and GCM predictors. Figure 2b shows a scatter plot of intersite correlations computed by the models versus actual correlations between the sites, for validation years

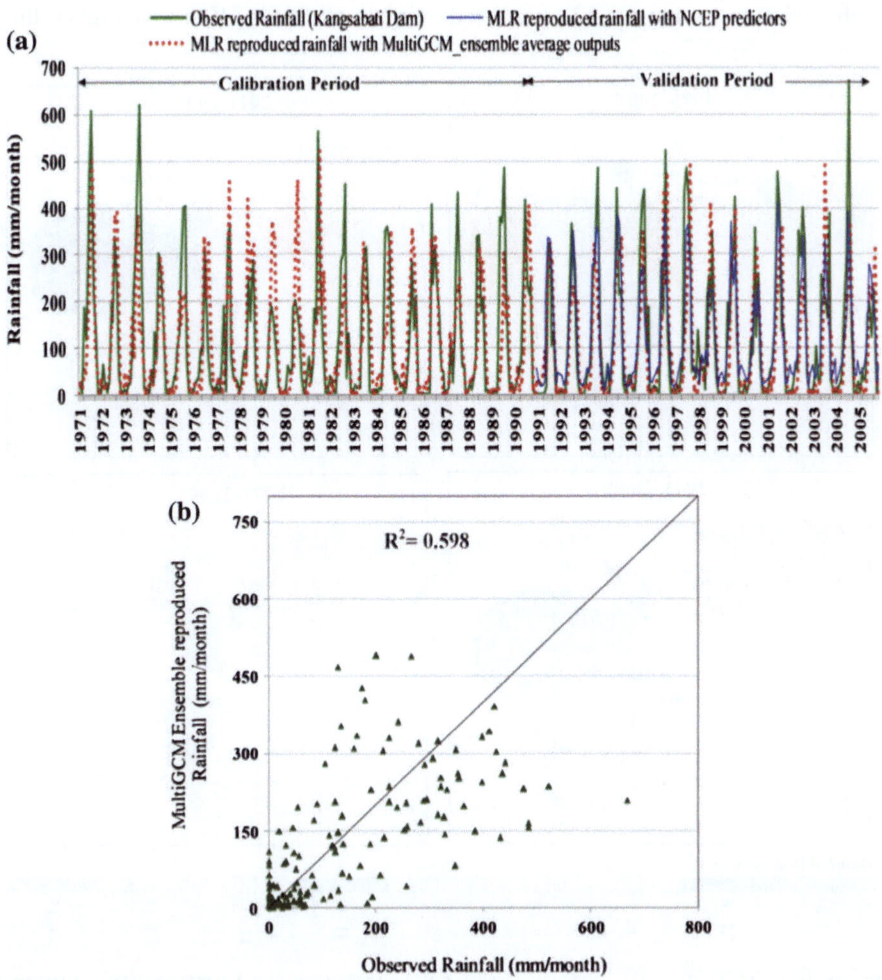

Fig. 2 **a** Rainfall reproduced by MLR statistical downscaling technique with NCEP and multi-GCM ensemble average at for calibration (1971–1990) and validation period (1991–2005), **b** scatter plot of multi-GCM ensemble reproduced and observed rainfall for validation period

1991–2005. It shows a capability of the MLR technique in simulating the daily rainfall for monsoon season using multi-GCM ensemble for Kangsabati reservoir catchment and command.

In the case of high emission RCP8.5 scenario, rainfall will be increasing at Jhargram and Bankura in command area and at Tusama station of catchment for monsoon season. The magnitude of increase in rainfall is observed to be more under RCP8.5 scenario than historical scenario. Figure 3 shows the spatial variation of mean daily rainfall for RCP8.5 scenario for 2011–2040, 2041–2070 and 2071–2100 periods. The increase in future monsoon rainfall in this study area is in line with studies of Kumar et al. (2013), Salvi et al. (2013) in northeastern part (West Bengal) of India under high emission scenario. Mishra et al. (2009) also found that under high emission scenario (A2) predicted rainfall of 2051–2100 was higher than

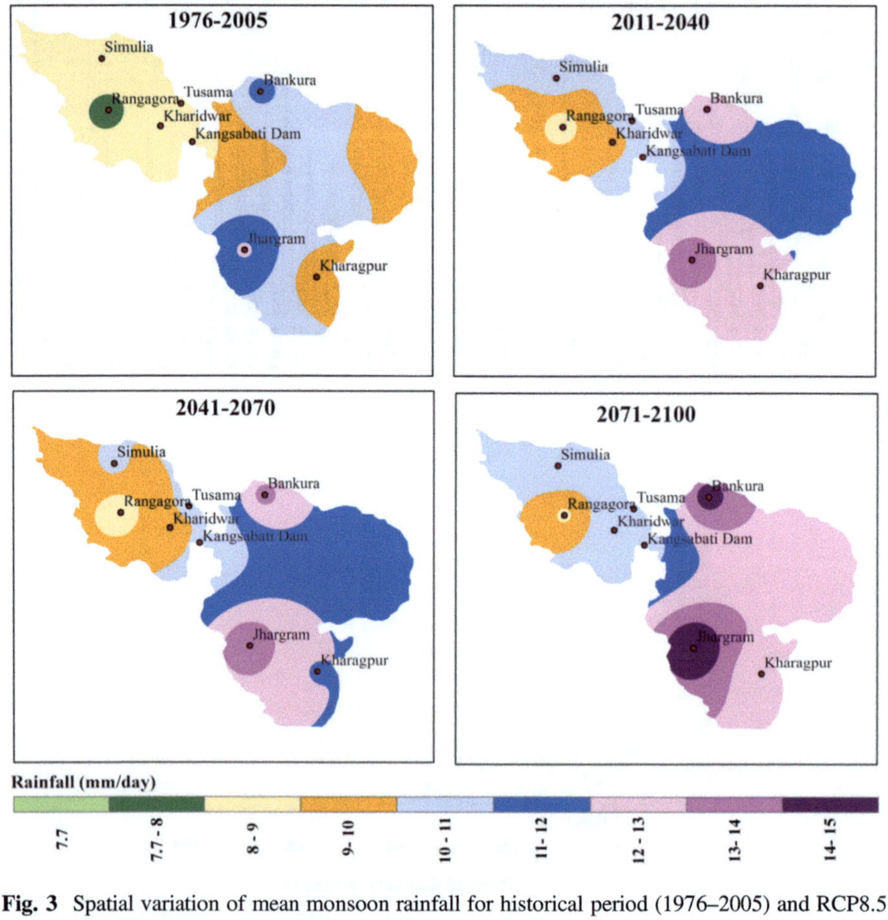

Fig. 3 Spatial variation of mean monsoon rainfall for historical period (1976–2005) and RCP8.5 scenario for 2011–2040, 2041–2070 and 2071–2100 periods

2011–2050 and the historical period for Kangsabati catchment study area. The heavy rainfall implies that in future will increase in these areas that may lead to flooding, crop damage problem in Kharif season.

Conclusions

In this study, the rainfall was predicted for RCP8.5 scenario using well-performed technique for Kangsabati reservoir catchment and command. The intercomparison of two statistical downscaling techniques viz., MLR and KR was carried out to identify the effective prediction technique for rainfall. In statistical downscaling, calibration and validation was performed using 11 NCEP predictors and observed rainfall predictand for the period of 1971–1990 and 1991–2005, respectively. The results of calibration and validation showed the MLR technique performed better than KR technique at all stations for rainfall. Therefore, future prediction of rainfall was carried out only by employing MLR technique using four GCM variables for RCP8.5 scenario. The results showed that the predicted rainfall will be increased over the observed reference period 1976–2005 and among the three time slices under high emission scenario. The use of multi-GCM while predicting future climate condition and ensuring impacts could help to capture key climate features are not well explained by one of the GCM. Hence, future attempts for region could rely on realizations from multi-GCMs.

References

Benioff R, Warren J, Warren J (1996) Steps in preparing climate change action plans: a handbook. US Country Studies Program

Kumar P, Wiltshire A, Mathison C, Asharaf S, Ahrens B, Lucas-Picher P, Christensen JH, Gobiet A, Saeed F, Hagemann S, Jacob D (2013) Downscaled climate change projections with uncertainty assessment over India using a high resolution multi-model approach. Sci Total Environ 468:S18–S30

Li H, Sheffield J, Eric F (2010) Bias correction of monthly precipitation and temperature fields from Intergovernmental panel on Climate change AR4 model using equidistant quantile matching. J Geophys Res 115(D10):1–20. doi:10.1029/2009JD012882

Mishra AK, Özger M, Singh VP (2009) Trend and persistence of precipitation under climate change scenarios for Kansabati basin, India. Hydrol Process 23(16):2345–2357

Nash JE, Sutcliffe JV (1970) River flow forecasting through conceptual models. Part 1: a discussion of principles. J Hydrol 10(3):282–290

Sachindra DA, Huang F, Barton AF, Perera BJC (2014) Multi-model ensemble approach for statistically downscaling general circulation model outputs to precipitation. Q J Roy Meteor Soc 140(681):1161–1178

Salvi K, Kannan S, Ghosh S (2013) High-resolution multisite daily rainfall projections in India with statistical downscaling for climate change impacts assessment. J Geophy Res-Atmos 118 (9):3557–3578

Wilby RL, Hay LE, Leavesley GH (1999) A comparison of downscaled and raw GCM output: Implications for climate change scenarios in the San Juan river basin. Colorado J Hydrol 225(1–2):67–91

Yun WT, Stefanova L, Mitra AK, Kumar TV, Dewar W, Krishnamurti TN (2005) A multi-model superensemble algorithm for seasonal climate prediction using DEMETER forecasts. Tellus A 57(3):280–289

Investigation of the Relationship Between Natural Aerosols and Indian Summer Monsoon Rainfall Using a Climate Model

Charu Singh, Dilip Ganguly and S.K. Dash

Abstract Using all forcing simulations of fully coupled climate model GFDL-CM3, an attempt has been made to represent the role of natural aerosols (such as dust) in modulating the summer monsoon rainfall over the Central Indian region. For this purpose, long-term data set of dust, winds and rainfall have been obtained from CMIP5 data portal for the past 54 years of time period. The spatial pattern of dust load from GFDL-CM3 is able to capture the potential dust sources such as Sahara Desert, Arabian Peninsula region. Further to this, it has been observed that the load of dust over the Arabian Sea and Arabian Peninsula is significantly correlated (significant at 1% significance level) with the rainfall over the central Indian region, suggesting an in-phase relationship between the two parameters.

Keywords Climate model · Natural aerosol · CMIP5 · Significant

Introduction

Atmospheric aerosols are the suspended liquid or solid particles in the air, which could be of different compositions, sizes, shapes and optical properties, and these aerosol properties depend upon their source of origin such as natural and anthropogenic. Aerosols through their optical properties modify the Earth's radiation budget. Aerosols could cut down the solar radiation before it reaches the Earth surface, and hence are responsible for the reduction in the land surface temperature thereby affecting the vegetation growth (Prasad et al. 2004, 2005).

C. Singh (✉)
Marine and Atmospheric Sciences Department, Indian Institute of Remote Sensing, ISRO, Dehradun, India
e-mail: singhcharu105@gmail.com

D. Ganguly · S.K. Dash
Centre for Atmospheric Sciences, Indian Institute of Technology Delhi, New Delhi, India

V.P. Singh et al. (eds.), *Climate Change Impacts*, Water Science and Technology Library 82, https://doi.org/10.1007/978-981-10-5714-4_11

The aerosols are considered as one of the major contributors in Earth's climate change; however, the exact role of aerosols in modifying Earth's climate is not yet quantified. Several previous studies documented the fact that the Indian summer monsoon rainfall is significantly related to the amount of the aerosols (Jin et al. 2014; Vinoj et al. 2014; Singh et al. 2015, 2017a,b). Aerosols modulate the monsoon in different ways. Ramanathan et al. (2005) proposed that the aerosols induce the cooling effect by absorption and scattering of the solar radiation, which could reduce the land–ocean temperature difference and in turn reduces the Indian summer monsoon rainfall. This theory is known as "solar dimming effect". Lau et al. (2006) documented the elevated heat pump theory, and elaborated the fact that the absorbing aerosols coupled with the Tibetan Plateau heating induce the strong land–ocean temperature gradient which is responsible for the early arrival and intensification of the monsoon rainfall. Recently, based on GFDL-CM3 climate model simulations, Bollasina et al. (2011) suggested that the anthropogenic aerosols are responsible for the reduction in the monsoon rainfall over the central Indian region. The fast and slow responses of the monsoon system to the anthropogenic aerosols have been discussed in detail by Ganguly et al. (2012a, b). Previous studies for example, Kuhlmann and Quaas (2010), Gautam et al. (2009) reported that the dust aerosols contribute more towards the aerosol loading over the Arabian Sea during pre-monsoon and monsoon period. Therefore, this study is designed to investigate the relationship between the amount of natural aerosols (i.e. dust) and the Indian summer monsoon rainfall using a long-term data set simulated by a climate model.

Data Set and Methods

The load of dust, surface winds and precipitation data sets corresponding to all forcing simulations of the climate model GFDL-CM3 have been obtained from CMIP5 (coupled model intercomparison project 5) data portal. Data set for the time period from 1951 to 2004 has been utilised in the present study. Based on the spatial pattern of dust, the regions with significant dust loading are identified and further to this, the relationship between the dust over the identified regions and the rainfall over the central Indian region have been studied using scatter plot and lead–lag relationship. The significance of the correlation is tested using t-test under the null hypothesis of $r = 0$ (i.e., no correlation exists between the variables). The test statistics of the t-test to estimate the significance of the correlation between N number of samples, may be expressed as follows:

$$\text{Test statistics} = r\sqrt{\frac{(N-2)}{(1-r^2)}} \tag{1}$$

Results

The climatology of the load of dust simulated by GFDL-CM3 for the past 50 years of the time period is shown in Fig. 1. TOM's aerosol index and GOCART dust optical depth are also illustrated in the top panel of Fig. 1 for the comparison purpose. It appears that the GFDL-CM3 is able to identify the dust sources such as Sahara Desert, Arabian Peninsula region and Gobi Desert, etc., and follows the GOCART dust optical depth pattern. Based on the previous studies (Vinoj et al. 2014; Jin et al. 2014; and references therein) and as per the spatial pattern of the dust load, four sub-regions which indicate sufficiently high dust loading are selected for the further study in this present work. Four regions are shown in the top panel of Fig. 1 (SD: Middle West part of Africa encompassing the Sahara Desert, AP: Arabian Peninsula, AS: The Arabian Sea and CI: central Indian).

Spatial plots illustrate the TOMS aerosol index, dust optical depth from GOCART model in the top panel and dust load from GFDL-CM3 in the bottom panel. The colour bars in the top panel represents the aerosol index and optical depth, respectively, and the colour bar in the bottom panel indicates dust load in mg/m^2. The four identified regions are shown by white text in the first figure.

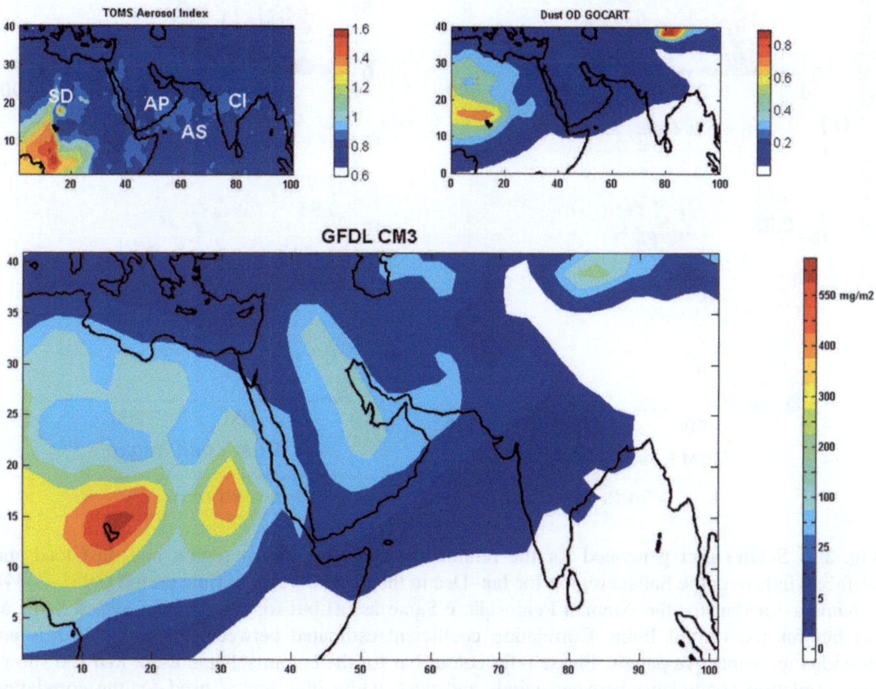

Fig. 1 Spatial distribution of the climatological Jan–Dec load of dust for the time period starting from 1951 to 2004

Further to this, the scatter plots are generated for load of dust and winds (surface winds) over the four identified regions and the rainfall over the central Indian region. For this purpose, monthly data sets of dust load, winds and rainfall from 1951 to 2004 have been utilised. The correlation coefficient has also been computed between the two parameters (i.e. winds and dust and dust and winds). Based on *t*-test, 0.2 is the significant correlation coefficient for 648 samples at 1% significance level. Estimated correlation coefficients are provided on respective panels. For illustration, please refer Fig. 2.

Rainfall over the central Indian region is significantly positively correlated with a load of dust over the distant sources (Fig. 2). Further to this, it appears that the correlation between the area averaged central India rainfall improves when one moves from the Sahara dust source to the Arabian Peninsula and then to the

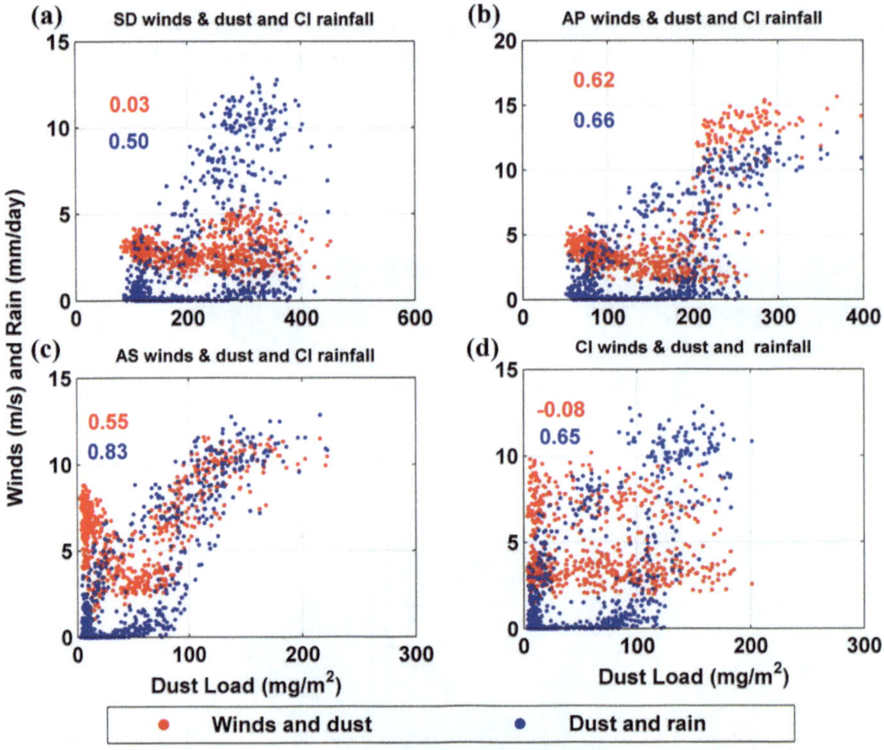

Fig. 2 **a** Scatter plot generated for the rainfall over Central Indian region and dust load and surface winds over the Sahara region for Jan–Dec in the past 50 years of time period (1951–2004). **b** Same as (**a**) but for the Arabian Peninsula. **c** Same as (**a**) but for the Arabian Sea. **d** Same as (**a**) but for the central India. Correlation coefficient estimated between the two data sets are provided on respective panels. Please refer colour bar for the colour scheme used. *Red text* shows the correlation coefficients between winds and dust, while *blue text* is used for the correlation coefficient between dust load and rainfall. *X*-axis represents dust load in mg/m^2 and *Y*-axis shows rainfall in mm/day and winds in m/s

Arabian Sea. However, the correlation coefficient reduces when the coefficient is estimated between the dust load and rainfall over the central India. The strongest relationship is observed between the central Indian rainfall and dust load over the Arabian Sea. Surface winds over the Sahara region are weakly positively correlated with the amount of dust, whereas the strong positive correlation between the surface winds and dust load is noted over the Arabian Peninsula and the Arabian Sea. Additionally, dust load over the central Indian region found to be negatively correlated with the surface winds. This is also worth to mention here that the correlation between the dust load and winds improves substantially over the Central India for 850 hPa winds (correlation coefficient ~0.67). However, the relationship between the winds and dust load weaken over the Arabian Peninsula (0.19) and Sahara region (−0.49) for 850 hPa winds (figures are not shown).

In this study, we also examined the lead-lag relationship between the load of dust and winds over the four identified regions and the rainfall over the central Indian region (Fig. 3). It appears from the figure that the dust load shows the

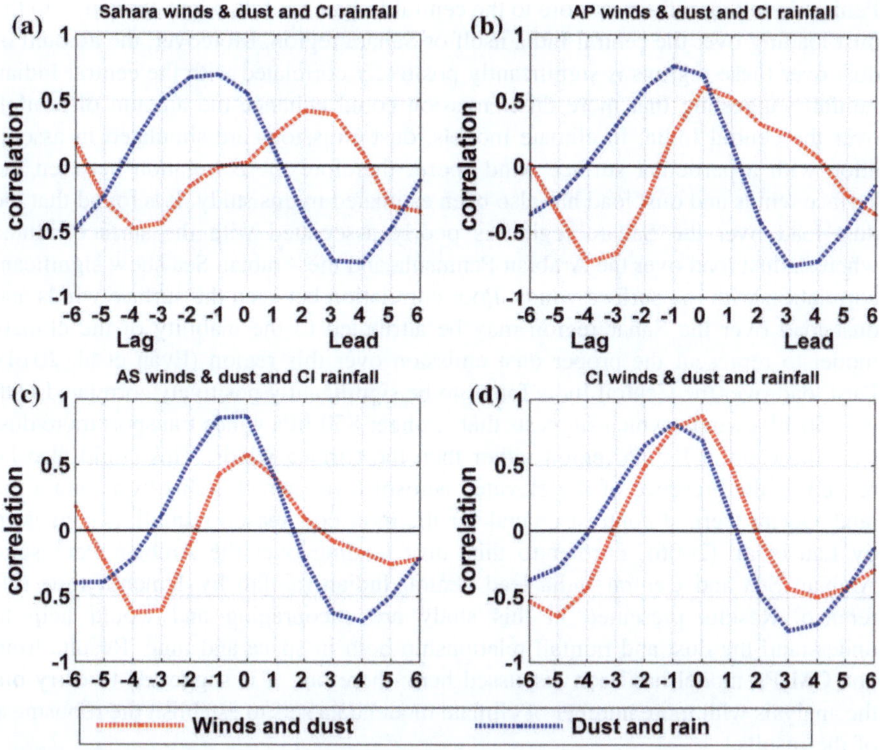

Fig. 3 **a** Cross correlation estimated between the precipitation over Central Indian region and dust load and surface winds over the Sahara region for Jan–Dec in the past 50 years of time period (1951–2004) and denoted as a function of lead and lag at monthly scale. **b** Same as (**a**) but for the Arabian Peninsula. **c** Same as (**a**) but for the Arabian Sea. **d** Same as (**a**) but for the central India. Load of dust is considered as a leading parameter for the computation. *X*-axis represents lead-lag in months and *Y*-axis shows correlation coefficients

maximum correlation with the surface winds for zero lag (i.e. for same month) over the Arabian Peninsula, Arabian Sea and Central India, however similar association is not seen in the Sahara region. Further, the dust load over the Sahara region seems to lead the central Indian rainfall by one to two months' time difference. On the other hand, the dust load over the Arabian Peninsula, the Arabian Sea and central India found to be leading ahead of the central Indian rainfall by one month, since the maximum correlation appears corresponding to the 1 month lead period for all these three regions.

Summary

Using historical simulations of GFDL-CM3 climate coupled model, the relationship between the amount of dust and the rainfall is examined in the present work. It is observed that the GFDL-CM3 represents the potential dust sources reasonably well. Our analysis reveals that the amount of dust over the Arabian Sea and Arabian Peninsula region influence more to the central Indian rainfall, when compared to the dust loading over the central India itself or Sahara region. Moreover, the amount of dust over these regions is significantly positively correlated with the central Indian rainfall suggesting that more dust emission could enhance the amount of rainfall over the central India. In climate models, dust emissions are simulated in association with a particular surface wind speed, therefore the correlation between the surface winds and dust load has also been estimated in this study. It is found that the dust load over the Sahara region is poorly associated with the surface winds, whereas dust load over the Arabian Peninsula and the Arabian Sea show significant correlation with the surface winds. Poor correlation between the surface winds and dust load over the Sahar region may be attributed to the inability of the climate model to represent the proper dust emission over this region (Evan et al. 2014). Dust load over the Central India found to be significantly positively correlated with the 850 hPa winds, which suggests that stronger 850 hPa winds transport more dust over the Central Indian region rather than the surface winds. This could also be related to the presence of the elevated aerosol layer, which helps to maintain the land–Ocean thermal contrast crucial for the monsoon season rainfall as suggested by Lau et al. (2006). Further to this, dust loading over the Arabian Peninsula, Arabian Sea and Central India lead central Indian rainfall by 1 month time difference. Results presented in this study are encouraging and would help to understand the dust and rainfall relationship both in space and time. Results from one CMIP5 model has been discussed here; therefore, it is suggested to carry out the analysis with more number of climate model data sets to establish the robustness of the results.

Acknowledgements First author wishes to thank Head MASD, Group Director ER and SS Group and Director, IIRS for providing the support and encouragement to carry out the present work. We acknowledge the World Climate Research Programme's Working Group on Coupled Modelling, which is responsible for CMIP5, and we thank the climate modelling groups (specifically GFDL-CM3) for producing and making available their model output. TOMS aerosol index and GOCART dust optical depth have been downloaded from http://disc.sci.gsfc.nasa.gov/data-hldings/PIP/aerosol_index.html and http://acd-ext.gsfc.nasa.gov/People/Chin/gocartinfo.html respectively.

References

Bollasina MA, Ming Y, Ramaswamy V (2011) Anthropogenic aerosols and the weakening of the South Asian summer monsoon. Science 502–505

Evan AT, Flamant C, Fiedler S, Doherty O (2014) An analysis of aeolian dust in climate models. Geophys Res Lett 41. doi:10.1002/2014GL060545

Ganguly D, Rasch PJ, Wang H, Yoon J-H (2012a) Fast and slow responses of the South Asian monsoon system to anthropogenic aerosols. Geophys Res Lett 39:L18804. doi:10.1029/2012GL053043

Ganguly D, Rasch PJ, Wang H, Yoon J-H (2012b) Climate response of the South Asian monsoon system to anthropogenic aerosols. J Geophys Res 117:D13209. doi:10.1029/2012JD017508

Gautam R, Hsu NC, Lau KM, Kafatos M (2009) Aerosol and rainfall variability over the Indian monsoon region: distributions, trends and coupling. Ann Geophys 27:3691–3703

Jin Q, Wei J, Yang Z-L (2014) Positive response of Indian summer rainfall to middle east dust. Geophys Res Lett. doi:10.1002/2014GL059980

Kuhlmann J, Quaas J (2010) How can aerosols affect the Asian summer monsoon? Assessment during three consecutive pre-monsoon seasons from CALIPSO satellite data. Atmos Chem Phys 10(10):4673–4688. doi:10.5194/acp-10-4673-2010

Lau KM, Kim MK, Kim KM (2006) Asian summer monsoon anomalies induced by aerosol direct forcing: the role of the Tibetan Plateau. Clim Dynam 26(7–8):855–864. doi:10.1007/s00382-006-0114-z.1

Prasad AK, Singh RP, Singh A (2004) Variability of aerosol optical depth over Indian subcontinent using MODIS data. J Indian Soc Remote Sens 32(4) (Dec 2004 Issue)

Prasad AK, Singh RP, Singh A, Kafatos M (2005) Seasonal variability of aerosol optical depth over Indian subcontinent. In: IEEE, pp 35–38

Ramanathan V et al (2005) Atmospheric brown clouds: impacts on South Asian climate and hydrological cycle. Proc Natl Acad Sci USA 102(15):5326–5333. doi:10.1073/pnas.0500656102

Singh C, Thomas L, Kumar KK (2015) Impact of aerosols and cloud parameters on Indian summer monsoon rain at intraseasonal scale: a diagnostic study. Theor Appl Climatol. doi:10.1007/s00704-015-1640-6

Singh C, Ganguly D, Dash SK (2017a) On the dust load and rainfall relationship in South Asia: an analysis from CMIP5. Clim Dyn. doi:10.1007/s00382-017-3617-x

Singh C, Ganguly D, Dash SK (2017b) Dust load and rainfall characteristics and their relationship over the South Asian monsoon region under various warming scenarios. J Geophys Res: Atmos 122. doi:10.1002/2017JD027451

Vinoj V et al (2014) Short-term modulation of Indian summer monsoon rainfall by West Asian dust. Nat Geosci 7:308–313. doi:10.1038/ngeo2107

Part III
Impacts of Climate Change

Change Point Analysis of Air Temperature in India

N.R. Chithra, Santosh G. Thampi, Dilber Shahul, Sankar Muralidhar, Upas Unnikrishnan and K. Akhil Rajendran

Abstract Change point analysis was performed on air temperature at different pressure levels in the Indian subcontinent to identify the time at which a major change in trend, if any, has occurred. Pettit test, a nonparametric test to identify change points in a time series was used for this purpose. It tests, the null hypothesis that the variable follows one or more distributions that have the same location parameter against the alternate hypothesis that a change point exists. The significance of the change point is determined and if it is greater than the considered level of 90% confidence, then, the change point is considered to be significant. The test was performed on the surface temperature data of the Indian subcontinent for the period 1949–2014, obtained from the NCEP/NCAR reanalysis data set at a resolution of 2.5°. The results of the test for the dry period indicate that the southern, northern and northeastern parts of India exhibited a significant change point in the nineteen seventies. During the wet season and the southwest monsoon season, a significant change was observed in the southern, central and eastern parts of India in the last decade. Analysis of the annual mean temperature revealed that a significant change point occurred in South India in the last decade.

Introduction

It has been observed in many studies that the global climate has taken a significant turn in the recent decades. The impact of climate change is projected to have different effects within a country and between countries. Information about such changes is required at global, regional and basin scales for a variety of purposes. According to the assessments by the Intergovernmental Panel on Climate Change (IPCC) 2001, increase in greenhouse gas concentrations caused an increase in the annual mean

N.R. Chithra (✉) · S.G. Thampi · D. Shahul · S. Muralidhar · U. Unnikrishnan · K. Akhil Rajendran
Department of Civil Engineering, National Institute of Technology Calicut, Calicut 673601, Kerala, India
e-mail: chithranr@nitc.ac.in

V.P. Singh et al. (eds.), *Climate Change Impacts*, Water Science and Technology Library 82, https://doi.org/10.1007/978-981-10-5714-4_12

global temperature by 0.6 ± 0.2 °C since the late nineteenth century (Houghton et al. 2001). According to the estimates by the IPCC (2007), the earth's linearly averaged surface temperature has increased by 0.74 °C during the period 1901–2005 (Pachauri and Reisinger 2007). Weather reports indicate that the global mean surface temperature has risen, approximately by 0.6 °C, since 1850. It is expected that by 2100, the increase in temperature could be in the range 1.4–5.8 °C (Singh et al. 2008).

The study by Srivastava et al. (1992) on decadal trends in climate over India gave the first indication that temperature trends in India are quite different from that observed over various parts of the globe. They observed that the maximum temperatures show much larger increasing trends than a minimum temperature, over a major part of the country and an overall slightly increasing trend of the order of 0.35 °C over the last 100 years. Rupa Kumar et al. (1994) have shown that the countrywide mean maximum temperature has risen by 0.6 °C. Lal et al. (1995) suggested that increase in the annual mean minimum and maximum surface air temperatures would be of the order of 0.7–1.0 °C in the 2040s, when compared to that in the 1980s. Tabari and Hosseinzadeh Talaee (2011) analysed temperature series from 29 stations in Iran for the period 1966–2005 using the Mann–Kendall and Mann–Whitney tests. Results indicated that the annual mean temperature increased at 25 out of 29 stations, of which 17 stations showed significant trends. The analysis also indicated that most of the positive significant change points occurred first in 1972 at all stations except the coastal stations.

Bisai et al. (2014) performed change point analysis for the Krishnanagar weather observatory, West Bengal, India by applying cumulative sum chart and bootstrapping test to the time series of temperature data. They concluded that the major change point in the annual mean temperatures occurred around the year 2001. In this study, change point analysis was performed using Pettit's test for air temperature to identify the time at which major changes have occurred in this during the Indian subcontinent.

Methodology

Study Area

The study area chosen for this study consists of the Indian subcontinent, between 8° 4′ and 37° 6′ north latitude and 68° 7′ and 97° 25′ east longitude. This area contains a variety of geographical features. The Indian subcontinent is surrounded by the Arabian Sea in the west, the Bay of Bengal in the east and the Indian Ocean in the south. South India is a peninsula with two coastal lines at the boundaries and a plateau in the centre. North India occurs in the valley of the Himalayas and northeastern India is mainly the foothills and peaks of the Himalayas. There exists a wide variation in geographical features and this could result in highly varying climatic conditions. The study area is presented in Fig. 1.

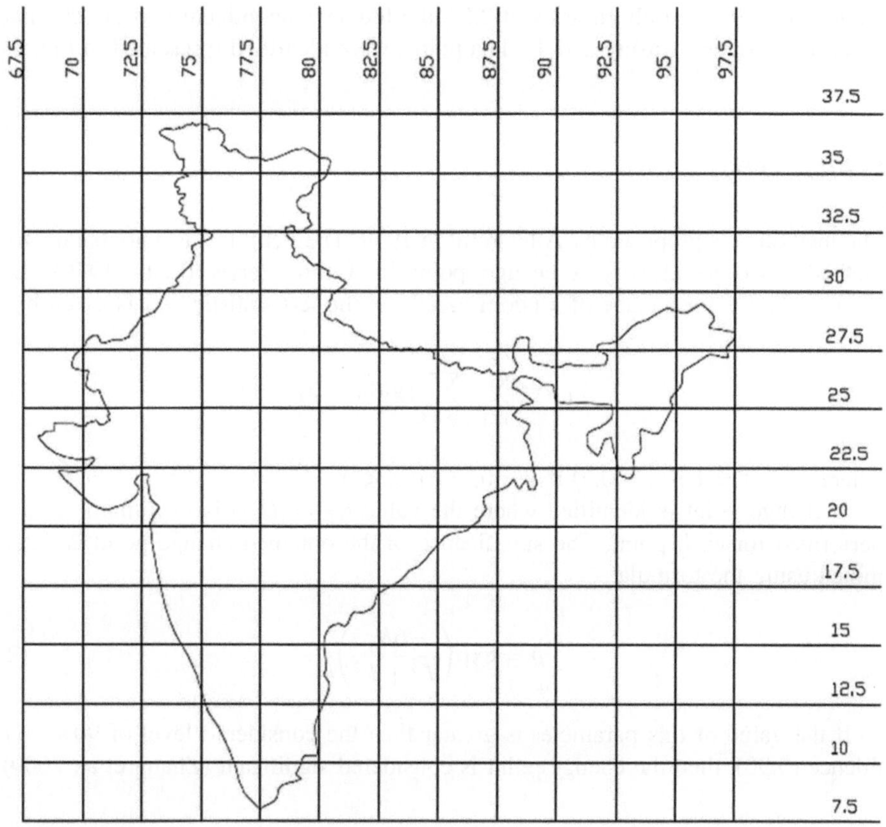

Fig. 1 Latitude and longitude of study area

Data Used

This study uses the National Center for Environmental Protection/National Center for Atmospheric Research (NCEP/NCAR) reanalysis data for the period of 1948–2014. This is the output of an offline run of the T62 operational model in 2.5° × 2.5° grid. As data assimilation has changed considerably in the satellite era, time-dependant in-homogeneities may be present. However, the NCEP/NCAR data is a reliable basis for analysis of the natural variability over the last several decades, especially in the Northern Hemisphere (Rudeva and Gulev 2011). Due to the lack of availability of observed meteorological data in extremes terrains like the north-eastern parts of India, the reanalysis data is considered most complete and a physically consistent data set (Simmonds and Keay 2000; Dell'Aquila et al. 2005). The data assimilation system uses a 3-D variational analysis scheme, with 28 sigma levels in the vertical and a triangular truncation of 62 waves which corresponds to a horizontal resolution of approximately 200 km (Kalnay et al. 1996). As the data points are available only in grids of 2.5°, the Indian subcontinent was divided into grids of the same measure and 47 data points were identified (presented in Fig. 1).

Pettit's Test

The method was proposed by A.N. Pettit in 1979. The Pettit test is a nonparametric method, used to identify a change point in a time series (Pettit 1979). Let $X_1, X_2 \ldots X_n$ be a sequence of random variables; the test statistic $U_{t,T}$ is given by

$$U_{t,T} = \sum_{i=1}^{t} \sum_{j=t+1}^{T} \operatorname{sgn}(x_i - x_j) \tag{1}$$

where $\operatorname{sgn}(x) = 1$ if $x > 0$, 0 if $x = 0$, -1 if $x < 0$.

A change point is identified where the value $K_T = |U_{t,T}|$ is maximum. This is performed for each point. The significance of the obtained change point is determined using the formula

$$\rho = \exp\left(\frac{-6K_T^2}{T^3 + T^2}\right) \tag{2}$$

If the value of this parameter is greater than the considered level of 90% confidence (0.90), then the change point is considered significant (Zhang et al. 2009).

Results and Discussions

Change Point Analysis

The Pettit's test was conducted for all grid points at various pressure levels in all seasons in the Indian subcontinent. In the dry season, at surface level, 14 points showed significant change as per Pettitt's test. 10N 77.5E, 12.5N 77.5E, 15N 75E, 25N 87.5E had a change point around the year 1976–'78. 25N 85E, 25N 92.5E, 27.5N 92.5E, 27.5N 95E had a change point around the period 1972–'73.35N 77.5E, 35N 80E had a change point in the year 2004. The other points are 27.5N 82.5E, 25.0N 82.5E, 22.5N 87.5E, 12.5N 80.0E which had a change point at 1970, 1966, 1988 and 1997, respectively. This indicates the presence of a change in trend during the period 1976–'78 in South India and 1972–'73 in northeastern India.

Similarly, a change in trend was detected in level 1000 mb in Southern and Central India during the past decade. Northern Plains had a change in the pattern of temperature during 1973–'75 at the level 850 mb. Almost all points had a change point in the period 2004–'06 at the level 500 mb. The same is illustrated in Fig. 2.

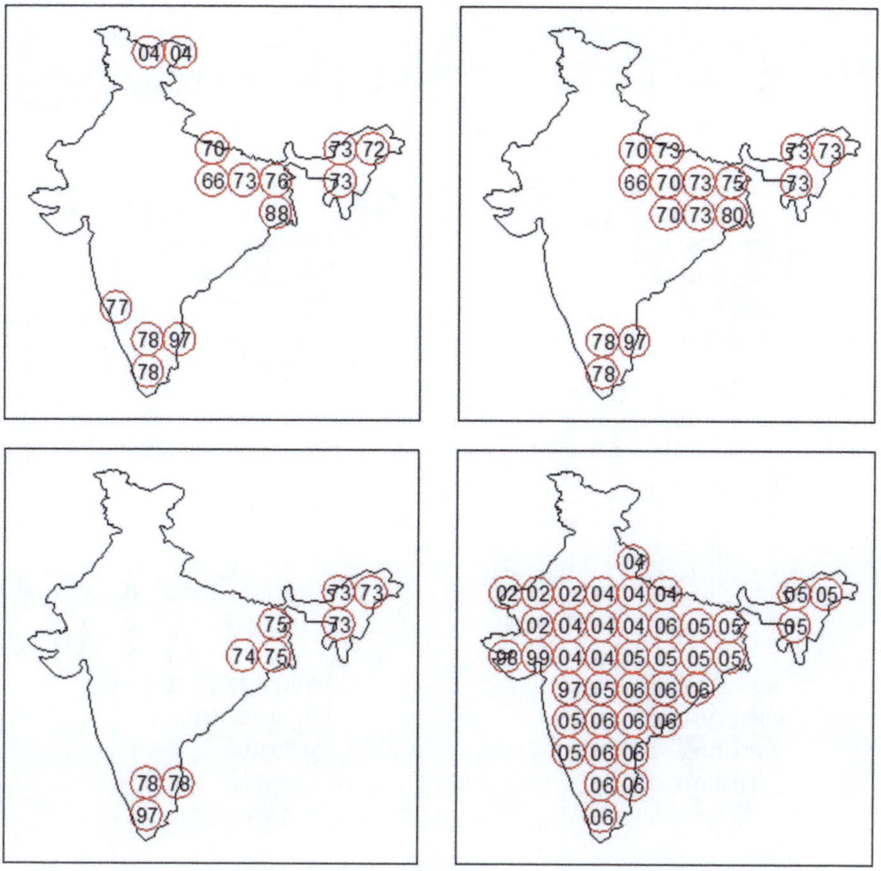

Fig. 2 Significant change points (year) in monthly mean temperature during the dry season estimated using Pettit's test (*clockwise from left top* level surface, level 1000, level 500, level 850)

For the wet season, at surface level, it can be concluded that all of southern India and eastern coastal area exhibit a major change in the trend of temperature during the past decade. At level 1000 mb, it is observed that southern India showed a significant change point in the last decade and Kashmir showed a significant change point during 1970–'71. Southern and northeastern India has a significant change point in 2004–'12 at 850 mb level. The region comprising the states of Kerala, Karnataka, Andhra Pradesh and Tamil Nadu has a change in pattern in the period 2006–'08 and the region consisting of Maharashtra, Madhya Pradesh and Orissa has a change in the period of 1976–'79 at 500 mb level (illustrated in Fig. 3).

During the southwest monsoon season, at surface level, the conclusion drawn is that all of south India, an eastern coastal region extending till West Bengal and northeast region show a significant change of pattern in the last decade. It is observed that, at 1000 mb level, a change in pattern exists in southern India, the

Fig. 3 Significant change points (year) in monthly mean temperature during the wet season estimated using Pettit's test (*clockwise from left top* level surface, level 1000, level 500, level 850)

eastern coastal region extending till West Bengal and northeast India during the last decade. Kashmir region showed a significant change point during the year 1971. At 850 mb level, southern India, eastern and western coastal regions and northeast India showed a significant change in the pattern of temperature during the period 2004–'12 (refer Fig. 4).

From these results, it can be observed that the major change points exist in the past decade. In order to identify any other significant pattern before this period, an annual analysis was performed till the year 1999. The monthly data was averaged to obtain annual data for the years and Pettit's test was performed. The result obtained indicates that, on a surface level, south India showed a significant change point in the period 1976–'84. Central India had a significant change of pattern in the period of 1960–'73. North and northeastern India showed a significant change point in the period 1970–'80. Similarly at 1000 mb pressure level, south India showed a

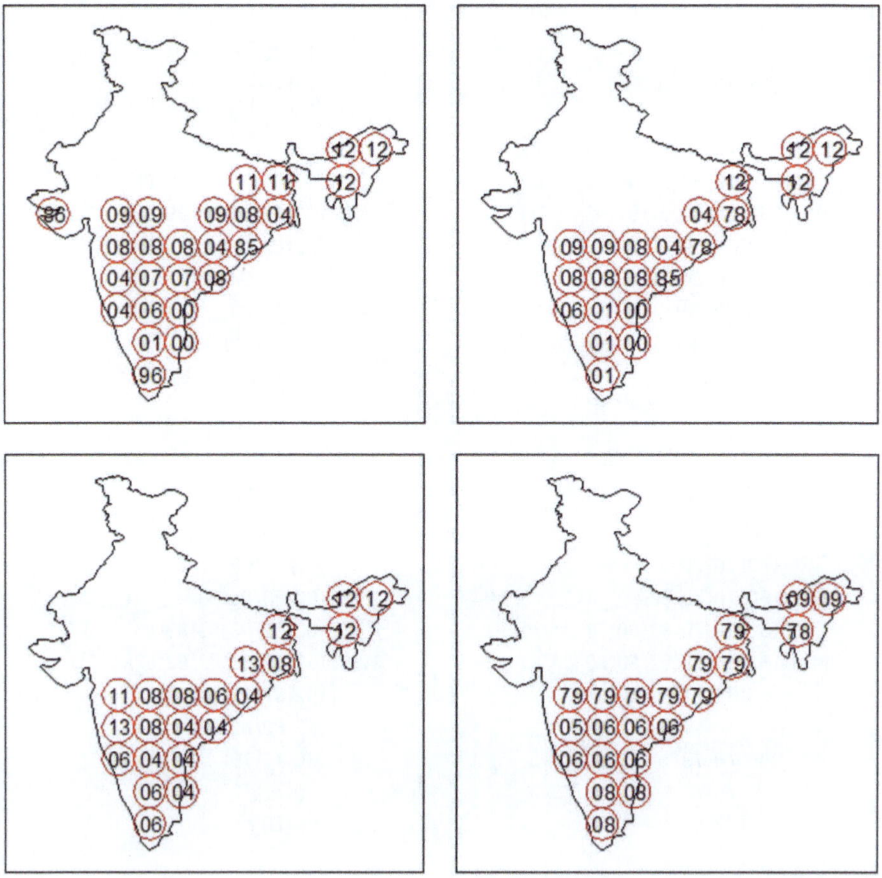

Fig. 4 Significant change points (year) in monthly mean temperature during the southwest monsoon season estimated using Pettit's test (*clockwise from left top* level surface, level 1000, level 500, level 850)

significant change point in 1976–'97, central India showed a change in pattern in the period of 1960–'74 and north and northeastern India showed a significant change point in 1970–'88. At 850 mb pressure level, south India had a change point during the period 1976–'97, central India showed a change in the period 1960–'74', whereas north and northeastern India showed a change point in the period 1970–'88. At 500 mb level, south India showed a change point in the year 1979, central India showed a change point during the period 1982–'83, north and northeastern India had a change point during the period 1957–'69. This is illustrated in Fig. 5.

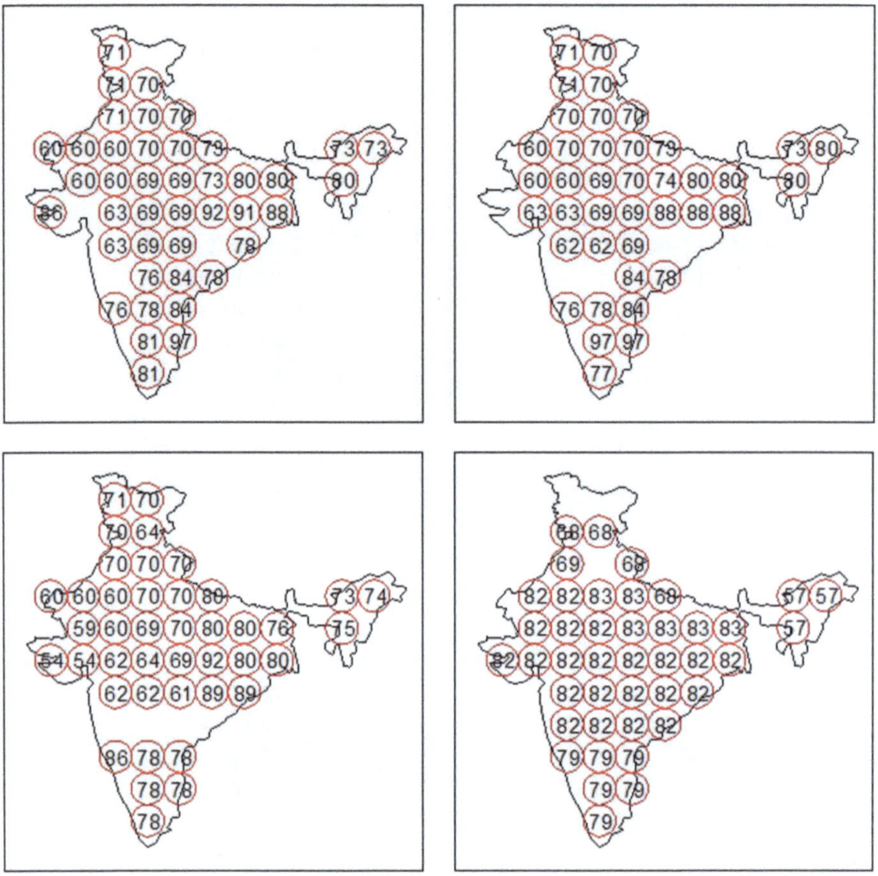

Fig. 5 Significant change points (year) in annual mean temperature (excluding 2000–2014) estimated using Pettit's test (*clockwise from left top* level surface, level 1000, level 500, level 850)

Summary and Conclusions

In this study, change point analysis using Pettit's test was performed on air temperature at different pressure levels. The test was first done on the seasonal subsets and then on the whole data taken annually.

The results indicated that except for the dry season, all other seasonal data showed a change point during the period of 2000–'14 in the southern peninsula and northeastern India. Results for the dry season showed a change point during the 1970s for this region. Further, on analysis of the data, excluding the data for the period since 2000, it is observed that South India had a change of pattern in 1974–'84, Central India experienced the same in the 1960s and North and Northeastern India experienced this during the 1970s.

References

Bisai D, Chatterjee S, Khan A, Barman NK (2014) Long term temperature trend and change point: a statistical approach. Open J Atmos Clim Chang 1:32–42

Dell'Aquila A, Lucarini V, Ruti PM, Calmanti S (2005) Hayashi spectra of the northern hemisphere mid-latitude atmospheric variability in the NCEP–NCAR and ECMWF reanalyses. Clim Dyn 25:639–652

Houghton J, Ding Y, Griggs DJ, Noguer M, Van der Linden PJ, Dai X, Maskell K, Johnson CA (2001) IPCC 2001: climate change 2001. The climate change contribution of working group I to the third assessment report of the intergovemmental panel on climate change, p 159

Kalnay E, Kanamitsu M, Kistler R, Collins W, Deaven D, Gandin L, Iredell M, Saha S, White G, Woollen J (1996) The NCEP/NCAR 40-year reanalysis project. Bull Am Meteor Soc 77: 437–471

Lal M, Cubasch U, Voss R, Waszkewitz J (1995) Effect of transient increase in greenhouse gases. Curr Sci 69

Pachauri RK, Reisinger A (2007) IPCC fourth assessment report. IPCC, Geneva

Pettit AN (1979) A non-parametric approach to the change-point detection. Appl Stat 28:126–135

Rudeva I, Gulev SK (2011) Composite analysis of North Atlantic extratropical cyclones in NCEP-NCAR reanalysis data. MWRv 139:1419–1446

Rupa Kumar R, Kumar KK, Pant GB (1994) Diurnal asymmetry of surface temperature trends over India. Geophys Res Lett 21:677–680

Simmonds I, Keay K (2000) Mean Southern Hemisphere extratropical cyclone behavior in the 40-year NCEP-NCAR reanalysis. J Clim 13:873–885

Singh P, Kumar V, Thomas T, Arora M (2008) Basin-wide assessment of temperature trends in northwest and central India. Hydrol Sci J 53:421–433

Srivastava HN, Dewan BN, Dikshit SK, Prakash Rao GS, Singh SS, Rao KR (1992) Decadal trends in climate over India. Mausam 43:7–20

Tabari H, HosseinzadehTalaee P (2011) Shift changes and monotonic trends in autocorrelated temperature series over Iran. Theor Appl Climatol 109(1–2):95–108

Zhang Q, Xu CY, Becker S, Zhang ZX, Chen YD, Coulibaly M (2009) Trends and abrupt changes of precipitation maxima in the Pearl River basin, China. AtScL 10:132–144

Greenhouse Gas Emissions from Sewage Treatment Plants Based on Sequential Batch Reactor in Maharashtra

Vipin Singh, Harish C. Phuleria and Munish K. Chandel

Abstract Wastewater treatment systems contribute significantly to anthropogenic greenhouse gas emissions. The main greenhouse gases emitted during the wastewater treatment processes are methane (CH_4), nitrous oxide (N_2O) and carbon dioxide (CO_2). Sequential batch reactor (SBR) is a type of an activated sludge process, and due to its high efficiency, currently, this is the preferred technology for the construction of new wastewater treatment plants (WWTPs). This study presents the estimation of greenhouse gas (GHG) emissions from SBR domestic wastewater treatment plants in Navi Mumbai, Maharashtra. We estimated direct emissions from wastewater treatment processes as well as indirect emissions due to energy usage during the treatment process. A total emission of ~35 kt CO_2-eq/year was estimated for six SBR-based WWTPs having combined treatment capacity of 474 MLD. All except one of these plants were well managed. In the case of not so well-managed SBR plants, significant methane production occurs during the treatment process. In the long run, if these plants are not well managed, the emission could increase by three to fourfolds for the same treatment capacity. In either case, major GHG emissions are due to CH_4 emission during the treatment process. The contribution of N_2O is negligible towards total GHG emissions.

Keywords Greenhouse gases · Methane emission · Nitrous oxide emission
Emission factor · Domestic wastewater · IPCC guidelines

Abbreviation

MLD	Million litre per day
GHG	Greenhouse gas
CO_2	Carbon dioxide
CO_2-eq	Carbon dioxide equivalent
WWTP	Wastewater Treatment Plant

V. Singh · H.C. Phuleria · M.K. Chandel (✉)
Centre for Environmental Science and Engineering, Indian Institute of Technology Bombay, Powai, Mumbai 400076, India
e-mail: munish.chandel@iitb.ac.in

V.P. Singh et al. (eds.), *Climate Change Impacts*, Water Science and Technology Library 82, https://doi.org/10.1007/978-981-10-5714-4_13

N_2O	Nitrous oxide
CH_4	Methane
GWP	Global warming potential

Introduction

Wastewater treatment systems contribute significantly to anthropogenic greenhouse gas emissions (USEPA 1997; Rittmann and McCarty 2001). The main greenhouse gases emitted during the wastewater treatment processes are methane (CH_4), nitrous oxide (N_2O) and carbon dioxide (CO_2). Carbon dioxide is formed as a result of aerobic microbial degradation and combustion of organic matter whereas methane is produced through the anaerobic degradation of organic matter. Nitrous oxide, on the other hand, is generated as a result of nitrification and denitrification processes.

This is observed that GHG emission and energy utility analysis for sewage treatment is the country wide challenge in India as their environmental impact, e.g. GHG emissions and global warming have not been assessed. According to USEPA (2014), Wastewater treatment plants (WWTPs) are the eight largest sectors that contribute to CH_4 and N_2O emissions. So in such case, the emission estimation from WWTPs could be one of the important sectors to consider along with treatment efficiency of the plant.

The sequencing batch reactor (SBR) is upgraded form of the activated sludge process (ASP). In SBR, the processes of filling, aeration, settling and decanting are consecutively carried out in the same tank. These phases constitute a cycle; during which, the fluid volume inside the reactor increases from a set working base water level. Mixed liquor is recycled into a selector from aeration zone during fill–aeration. End of the air circulation permits the biomass to flocculate and settle under at a predetermined time of the cycle. The treated supernatant is tapped after a particular setting period. The solids are isolated from the reactor amid the emptying stage. This system does not use secondary clarifier system for concentrating the sludge. The activated sludge is recycled, whereas the surplus is discarded from the basin. The BOD removal efficiency is higher in this type of treatment systems, which can be achieved up to 98% in case of BOD removal through SBR-based WWTPs (Compendium of Sewage Treatment Technologies 2009). Another study (Wakode and Sayyad 2014) in Mumbai has shown the average biochemical oxygen demand (BOD) removal of ~95–97% for SBR wastewater treatment plant. According to Loganathan et al. (2012) the removal efficiency of carbonaceous fractions (e.g. BOD, COD and TSS) was achieved more than 90%, similar removal efficiencies (80–90%) were shown in another study by Iaconi et al. (2008), whereas the BOD removal efficiencies for ASP-based WWTPs were seen in the range of 77–95% in Delhi (Jamwal et al. 2009) and for UASB based WWTPs 67% in Surat (Nair and Ahammed 2015) and Seghezzo et al. (1998) reported in the range of

50–70% in Netherland for UASB treatment plant. The COD removal was reported 43–56% and 68–85% for ASP and UASB, respectively (Sperling et al. 2001).

The sequencing batch reactor (SBR) is a kind of recent wastewater treatment technology, and there are not many studies found regarding GHG emissions from these type of WWTPs. A study in Noida, India (Gupta and Singh 2012) on a 33 MLD WWTP based on SBR technology using IPCC 2006 and population equivalent data found 3027.8 CO_2-eq/year.

Due to the latest technology for wastewater treatment because of its great removal efficiency, SBR could be one of the choices of selectors for the establishment of new WWTPs in India. Considering this, the WWTPs were selected for the present study of Navi Mumbai, Maharashtra state of India, as a part of estimation of GHG emissions contributed by WWTPs sectors interestingly all plants are based upon SBR technology.

Materials and Methodology

Six differently located SBR-based WWTPs for 474 MLD capacity were selected for our research study. All WWTPs have similar design and consist of preliminary, secondary and advance treatment processes within. A schematic diagram of SBR wastewater treatment plant in Navi Mumbai has been shown in Fig. 1.

The personal visits were made to individual WWTP and required data such as Treatment Capacity and BOD were collected from individual WWTPs in Navi Mumbai. For these WWTPs the range of BOD (raw sewage) was found to be 118–164 mg/l, with an average value of 138.7 mg/l (Table 1). The data for per capita intake of protein per day was taken 52.7 gm for Maharashtra state (NSSO 2014), which is required for N_2O emission calculation. On the basis of average wastewater generation in India, the per capita generation of wastewater is taken 121 l/capita/day, which is based on the studies carried out by Central Pollution Control Board (CPCB 2009). The global warming potential (GWP) is taken as 25 and 298 for CH_4 and N_2O respectively (IPCC 2006).

Setting system boundary: The system boundary is sewage treatment plant which includes the CH_4 and N_2O emissions from sewage treatment process and

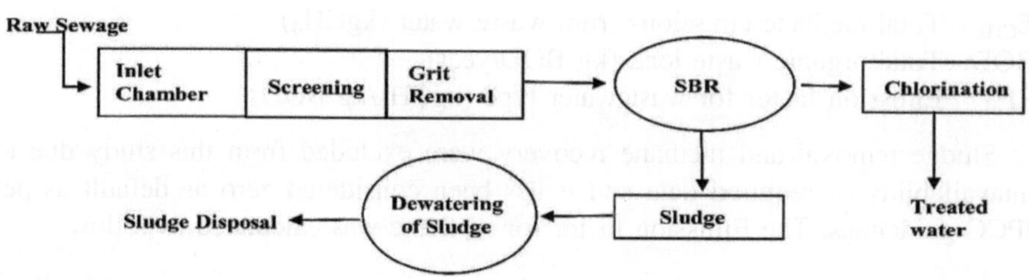

Fig. 1 Process flow diagram for SBR WWTP in Navi Mumbai

Table 1 Detail data of WWTPs in Navi Mumbai

WWTPs code	WWTP capacity in MLD	Total electricity use (KWh/month)	BOD (inlet) mg/l	Wastewater generation by population
WWTP1	100	236,170	128	723,140
WWTP2[a]	87.5	227,220	140	826,446
WWTP3	100	235,500	164	826,446
WWTP4	87.5	226,880	132	723,140
WWTP5	19	45,420	118	157,025
WWTP6	80	198,400	150	661,157
Total	474	1,169,590		3,917,355

[a]WWTP with improper management

CO_2 emissions due to use of electricity in the WWTPs. The emissions from the sludge treatment were not considered in this study due to data constraints CO_2 emissions of biogenic origin were not considered in the treatment process.

Direct and Indirect GHG emissions: There are three greenhouse gases calculated in the scope of present study, i.e. CO_2, N_2O and CH_4 using IPCC guidelines, 2006. The direct GHG emission was calculated from the generation of CO_2 by the breakdown of the organic component during the treatment process. The amount CH_4 is generated in case of improper management of WWTP which could be specific in aeration basin as there is no primary clarifier in the WWTPs considered. The N_2O emission was calculated on the basis of protein intake per capita. The Indirect GHG emission was calculated through power consumed in WWTPs. The total electrical power use in all six WWTPs was 1169.6 MWh/month for the capacity of 474 MLD wastewater treatment. The details of power uses for the operation of individual WWTP has been shown in Table 1, which was collected by the visit of WWTPs in Navi Mumbai.

The methodology for the estimation of emissions is based on IPCC guidelines for national greenhouse gas inventories (2006). Estimation of methane emissions is done using the following equation:

$$E_{CH_4} = (\text{kg BOD} \times \text{EF} \times 365) - S - R \tag{1}$$

where

E_{CH_4} Total methane emissions from waste water ($kgCH_4$)
BOD Total organic waste load (kg BOD/year)
EF Emission factor for wastewater type ($kgCH_4$/kg BOD)

Sludge removal and methane recovery were excluded from this study due to unavailability of required data and it has been considered zero as default as per IPCC guidelines. The Emission factor for methane was calculated as follows:

$$EF = B_0 \times MCF \tag{2}$$

where

B_0 Maximum methane producing capacity from raw sewage (default value for B_0 is 0.6 $kgCH_4/kgBOD$, IPCC 2006)
MCF Methane correction factor (for SBR the default value of well-managed WWTP is 0 whereas it is 0.1 in case of not well-managed Pant)

Calculation of CO_2 emissions due to electricity use (IPCC 2006):

$$CO_2\ \text{Emissions from electricity use} = \text{Total Use of electricity in MWh} \times \text{EFGE} \tag{3}$$

where

MWh Electricity consumption in megawatt hour.
EFGE Emission factor for generation of electricity in Maharashtra was taken 0.88 t CO_2/MWh from Central Electricity Authority of India (CEA 2011).

Estimation of nitrous oxide emissions is done using the following equation (IPCC 2006):

$$\text{Total}_{N_2O} = \text{Emission Factor} \times \text{Conversion Factor of kg } N_2O - N \text{ to kg } N_2O \times N_{\text{Effluent}} \tag{4}$$

where

$$N_{\text{Effluent}} = HP \times P \times F_{NPR} \times F_{NON-CON} \times F_{IND-COM} \tag{5}$$

N_{Effluent} Total amount of nitrogen in the wastewater effluent annually
HP Population of Human being contributing to the generation of wastewater generation
Protein Per capita protein consumption annually (kg/person/year)
F_{NPR} Nitrogen's fraction in protein (the default value = 0.16, kg N/kg protein)
$F_{NON\text{-}CON}$ Factor for non-consumed protein added to the wastewater (the value for developing countries is 1.1)
$F_{IND\text{-}COM}$ Factor of commercial and industrial co-discharged protein into the sewerage system (the default value is 1.25)

N_2O–N to kgN_2O = 1.57
Emission Factor = 0.0005 kgN_2O–N/KgN

Results and Discussion

The total estimated GHG emissions for the six SBR-based WWTPs was ~35 kt CO_2-eq/year in current operation of WWTPs, whereas 83.33% WWTPs were properly managed. Where as if all plants are partially managed, the emissions will be ~127 kt CO_2-eq/year (Fig. 2).

Since the SBR is an aerobic process, methane emissions are negligible. However, if the WWTP is not functioning properly, methane could be generated during the treatment process. Hence, in case of partially managed WWTPs and ~87% of total GHG emission is due to the production of CH_4. In the current working conditions, where all plants, except one are well-managed, the methane emissions are estimated to ~53%.

The total N_2O emissions were ~3.9 kt CO_2-eq/year from all WWTPs; out of which the N_2O contribution to total GHG emissions is 11.2% in current working conditions of WWTPs and will reduce to 3.1% if WWTPs do not work well. The total emissions due to electricity consumption were 12.4 kt CO_2/year which is 35.7% (of the total GHG emissions) in the current situation and 9.7% in not well-managed WWTPs (Fig. 3).

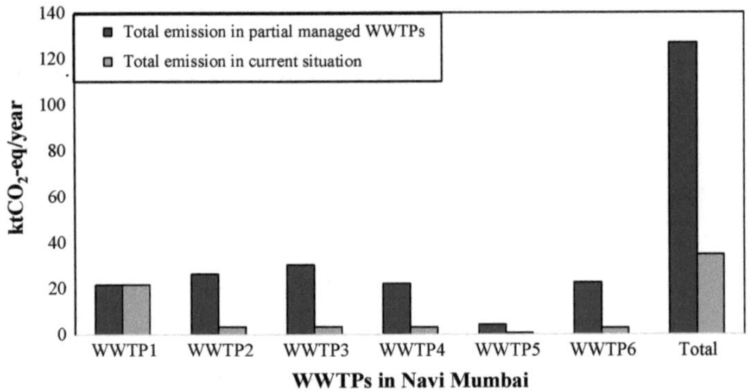

Fig. 2 Total GHG emissions from WWTPs in Navi Mumbai, Maharashtra

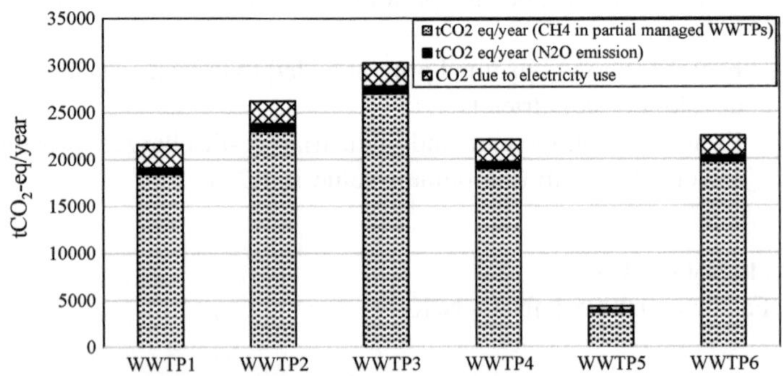

Fig. 3 GHG emissions from WWTPs in Navi Mumbai, Maharashtra (the individual emission of N_2O, CH_4 and CO_2 from the WWTPs has been shown in terms of CO_2-eq)

The GHG emissions data from WWTPs is very sparse in India. Gupta and Singh (2012), estimated GHG emissions from a SBR-based WWTP in Noida, India with a treatment capacity of 33 MLD. They have reported ~90 t CO_2/year emission from the WWTP, whereas we estimate ~75 t CO_2/year GHG emission for per MLD treatment of wastewater in Navi Mumbai. The slight difference in emissions could be due to the working condition of the WWTPs and efficiency of the individual units, perhaps also due to climatic conditions.

In most of the cases, it has been seen the methane generated in all WWTPs is discharged in the atmosphere and in some cases it is flared which adds to GHG concentration. The WWTPs also produce sludge during the treatment of wastewater in these WWTPs but it has been observed that there is no proper utilization or treatment of generated sludge as well as WWTPs were not equipped for capturing of generated methane. Bianchini et al. (2015) reported the energy content in sludge in the range of 12.7–15.5 MJ/kg of dry sludge. So if the produced sludge can be further utilized such as for biogas generation, then the energy could be generated and utilized in respective WWTPs. Similarly, if the generated methane in the WWTPs could be captured and utilized for energy generation. Thus, the overall GHG emission could be reduced as there will be less requirement of energy from external sources.

Conclusion

The study has estimated greenhouse gas emissions from SBR-based WWTPs in Navi Mumbai, Maharashtra for current operation as well as in case the WWTPs are not managed properly. The GHG emissions are primarily due to the treatment of sewage water and energy used associated in mechanical or electrical part of the WWTPs. Methane emissions contribute majorly in not so well-managed WWTPs, whereas the generation of N_2O and CO_2 emissions are same in both the situations, i.e. if the WWTPs functioning well or not. So if all WWTPs are partially managed, GHG emissions are estimated to be ~127 kt CO_2-eq/year for the whole wastewater treatment capacity available in Navi Mumbai. In this current situation, where most of the WWTPs are in proper working conditions the GHG emission is found to be ~35 kt CO_2-eq/year.

The recommendations could be proper management and utilization of sludge and, capturing of methane if it is produced during treatment of wastewater for generation of energy, which could be used within respective WWTP. That way, it may likely reduce energy dependency of WWTP and plants can be self-sustained to some extent. It would also help in reduction of environmental burden in terms of overall GHG emissions.

References

Bianchini A, Bonfiglioli L, Pellegrini M, Saccani C (2015) Sewage sludge drying process integration with a waste-to-energy power plant. Waste Manag 42:159–165

CEA (2011) CO_2 Baseline Database for the Indian Power Sector, Central Electricity Authority, Ministry of Power, Govt of India. Retrieved 03 October 2015 from http://www.cea.nic.in/reports/others/thermal/tpece/cdm_co2/user_guide_ver6.pdf

Compendium of Sewage Treatment Technologies (2009) Retrieved 10 October 2015 from https://nmcg.nic.in/writereaddata/fileupload/15_Technologies%20Involved.pdf

CPCB (2009) Status of water supply, wastewater generation and treatment in class-I cities and class-II towns of India. Retrieved 15 March 2015 from http://cpcb.nic.in/upload/NewItems/NewItem_153_Foreword.pdf

Gupta D, Singh SK (2012) Greenhouse gas emissions from wastewater treatment plants: a case study of Noida. 2(2)

Iaconi CD, Sanctis MD, Rossetti S, Ramadori R (2008) Technological transfer to demonstrative scale of sequencing batch bio-filter granular reactor (SBBGR) technology for municipal and industrial wastewater treatment. Water Sci Technol 58(2):367–372

IPCC (2006) Guidelines for national greenhouse gas inventories. Retrieved 10 March 2015 from http://www.ipcc-nggip.iges.or.jp/public/2006gl/pdf/Vol5/Ch6_Wastewater.pdf

Jamwal P, Mittal AK, Mouchel JM (2009) Efficiency evaluation of sewage treatment plants with different technologies in Delhi (India). Environ Monit Assess 153(1–4):293–305

Loganathan R, Rasappan K, Jebamani IM, Kumar JNK (2012) Biological treatment of domestic wastewater using sequential batch reactor (SBR). Indian J Environ Prot 32(7):554–559

Nair AT, Ahammed MM (2015) Water treatment sludge for phosphate removal from the effluent of UASB reactor treating municipal wastewater. Process Saf Environ Prot 94(C):105–112

National Sample Survey Office (2014) Nutritional Intake in India, 2011–12, published in Oct 2014. Retrieved 15 March 2015 from http://mospi.nic.in/Mospi_New/upload/nss_report_560_19dec14.pdf

Rittmann BE, McCarty PL (2001) Environmental biotechnology: principles and applications. McGraw-Hill, New York

Seghezzo L, Zeeman G, Van Lier JB, Hamelers HVM, Lettinga G (1998) A review: the anaerobic treatment of sewage in UASB and EGSB reactors. Biores Technol 65(3):175–190

Sperling VM, Freire VH, Chernicharo CD (2001) Performance evaluation of a UASB-activated sludge system treating municipal wastewater. Water Sci Technol 43(11):323–328

USEPA (1997) Estimates of global greenhouse gas emissions from industrial and domestic wastewater treatment. Office of Policy, Planning and Evaluation, Washington, DC, EPA-600/R-97-091

USEPA (2014) Inventory of U.S. greenhouse gas emissions and sinks 1992–2012, published in 2014. Retrieved 10 Dec 2015 from http://www3.epa.gov/climatechange/downloads/Framework-for-Assessing-Biogenic-CO2-Emissions.pdf

Wakode PN, Sayyad SU (2014) Performance evaluation of 25 MLD sewage treatment plant at Kalyan. Am J Eng Res (AJER) 03(03):310–316 (2320-0936)

Study of Climate Change in Uttarakhand Himalayas: Changing Patterns of Historical Rainfall

Archana Sarkar and Vaibhav Garg

Abstract The subject area of climate change is vast, but the changing pattern of rainfall is a topic within this field that deserves urgent and systematic attention since it affects the availability of freshwater, food production and the occurrence of water related disasters triggered by extreme events. The detection of trends in rainfall is essential for the assessment of the impacts of climate variability and change on the water resources of a region. In June 2013, several days of extremely heavy rain caused devastating floods in the region, resulting in more than 5000 people missing and presumed dead. The present study aims to determine trends in the annual, seasonal, and monthly rainfall over Uttarakhand State. Long-term (1901–2013) gridded daily rainfall data at 0.25° grid have been used. Daily rainfall data at ten grid center locations (five each in Garhwal and Kumaon divisions) in the vicinity of Haridwar, Tehri, Uttarkashi, Rudraprayag, Joshimath, Almora, Bageshwar, Munsiyari, Pithoragarh, and Rudrapur have been processed and analyzed for a period of 113 years (1901–2013). Historical trends in daily rainfall have been examined using parametric (regression analysis) and non-parametric (Mann–Kendall (MK) statistics). On the basis of regression and MK test, rising and falling trend in rainfall and anomalies at various stations have been analyzed. The result shows that many of these variables demonstrate statistically significant changes occurred in last eleven decades. Statistically significant increasing trends of annual as well as monsoon rainfall have been observed at Haridwar, Rudraprayag, Joshimath, Almora, and Munsiyari whereas statistically significant decreasing trends of monthly rainfall (August, September, and October) have been observed at Uttarkashi and Tehri stations.

A. Sarkar (✉)
National Institute of Hydrology, Roorkee 247667, Uttarakhand, India
e-mail: archana.nihr@gov.in

V. Garg
Indian Institute of Remote Sensing, Dehradun, Uttarakhand, India
e-mail: vaibhav@iirs.gov.in

V.P. Singh et al. (eds.), *Climate Change Impacts*, Water Science and Technology Library 82, https://doi.org/10.1007/978-981-10-5714-4_14

Introduction

The climate of earth was never stable for any extended period. Potential climate change and its impacts on hydrologic systems pose a threat to water resources throughout the world. There are a number of natural causes of climate variability, namely variations in the amount of energy emitted by the Sun, changes in the distance between the Earth and the Sun, the presence of volcanic pollution in the upper atmosphere and presence of green house gases, etc. (Scafetta and West 2005). Natural and human influences, called "forcings" in the climate-science literature alter the flow of radiant energy in the atmosphere, cooling and warming Earth by perturbing its energy balance. Positive forcings warm the planet while negative ones cool it. One of these forcings is induced by the greenhouse gases, which alter the planet's energy balance by absorbing infrared radiation that would otherwise escape to space. The major greenhouse gases include CO_2, methane, nitrous oxide, tropospheric ozone, chlorofluorocarbons (CFCs), and water vapor. With the exception of water vapor, the concentrations of all the greenhouse gases are more or less directly dependent on human activities. (Water vapor levels depend on Earth's temperature and the availability of liquid water, and thus are indirectly affected by humans). Other forcings include reflective aerosols (mostly sulfate particles from burning of fossil fuel), black carbon particles (soot), land-cover changes, variations in solar output, and cloud-cover changes resulting from global temperature variations and aerosols (IPCC 2013).

The important climatic variables that influence the ecosystem are precipitation, radiation, temperature and stream flow. It is a challenge to the scientific community to understand the complicated processes involved in climate change and alert the society to tackle the problem. The changing pattern of precipitation deserves urgent and systematic attention as it will affect the availability of food supply (Dore 2005) and the occurrence of water related disasters triggered by extreme events. Precipitation is the major driving force of the land phase of the hydrologic system, and changes in its pattern could have direct impacts on water resources. A higher or lower rainfall or changes in its distribution would influence the spatial and temporal distribution of runoff, soil moisture, groundwater reserves, and would alter the frequency of droughts and floods.

The southwest monsoon, which brings about 80% of the total precipitation over the country, is critical for the availability of fresh water for drinking and irrigation. Changes in climate over Indian region, particularly the southwest monsoon, would have a significant impact on agriculture production, water resources management, and overall economy of the country. According to IPCC (2013), future climate change is likely to affect agriculture, increase the risk of hunger and water scarcity, and would lead to more rapid melting of glaciers. Freshwater availability in many river basins in Asia is likely to decrease due to climate change. This reduction along with population growth and rising living standards could adversely affect more than a billion people in Asia by the 2050s. Accelerated glacier melt is likely to cause an increase in the number and severity of glacier melt related floods, slope destabilization and a decrease in river flows as glaciers recede (IPCC 2013). Lal (2001) has discussed implications of climate change on Indian water resources. Gosain et al. (2006) have quantified the impact of climate change on the water resources of Indian River systems.

Global averaged precipitation is projected to increase, but both increases and decreases are expected at the regional and continental scales (IPCC 2007). Similar trends were reported in rainfall by various authors in India (Thapliyal and Kulshrestha 1991; Kumar et al. 1992; Sinha Ray and De 2003; Singh et al. 2008). Though the monsoon rainfall in India is found to be trendless over a long period of time, particularly on the all India scale, pockets of significant long-term rainfall changes have been identified (Srivastava et al. 1998). Climate change projections using various global climate models (GCMs) and regional climate models (RCMs) showed increasing temperature and changing patterns in rainfall during the twentyfirst Century over India (Kumar et al. 2006; Rajendran and Kitoh 2008).

In last few decades, several individual and collaborative researches were undertaken to study climate change. The linear relationship is one of the most common methods used for detecting rainfall trends (Hameed et al. 1997). Both parametric and non-parametric tests are widely used for trend study. The advantage with a non-parametric test is that it only requires data to be independent and can tolerate outliers in the data (Hameed and Rao 1998). One of the popular non-parametric tests widely used for detecting trends in the time series is the Mann–Kendall test (Mann 1945; Kendall 1955). The two important parameters of this test are the significance level that indicates the trend strength and the slope magnitude that indicates the direction as well as the magnitude of the trend (Burn and Elnur 2002). The advantage of the test is that it is distribution-free, robust against outliers and has a higher power than many other commonly used tests (Hess et al. 2001). Many climate studies applying Mann–Kendall test have been carried out in the last decade. Modarresa and Silva (2007) studied the rainfall trend in Iran; Birsan et al. (2005) used the test to study the stream flow trend in Switzerland; Shan Yu et al. (2002) studied the impact of climate change on water resources in Taiwan; Hesse et al. (2005) studied the temperature trends over India; Zhang et al. (2005) analyzed the trend of precipitation, temperature, and runoff in the Yangtze basin China. Mcbean and Rovers (1998) examined historical trends in precipitation, temperature, and stream flows in the Great Lakes using regression analysis and Mann–Kendall statistics.

Keeping in view the above back ground, the present study has been carried to evaluate the trend of rainfall in the State of Uttarakhand, India. The Uttarakhand State is located in the Himalayan region has been selected for the study. Several major and minor hydro-electric power stations are being built in over the tributaries of Ganga River in the mountainous part of the State several more are under consideration. Hence, it is very important to understand the impact of climate change on the hydrology of this Hilly State for proper planning and management of the water resources. The major objective of the study was to observe the trend of rainfall in the State in the last 113 years. Trend analysis has been carried out using linear regression method and Mann–Kendall test.

Study Area and Data Used

Uttarakhand is a state in the northern part of India. It is often referred to as the "Land of the Gods" due to the many holy Hindu temples and pilgrimage centers found throughout the state. Uttarakhand is known for its natural beauty of the Himalayas, the Bhabhar, and the Terai. It borders the Tibet Autonomous Region on the north; the Mahakali Zone of the Far-Western Region, Nepal on the east; and the Indian states of Uttar Pradesh to the south and Himachal Pradesh to the northwest. The state is divided into two divisions, Garhwal and Kumaon, with a total of 13 districts. Two of the most important rivers in Hinduism originate in the region, the Ganga at Gangotri and the Yamuna at Yamunotri.

Uttarakhand has a total area of 53,484 km^2, of which 93% is mountainous and 65% is covered by forest. Most of the northern part of the state is covered by high Himalayan peaks and glaciers. Uttarakhand lies on the southern slope of the Himalaya range, and the climate and vegetation vary greatly with elevation, from glaciers at the highest elevations to subtropical forests at the lower elevations. The highest elevations are covered by ice and bare rock. Below them, between 3,000 and 5,000 m (9,800 and 16,400 ft.) are the western Himalayan alpine shrub and meadows. The temperate western Himalayan sub-alpine conifer forests grow just below the tree line. At 3,000–2,600 m (9,800–8,500 ft.) elevation they transition to the temperate western Himalayan broadleaf forests, which lie in a belt from 2,600 to 1,500 m (8,500 to 4,900 ft.) elevation. Below 1,500 m (4,900 ft.) elevation lie the Himalayan subtropical pine forests. The Upper Gangetic Plains moist deciduous forests and the drier Terai-Duar savanna and grasslands cover the lowlands along the Uttar Pradesh border in a belt locally known as Bhabhar. These lowland forests have mostly been cleared for agriculture, but a few pockets remain. In June 2013, several days of extremely heavy rain caused devastating floods in the region, resulting in more than 5000 people missing and presumed dead. The flooding was referred to in the Indian media as a "Himalayan Tsunami."

Daily rainfall data at ten grid center locations (five each in Garhwal and Kumaon divisions) in the vicinity of Haridwar, Tehri, Uttarkashi, Rudraprayag, Joshimath, Almora, Bageshwar, Munsiyari, Pithoragarh, and Rudrapur procured from India Meteorological Department (IMD) have been used in this study. These stations are shown in Fig. 1 which represents the study area, i.e., the Uttarakhand State. Analysis has been performed for a period from 1901 to 2013, i.e., 113 years.

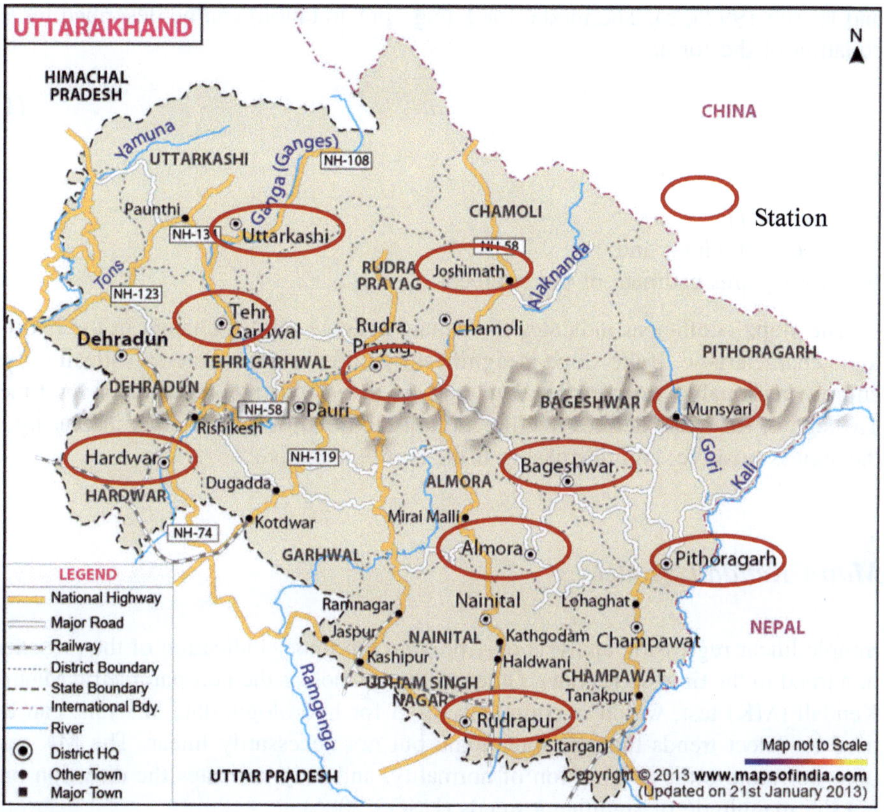

Fig. 1 Study area

Methodology

Trend Analysis

In the present study, two methods viz. regression and MK test have been used. These are described in the following sections.

Regression Model

One of the most useful parametric models to detect the trend is the "Simple Linear Regression" model. The method of linear regression requires the assumptions of normality of residuals, constant variance, and true linearity of relationship (Helsel and Hirsch 1992a, b). The model for Y (e.g., precipitation) can be described by an equation of the form:

$$Y = at + b \tag{1}$$

where,

t time (year)
a slope coefficient; and
b least-squares estimate of the intercept

The slope coefficient indicates the annual average rate of change in the hydrologic characteristic. If the slope is significantly different from zero statistically, it is entirely reasonable to interpret that there is a real change occurring over time. The sign of the slope defines the direction of the trend of the variable: increasing if the sign is positive, and decreasing if the sign is negative.

Mann Kendall Model

Simple linear regression analysis may provide a primary indication of the presence of a trend in the time-series data. Other methods, such as the non-parametric Mann–Kendall (MK) test, which is commonly used for hydrologic data analysis, can be used to detect trends that are monotonic but not necessarily linear. The MK test does not require the assumption of normality, and only indicates the direction but not the magnitude of significant trends (USGS 2005).

The trend in the data if any was quantified using Mann–Kendall's S-statistic (Mann 1945; Kendall 1955). The MK method assumes that the time series under research is stable, independent, and random with equal probability distribution (Zhang et al. 2005). The MK test is applied to uncorrelated data because it was reported that the presence of serial correlation might lead to an erroneous rejection of the null hypothesis (Helsel and Hirsch 1992a, b; Kulkarni and von Storch 1995; Yue et al. 2002; Yue and Wang 2002; Yue and Pilon 2003).

The computational procedure for the MK test is described in Adamowski and Bougadis (2003). Let the time series consist of n data points and T_i and T_j be two sub—sets of data where $i = 1, 2, 3\ldots, n - 1$ and $j = i + 1, i + 2, i + 3,\ldots n$. Each data point T is used as a reference point and is compared with all the T_j data points such that:

$$\text{Sign}(T) = \begin{cases} 1 \text{ for } T_j > T_i \\ 0 \text{ for } T_j = T_i \\ -1 \text{ for } T_j < T_i \end{cases} \tag{2}$$

The MK test used in the present study is based on the test statistic, S, defined as follows:

$$S = \sum_{i=1}^{n-1} \sum_{j=i+1}^{n} \text{sign}(T_j - T_i) \quad (3)$$

The variance for the S-statistic is defined by:

$$\sigma^2 = \frac{n(n-1)(2n+5) - \sum_{i=1}^{n} t_i(i)(i-1)(2i+5)}{18} \quad (4)$$

where t_i denotes the number of ties to extent i. The summation term in Eq. (4) is only used if data series contains the "tied" values. The test statistic, Z_s, can be calculated as

$$Z_s = \begin{cases} (S-1)/\sigma \text{ for } S > 0 \\ 0 \text{ for } S = 0 \\ (S+1)/\sigma \text{ for } S < 0 \end{cases} \quad (5)$$

In which, Z_s follows a standard normal distribution. Equation (5) is useful for record lengths greater than 10 and if the number of tied data is low. The test statistic, Z_s is used as a measure of the significance of the trend. In fact, this test statistic is used to test the null hypothesis, H_0: There is no monotonic trend in the data. If $|Z_s|$ is greater than $Z_{\alpha/2}$ where α represents the chosen significance level (usually 5%, with $Z_{0.025}$ = 1.96), then the null hypothesis is rejected, meaning that the trend is significant. For this study, the simple regression analysis technique was used to test the slopes of the trend lines for statistical significance at 5% level. The Mann–Kendall trend test procedure is applied to further verify the outcomes of regression analysis for the hydrological variables considered.

MK test holds well in the case of non-auto-correlated time-series data. For auto-correlated data, modified Mann–Kendall test proposed by Rao and Hamed (1997) was used, which is robust in presence of autocorrelation. It is based on the modified variance of S given by Eq. (5).

$$V*(S) = \text{var}(S)\frac{n}{n_s^*} = \frac{n(n-1)(2n+5)}{18}\frac{n}{n_s^*} \quad (6)$$

The recommended approximate value of n/n_s^* n/n_2 is given by the Eq. (7)

$$\frac{n}{n_s^*} = 1 + \frac{2}{n(n-1)(n-2)} \sum_{i=1}^{n-1} (n-i)(n-i-1)(n-i-2)\rho_s(i) \quad (7)$$

where n is the actual number of observations and $\rho_s(i)$ is the autocorrelation function of the ranks of the observations. The accuracy of the approximation given by the Eq. (6) was found to improve as n increases. The autocorrelation between ranks of observations $\rho_s(i)$ is first evaluated. The value of ranks of observations $\rho_s(i)$, however, must be calculated after subtracting a suitable non-parametric trend estimator (Sen 1968). Due to the nature of calculation in Eq. (5), which involves a large number of terms, it was found that insignificant values of $\rho_s(i)$ will have an adverse effect on the accuracy of the estimated variance of S. Therefore, only significant values of $\rho_s(i)$ are used in Eq. (6). This is achieved by requiring a suitable preset significance level for the autocorrelation to be included in the calculations, which can be taken equal to that of the rest.

In the present study, non-parametric Mann–Kendall trend test and modified Mann–Kendall test as proposed by Rao and Hamed (1997) were applied to study the historical trend in annual and monthly rainfall described in the following sections.

Data Preparation

The daily hydro-meteorological data available for above cited period at different grids/stations have been used to prepare annual, seasonal, and monthly series of rainfall as described below:

(1) *Annual*: Using the daily data, series of average annual data have been prepared for trend analysis for all the variables at various grids/stations
(2) *Seasonal*: Four types of average seasonal data series have been prepared. Data were divided into four seasons, namely pre-monsoon (March–May), monsoon (June–August), post-monsoon (September–November), and winter (December–February) based on the prevailing climate of India.
(3) *Monthly*: Using the daily data, series of average monthly data have been prepared for trend analysis for all the variables at various grids/stations.

Trend Analysis and Results

Using the gridded data of daily rainfall from IMD for the ten grid center locations (referred to as stations hereafter) under the present study, monthly, seasonal and annual rainfall series were generated for all the ten stations. Then, the time series were checked for randomness, autocorrelation, and long-term persistence before conducting the Mann–Kendall test. Time series having no autocorrelation were analyzed with Mann–Kendall's Test for the detection of trend and if significant autocorrelation was found in the data, the time series was tested with modified Mann–Kendall test as suggested by Rao and Hamed (1997). In this study, monthly, seasonal and annual rainfall for ten grid points were analyzed as given in Table 1 and Table 2 for Garhwal and Kumaon region, respectively. The selected stations

Table 1 Trend in different season for rainfall data (Garhwal Region)

Period	MKZ statistics				
	Joshimath	Rudraprayag	Uttarkashi	Tehri	Haridwar
January	−0.38	−0.93	−1.22	−0.75	−1.09
February	+0.76	+1.17	+1.34	+1.46	+0.96
March	+1.04	+0.53	+1.21	+1.27	−0.39
April	−0.28	+1.08	+1.49	+1.87	+0.72
May	+1.87	+1.85	**+2.43**	**+2.05**	+0.85
June	+1.37	+1.44	+0.78	+0.47	**+3.48**
July	**+2.85**	**+2.84**	−1.66	−1.73	**+2.75**
August	+1.49	+0.87	**−2.57**	**−2.50**	+1.17
September	−1.14	−0.99	**−5.12**	**−5.15**	−1.28
October	−1.39	−1.62	**−2.43**	**−2.26**	−0.94
November	−0.37	−1.20	−1.18	−0.83	−1.87
December	−0.01	+0.58	+0.06	+0.21	+1.35
Pre-mon	−1.01	−0.50	−0.26	−0.19	−0.33
Monsoon	**+2.92**	**+2.47**	−0.32	−0.59	**+2.81**
Post-mon	+0.72	+0.17	−0.10	−0.05	+0.50
Winter	−0.36	+0.62	−0.87	−0.07	**−2.25**
Annual	**+2.27**	**+3.05**	−0.07	−0.37	**+2.78**

Table 2 Trend in different season for rainfall data (Kumaon region)

Period	MKZ statistics				
	Almora	Bageshwar	Pithoragarh	Rudrapur	Munsiyari
January	−1.25	+0.30	−0.41	+0.24	+0.28
February	+0.21	+0.61	+0.97	+0.38	+1.11
March	−0.62	+0.24	−0.09	−0.88	+0.58
April	−0.05	+0.13	+0.26	+0.43	+0.33
May	+0.17	+0.70	+0.88	**+2.71**	+0.86
June	+1.11	+0.09	−0.18	**+2.60**	+0.20
July	+0.46	+0.54	−0.74	+0.29	+1.75
August	+0.22	−0.22	−1.20	−0.77	+0.64
September	−1.83	−1.66	−1.95	−1.89	−1.08
October	−1.23	−1.59	−1.75	−0.91	−1.77
November	−1.40	−0.57	−1.45	−0.86	−0.43
December	−0.03	−0.23	−0.41	+0.85	−0.29
Pre-mon	−0.62	−0.75	+0.52	+0.17	−0.22
Monsoon	**+2.00**	+1.35	+0.38	+1.53	**+2.27**
Post-mon	−0.41	+0.08	−0.02	+1.09	+0.90
Winter	−1.65	−0.06	−0.22	−1.67	+1.12
Annual	**+2.05**	+1.73	+0.58	+1.73	**+2.63**

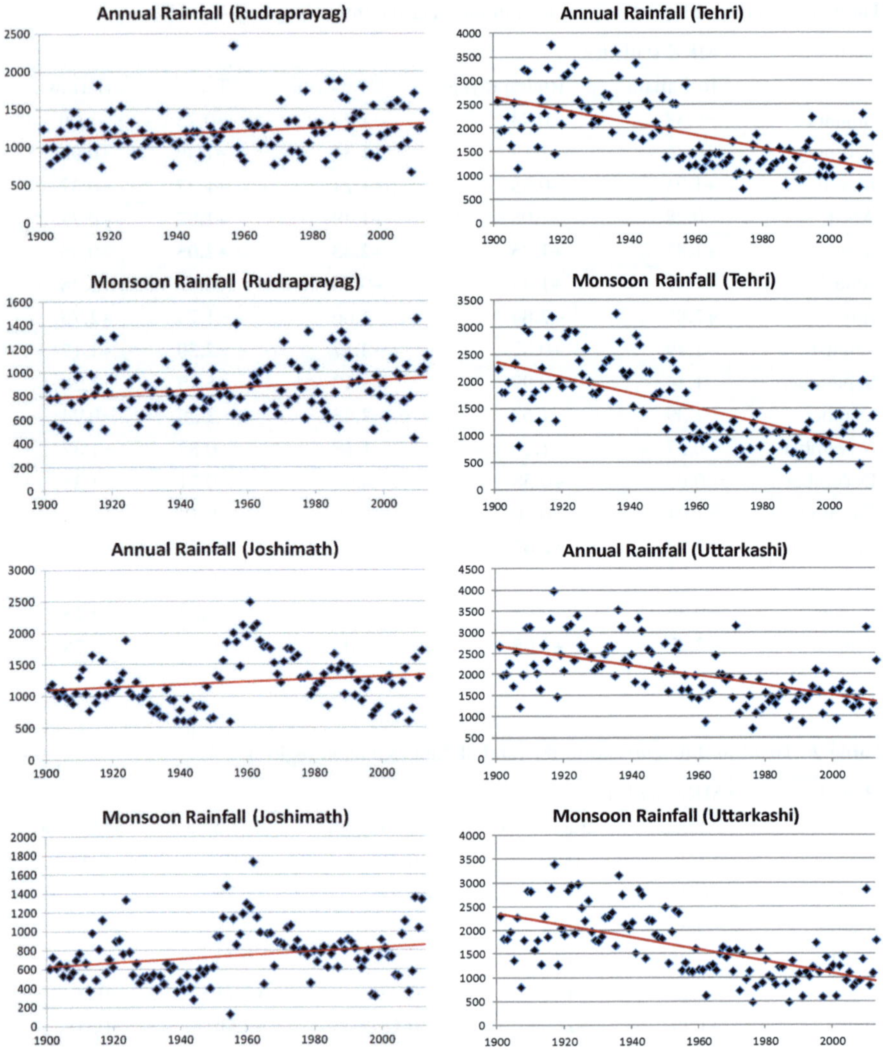

Fig. 2 Trends in annual and monsoon rainfall for all the stations

have 113 years of record length each. Linear regression analysis has been carried out for all the stations. Graphical presentations are shown in Fig. 2 for trends in annual and monsoon rainfall for all the stations. Annual and seasonal spatial trends in the Uttarakhand state are shown in Fig. 3.

Annual analysis of rainfall trends (Tables 1 and 2) shows a significantly increasing trend (at 5% significance level) at five rainfall stations, three in the Garhwal region and two in the Kumaon region. Monsoon rainfall also follows similar trends. However, the monthly rainfall analysis shows significantly

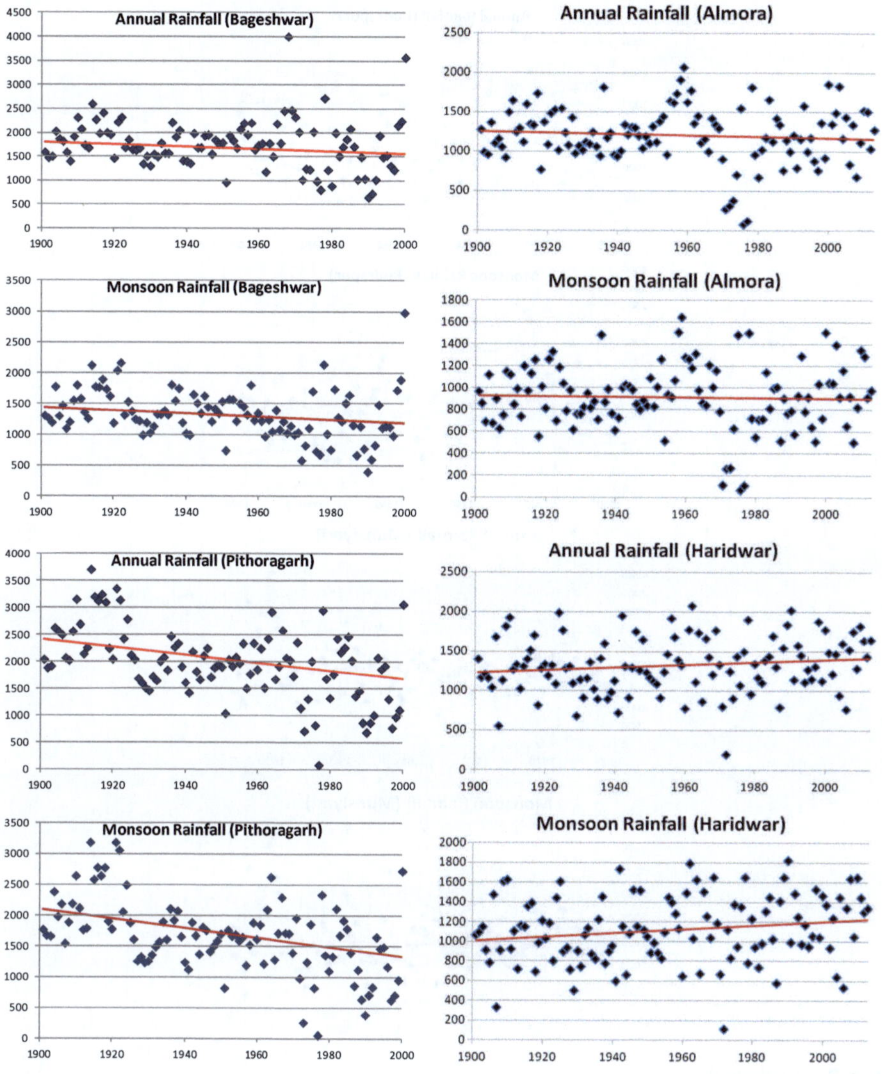

Fig. 2 (continued)

increasing trend of rainfall in the month of April over Bageshwar and Pithoragarh stations in the Kumaon region of the State. Statistically significant increasing rainfall trend has also been observed over Chamoli station of the Garhwal region during the month of August. Besides the above trends, the regression analysis shows increasing as well as decreasing trend of rainfall in all the stations during the different time period, but such trend is not statistically significant.

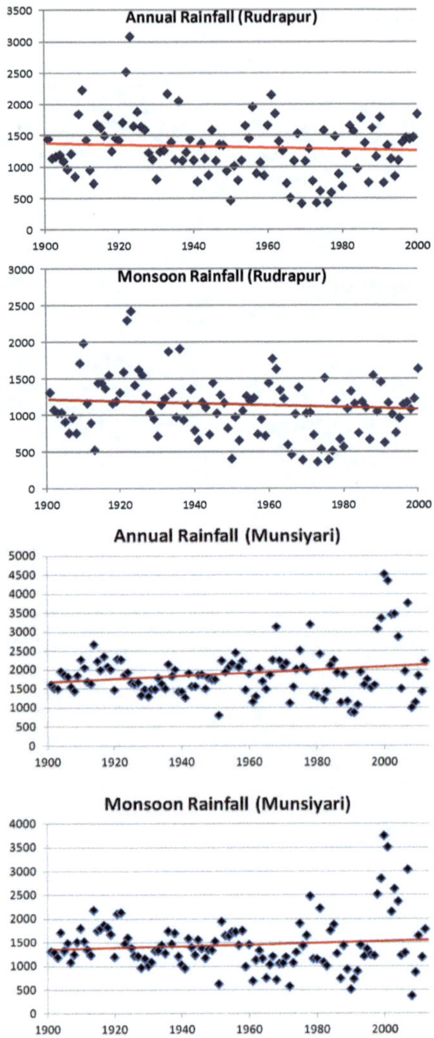

Fig. 2 (continued)

Conclusions

The Himalayas are highly sensitive to climate change. Any change in rainfall highly influences stream flow downstream. The Himalayan Rivers have witnessed a steep rise in glacial retreat in recent past. These events are a strong indication of climate change in the Uttarakhand State. However, the influence of anthropogenic factors cannot be rejected. The present study is based on the analysis daily rainfall data using simple linear regression and non-parametric Mann–Kendall trend test.

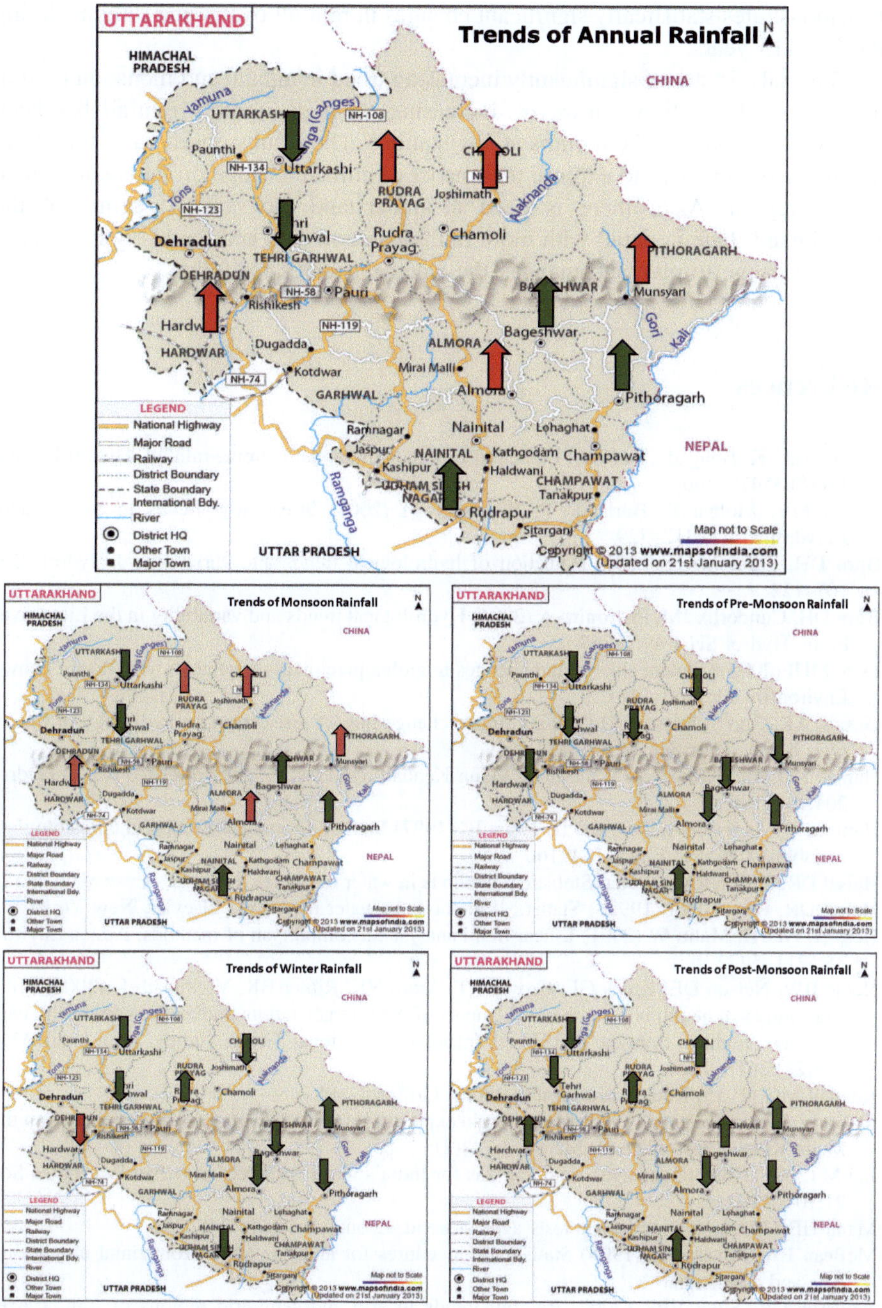

Fig. 3 Trend of annual rainfall over 113 Years (1901–2013)

It demonstrates statistically significant changes in rainfall over some stations during the last 113 years.

The analysis shows significantly increasing trend in annual and monsoon rainfall at some of the stations, however, increasing trend in monthly rainfall has been observed for some of the months at few stations. The results observed in this study are an encouragement to explore the impact of climate change on the local climate of the region. Also, there is need to understand the rainfall regime of the Uttarakhand State in detail with more rainfall station data and the likely impacts of climate change to plan and manage the water resources for future.

References

Adamowski K, Bougadis J (2003) Detection of trends in annual extreme rainfall. Hydrol Process 17(18):3547–3560

Birsan MV, Molnar P, Burlando P, Pfaundler M (2005) Streamflow trends in Switzerland. J Hydrol 314(1):312–329

Burn DH, Elnur MAH (2002) Detection of hydrological trends and variability. J Hydrol 255: 107–122

Burn DH, Cunderlik JM, Pietroniro A (2004) Hydrological trends and variability in the Liard river basin. Hydrol Sci J 49:53–67

Dore MHI (2005) Climate change and changes in global precipitation patterns: what do we know? Environ Int 31:1167–1181

Gosain AK, Rao S, Basuray D (2006) Climate change impact assessment on hydrology of Indian river basins. Curr Sci 90:346–353

Hamed KH, Rao AR (1998) A modified Mann-Kendall trend test for autocorrelated data. J Hydrol 204:182–196

Hameed T, Marino MA, De Vries JJ, Tracy JC (1997) Method for trend detection in climatological variables. J Hydrol Eng 4:154–160

Helsel DR, Hirsch RM (1992a) Statistical methods in water resources. Elsevier, Amsterdam, p 522

Helsel DR, Hirsch RM (1992b) Statistical methods in water resources. Elsevier, New York

Hess A, Iyer H, Malm W (2001) Linear trend analysis: a comparison of methods. Atmos Environ 35:5211–5222

Hesse BW, Nelson DE, Kreps GL, Croyle RT, Arora NK, Rimer BK, Viswanath K (2005) Trust and sources of health information: the impact of the Internet and its implications for health care providers: findings from the first Health Information National Trends Survey. Arch Intern Med 165(22):2618–2624

Kendall MG (1955) Rank correlation methods. Griffin, London

Kulkarni A, von Storch H (1995) Monte Carlo experiments on the effect of serial correlation on the Mann-Kendall test of trend. Meteorol Z 4(2):82–85

Lal M (2001) Climatic change—implications for India's water resources. J Indian Water Res Soc 21:101–119

Mann HB (1945) Nonparametric tests against trend. Econometrica 13:245–259

McBean EA, Rovers FA (1998) Statistical procedures for analysis of environmental monitoring data and assessmant

Modarresa R, Silva VPR (2007) Rainfall trends in arid and semi-arid regions of Iran. J Arid Environ 70:344–355

Rajendran K, Kitoh A (2008) Indian summer monsoon in future climate projection by a super high-resolution global model. Curr Sci 95:1560–1569

Rao AR, Hamed KH (1997) Regional frequency analysis of Wabash River flood data by L-moments. J Hydrol Eng 2(4):169–179

Rupa Kumar K, Pant GB, Parthasarathy B, Sontakke NA (1992) Spatial and sub seasonal patterns of the long term trends of Indian summer monsoon rainfall. Int J Climatol 12:257–268

Rupa Kumar K, Sahai AK, Kumar KK, Patwardhan SK, Mishra PK, Revadekar JV, Kamala K, Pant GB (2006) High-resolution climate change scenarios for India for the 21st century. Curr Sci 90:334–345

Scafetta N, West BJ (2005) Estimated solar contribution to the global surface warming using the ACRIM TSI satellite composite. Geophys Res Lett 32:L18713. doi:10.1029/2005GL023849

Sen PK (1968) Estimates of the regression coefficient based on Kendall's tau. J Am Stat Assoc 63:1379–1389

Shan Yu P, Yang TC, Wu CK (2002) Impact of climate change on water resources in southern Taiwan. J Hydrol 260:161–175

Singh P, Kumar V, Thomas T, Arora M (2008) Changes in rainfall and relative humidity in different river basins in the northwest and central India. Hydrol Process 22:2982–2992

Sinha Ray KC, De US (2003) Climate change in India as evidenced from instrumental records. WMO Bulletin 52:53–58

Srivastava HN, Sinha Ray KC, Dikshit SK, Mukhopadhaya RK (1998). Trends in rainfall and radiation over India. Vayu Mandal 41–45

Thapliyal V, Kulshreshtha SM (1991) Climate changes and trends over India. Mausam 42:333–338

Yue S Wang CY (2002) Applicability of prewhitening to eliminate the influence of serial correlation on the Mann-Kendall test, Water Resour Res 38(6). doi:10.1029/2001WR000861

Yue S, Pilon P (2003) Interaction between deterministic trend and autoregressive process. Water Resour Res 39(4). doi:10.1029/2001WR001210

Yue S, Pilon P, Cavadias G (2002) Power of the Mann-Kendall and Spearman's rho tests for detecting monotonic trends in hydrological series. J Hydrol 259:254–271

Zhang Q, Jiang T, Gemmer M, Becker S (2005) Precipitation, temperature and runoff analysis from 1950 to 2002 in the Yangtze basin, China. Hydrol Sci 50(1):65–79

The Impact of Climate Change on Rainfall Variability: A Study in Central Himalayas

L.N. Thakural, Sanjay Kumar, Sanjay K. Jain and Tanvear Ahmad

Abstract The impact of anthropogenic activities on the earth atmosphere and its systems has been an area of extensive research and engaging the attention of planners, governments, and politicians worldwide. The countries across the world are engaged in working out the impacts and associated vulnerabilities to the projected climate change. In India, the meteorological records indicate rise in the mean annual surface air temperature by 0.4 °C with not much variations in absolute rainfall. However, the rates of change in temperatures and precipitation have been found to be varying across the region. The intensity and frequency of heavy precipitation events have increased in the last 50 years. No change in the total quantity of rainfall is expected, however, the spatial pattern of the rainfall is likely to change, with rise in number and intensity of extreme rainfall events. The continuous warming and resulting change in rainfall pattern over different regions of Indian may adversely impact the natural resources on which majority of the population is dependent. Thus there is an ever increasing recognition of the need for micro and macro level assessments for greater understanding about the impact and implication of the current climate variability especially in eco-sensitive Himalayan region, where such studies are limited due to lack of adequate observational sites and related data. This study investigates the rainfall variability trends on annual and seasonal scales in the central Himalayan region. Eight observational sites at Almora, Nainital, Ranikhet, HawalBagh, Mukteshwar, Mukhim, Dehradun, and Mussoorie situated at different altitude in the central Himalaya region are considered for the study. These sites are situated within the altitude range of 682–2311 m. Statistical tools such as Mann–Kendall test, Sens's estimator of slope method (nonparametric) and regression (parametric), have been applied to analyze the rainfall variability trends at these observational sites. The significance of these trends has been tested at the 95% confidence level. The results indicate that there is decreasing trend at all the stations annually, however; only two stations show significantly decreasing trends at 95% confidence level. At seasonal scale, monsoon rainfall indicates decreasing trends at most of the stations in the region, whereas

L.N. Thakural (✉) · S. Kumar · S.K. Jain · T. Ahmad
National Institute of Hydrology, Roorkee 247667, Uttarakhand, India
e-mail: lnt@nih.ernet.in

V.P. Singh et al. (eds.), *Climate Change Impacts*, Water Science and Technology Library 82, https://doi.org/10.1007/978-981-10-5714-4_15

winter rainfall also shows the similar patterns. These temporal and spatial patterns of rainfall variability in central Himalayas may provide useful insights for the long-term planning and management of water resources and also useful for climatic studies in this region.

Keywords Climate change · Mann–kendal test · Regression analysis
Sen's estimator

Introduction

Water is of paramount importance and precious for the livelihood, from crucial drinking water to food production, production of energy to development of industries, and from the management of natural resources to environment conservation. The scarcity of water resources and its tremendous increasing demand, which is the outcome of growing population, intensification of agriculture sector, industrial and urban expansion, has necessitated its proper planning and management. Moreover, global warming and climate change have further added more intricacy. Himalayan region aptly called the water tower of Asia is to be greatly affected due to change in climatic conditions and impact the streamflow of snow-fed rivers originating from the region. The Himalaya is the youngest, highest and one of the most unstable regions of the world for which, the ecosystem is particularly fragile and more susceptible to the impacts of rapid climate change.

The changes in temperature, precipitation, and other climatic variables are likely to influence the amount and distribution of runoff in all river systems globally. Rainfall is an important factor in shaping the hydrology, water quantity and quality and the vegetal cover throughout the earth. A higher or lower rainfall or changes in its distribution would influence the spatial and temporal distribution of runoff, soil moisture, and ground water reserves and would increase the frequency of droughts and floods. The detection of trends in hydro-climatic data, particularly temperature, precipitation, and streamflow is essential for the assessment of impacts of climatic variability and its change on the water resources of a region. Trend analysis may thus focus on the overall pattern of change over time, help temporal and spatial comparisons for deriving future projections. Estimates of rainfall and temperature anomaly were better estimated using long-term series data. Several statistical methods apply parametric and nonparametric approach for the detection of trends.

Several studies in India have been carried out to determine the changes in temperature and rainfall and its association with climate change. Long-term trends in the maximum, minimum, and mean temperatures over the north-western Himalaya during the twentieth century (Bhutiyani et al. 2007) suggest a significant rise in air temperature in the north-western Himalaya, with winter warming occurring at a faster rate. Dimri and Ganju (2007) simulated the winter temperature and precipitation over the western Himalaya and found that temperature is underestimated and precipitation is overestimated in Himalaya. The changing trends of temperature and precipitation over the western Himalaya were examined and it was found that there was an increasing trend in temperature and decreasing trend in precipitation at some specific locations. Sharma et al. (2000) found an increasing trend in rainfall at some stations and a decreasing trend at other stations in Koshi basin in eastern Nepal and Southern Tibet. Similar trends in rainfall were found by Kumar et al. (2005) for the state of Himachal Pradesh. Basistha et al. (2009) investigated changes in the rainfall pattern for 30 stations during the twentieth century in the Indian Himalayas. They found that there was an increasing trend up to 1964 after which trend decreased during 1965–1980 for this region and changes were most explicit over the Shivaliks and southern part of Lesser Himalayas. Kumar and Jain (2010) analyzed rainfall and rainy days time series at five stations in Kashmir valley of India. They observe decreasing rainfall at four stations and increasing rainfall at one station, but none of the observed trends in annual rainfall were statistically significant. Choudhury et al. (2012) analyzed the long-term data (1983–2010) to detect a trend in the in Umiam located at mid altitude in Meghalaya. The results of the study indicated that there was a nonsignificant increasing trend (3.72 mm/year) in the total annual rainfall. Rai et al. (2010) investigated the persistence, trend, and periodicity in hydro-climatic variables in Yamuna river basin. The results indicated a significant difference in the patterns of monsoon and non-monsoon rainfall in terms of persistence and periodicity and about 20% of rainfall time series indicated the presence of persistence. They also observed an overall declining trend in the annual and monsoon rainfall, annual and monsoon rainy days and aridity index along with a rising trend in the onset of effective monsoon.

However, in Himalaya due to the poor and inadequate network of hydro-meteorological observations owing to inaccessibility for being rugged, dangerous with harsh climatic conditions, there is absence of relatively long-term consistent data. In the present paper, an attempt is made to analyze the trend of rainfall on seasonal and annual scales in the central Himalaya region.

The Study Area

The study is conducted for the eight stations of which five lie in the Kumaon region while other three stations lie in the Garhwal region of Uttarakhand state. Spatial distribution of the metrological stations is shown in Fig. 1. Uttarakhand state is located in the fragile region of Central Himalaya, India. The state has a total geographical area of 53,484 km^2 out of which 93% is mountainous, of which 65% is covered by forest. The state lies on the southern slope of the Himalaya range and has a highly varied topography with snow covered peaks, glaciers. Most of the northern part of the state is covered by high Himalayan peaks and glaciers. Two Indian largest rivers namely the Ganges and the Yamuna originates from the glaciers of Uttarakhand and fed by numerous lakes, glacial melts, and streams. It lies in the Northern part of India between the latitudes 28° 43′–31° 27′N and longitudes

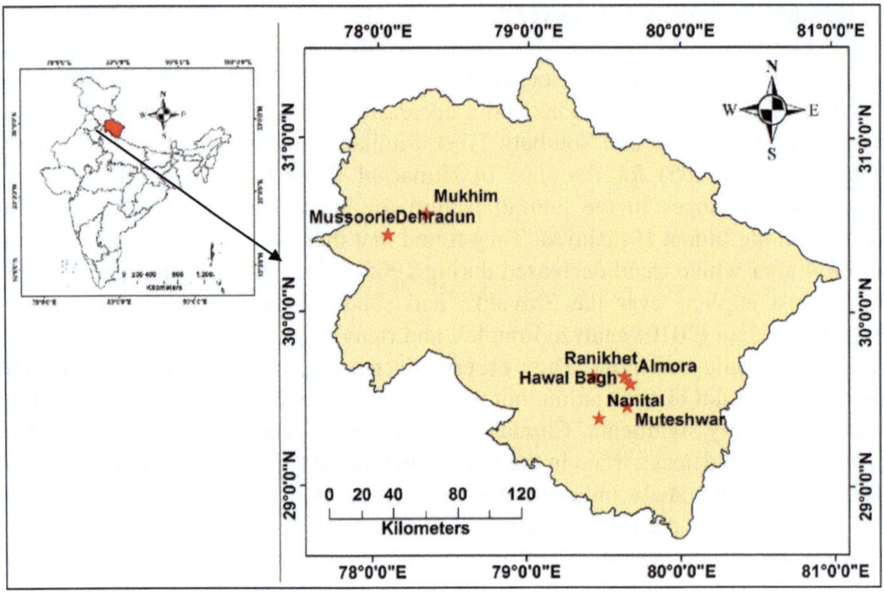

Fig. 1 Location of the metrological stations over central Himalaya

77° 34′–81° 02′E having a maximum dimension of east-west 310 and 255 km north–south covering an area of 53,484 km^2 with the elevation ranging from 210 to 7817 m above msl. The climate and the vegetation vary greatly with elevation, from glaciers at the highest elevations to subtropical forests at the lower elevations. The highest elevations are covered by ice and bare rock. Below them, between 3,000 and 5,000 m are the western Himalayan alpine shrub and meadows.

Data Used

For the present study, records on rainfall data have been collected from the IMD, Pune and Vivekananda Parvatiya Krishi Anusandhan Sansthan (VPKAS), Almora, and were further used for analysis. Moreover, the rainfall records at these stations were selected for different periods having continuous rainfall data. The meteorological stations used in the study and their details are shown in Table 1. For investigation of changes in rainfall at different time scales, a year was divided into four principal seasons:

1. Pre-monsoon season prevailing from March to May
2. Monsoon season prevailing from June to September
3. Post-monsoon season prevailing from October to November
4. Winter season prevailing from December to February

Table 1 Details of the meteorological stations located in the study area

Station	Longitude (dms)	Latitude (dms)	Altitude (m)	Period
Almora	79.67	29.6	1585	1980–2012
Dehradun	78.08	30.45	682	1957–2007
HawalBagh	79.634	29.642	1242	1981–2009
Mukhim	78.33	30.57	1945	1957–2007
Mukteshwar	79.65	29.47	2311	1901–1950
Mussoorie	78.08	30.45	1988	1930–1986
Nainital	79.47	29.4	1953	1953–1978
Ranikhet	79.43	29.643	1778	1974–2013

For evaluation of trend in rainfall, daily data have been used to form monthly totals. Monthly data of rainfall were further used to compute the seasonal and annual time series, which were in turn used for the investigation of trend on seasonal and annual time scale.

Methodology

Trends in data can be identified by using either parametric or nonparametric methods, and both the methods are widely used. The nonparametric methods do not require normality of time series and also are less sensitive to outliers and missing values. The nonparametric methods are extensively used for analyzing the trends in several hydrologic series namely rainfall, temperature, pan evaporation, wind speed etc. (Chattopadhyay et al. 2011; Dinpashoh et al. 2011; Fu et al. 2004; Hirsch et al. 1982; Jhajharia and Singh 2011; Jhajharia et al. 2009, 2011; Tebakari et al. 2005; Yu et al. 1993).

The present study analyzes the trends of rainfall series of each individual station using simple regression (parametric), Mann–Kendall test and Sens's estimator of slope (nonparametric).

Determination of Anomalies

For a better understanding of the observed trends, first of all, seasonal and annual anomalies of rainfall for each station were computed with reference to the mean of the respective variable for the available records. Further, these anomalies were plotted against time and the trend was examined by fitting the linear regression line. The linear trend value represented by the slope of the simple least square regression provided the rate of rise or fall in the variable.

Regression Model

One of the most useful parametric models to detect the trend is the "Simple Linear Regression" model. The method of linear regression requires the assumptions of normality of residuals, constant variance, and true linearity of relationship (Helsel and Hirsch 1992). The model for Y (e.g. precipitation) can be described by an equation of the form:

$$Y = at + b \tag{1}$$

where,

- t time (year)
- a slope coefficient; and
- b least-squares estimate of the intercept

The slope coefficient indicates the annual average rate of change in the hydrologic characteristic. If the slope is significantly different from zero, statistically, it is reasonable to interpret that there is a real change occurring over time. The sign of the slope defines the direction of the trend of the variable: increasing if the sign is positive, and decreasing if the sign is negative.

Magnitude of Trend

The magnitude of trend in a time series was determined using a nonparametric method known as Sen's estimator (Sen 1968). This method assumes a linear trend in the time series and has been widely used for determining the magnitude of trend in hydro-meteorological time series (Lettenmaier et al. 1994; Yue and Hashino 2003; Partal and Kahya 2006). In this method, the slopes (T_i) of all data pairs are first calculated by the following:

$$T_i = \frac{x_j - x_k}{j - k} \quad \text{for i} = 1, 2, \ldots, N \tag{2}$$

where x_j and x_k are data values at time j and k ($j > k$) respectively. The median of these N values of T_i is Sen's estimator of slope, which is calculated as follows:

$$\beta = \begin{cases} T_{\frac{N+1}{2}} & \text{if } N \text{ is odd,} \\ \frac{1}{2}\left(T_{\frac{N}{2}} + T_{\frac{N+2}{2}}\right) & \text{if } N \text{ is even.} \end{cases} \tag{3}$$

A positive value of β indicates an upwards (increasing) trend and a negative value indicates a downwards (decreasing) trend in the time series.

Significance of Trend

To ascertain the presence of a statistically significant trend in hydrologic climatic variables such as temperature, precipitation, and streamflow with reference to climate change, the nonparametric Mann–Kendall (MK) test has been employed by a number of researchers (Yu et al. 1993; Douglas et al. 2000; Burn et al. 2004; Singh et al. 2008a, b). The MK method searches for a trend in a time series without specifying whether the trend is linear or nonlinear. The MK test was also applied in the present study. The MK test checks the null hypothesis of no trend versus the alternative hypothesis of the existence of an increasing or decreasing trend. Following Bayazit and Onoz (2007), no pre-whitening of the data series was carried out as the sample size is large ($n \geq 50$) and slope of the trend was high (>0.01).

The statistic S is defined as (Salas 1993):

$$S = \sum_{i=1}^{N-1} \sum_{j=i+1}^{N} \operatorname{sgn}(x_j - x_i) \tag{4}$$

where N is the number of data points. Assuming $(x_j - x_i) = \theta$, the value of sgn (θ) is computed as follows:

$$\operatorname{sgn}(\theta) = \begin{cases} 1 & \text{if} \quad \theta > 0, \\ 0 & \text{if} \quad \theta = 0, \\ -1 & \text{if} \quad \theta < 0. \end{cases} \tag{5}$$

This statistic represents the number of positive differences minus the number of negative differences for all the differences considered. For large samples ($N > 10$), the test is conducted using a normal distribution (Helsel and Hirsch 1992) with the mean and the variance as follows:

$$E[S] = 0$$

$$\operatorname{Var}(S) = \frac{N(N-1)(2N+5) - \sum_{k=1}^{n} t_k(t_k - 1)(2t_k + 5)}{18} \tag{6}$$

where n is the number of tied (zero difference between compared values) groups and t_k is the number of data points in the kth tied group. The standard normal deviate (Z-statistics) is then computed as (Hirsch et al. 1993):

$$Z = \begin{cases} \frac{S-1}{\sqrt{\operatorname{Var}(S)}} & \text{if} \quad S > 0 \\ 0 & \text{if} \quad S = 0 \\ \frac{S+1}{\sqrt{\operatorname{Var}(S)}} & \text{if} \quad S < 0. \end{cases} \tag{7}$$

If the computed value of $\mid Z \mid > z_{\alpha/2}$, the null hypothesis H_0 is rejected at the α level of significance in a two-sided test. In this analysis, the null hypothesis was tested at 95% confidence level.

Results and Discussion

For better comprehension and visual interpretation of the observed trends, first of all, seasonal and annual anomalies of rainfall for each station were computed with reference to the mean of the respective variable for the available records. Further, these anomalies were plotted against time and the trend was examined by fitting the linear regression line. The linear trend value represented by the slope of the simple least square regression provided the rate of rise/fall in the variable. Thereafter, Mann–Kendall (MK) test has been used for identification and to test the statistical significance of trend at a confidence interval of 95%. Prior to which data series of all the variables were checked for the presence of auto-correlation. The Sen's estimator of slope (SE) was then applied to estimate the magnitude of the trend over the study period. The SE was applied to verify the outcomes of simple regression analysis. The outcomes of the analysis are shown in the form of Table and/or graph.

The anomalies of rainfall and their trends were determined for all the stations considered in the study. Anomalies in annual rainfall and their trends for the all the meteorological stations within the study area are shown in Fig. 2.

The figure shows the outcomes of the parametric approach which shows that there are a decreasing trend in almost all the stations. Further analysis using the

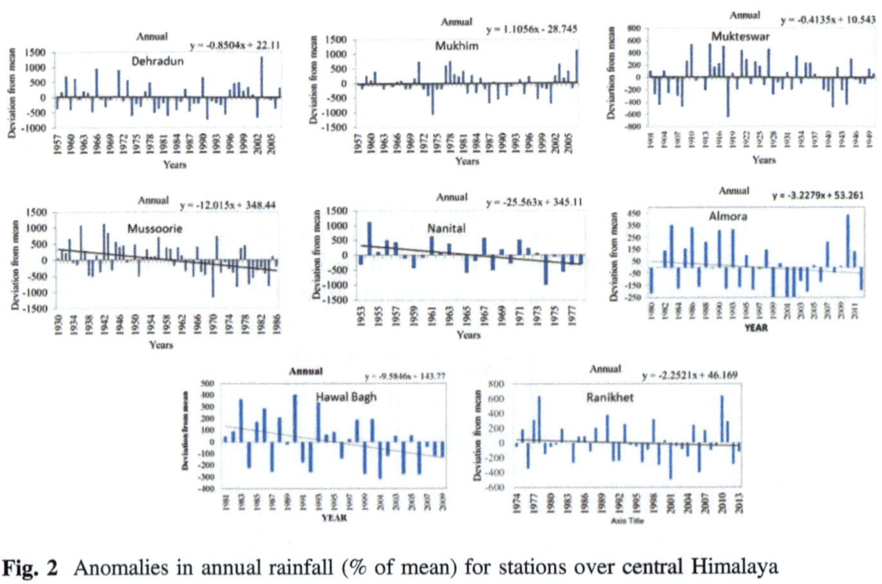

Fig. 2 Anomalies in annual rainfall (% of mean) for stations over central Himalaya

Table 2 Sen estimator of slope (mm/year) for seasonal and annual rainfall

Station	Pre monsoon		Monsoon		Post monsoon		Winter		Annual	
	Z	Sen slope	*Z*	Sen slope	*Z*	Sen slope	*Z*	Sen slope	*Z*	Sen slope
Almora	−1.41	−2.06	0.63	2.08	−0.84	−0.46	−3	**−3.53***	−0.66	−2.13
Dehradun	1.77	1.35	−0.48	−2.11	−1.86	−0.543	0.17	0.122	−0.29	−1.81
HawalBagh	−1.26	−2.69	−0.77	−2.81	−0.58	−0.25	−2.85	**−3.37***	−1.97	**−9.96***
Mukhim	1.96	2.268	−0.25	−1.837	−2.15	**0.723***	−0.25	−0.214	−0.12	−0.262
Mukteshwar	−0.89	−0.497	0.33	0.533	−0.45	−0.098	0.63	0.475	−0.18	−0.385
Mussoorie	0.93	0.68	−3.35	**−12.9***	0.66	0.398	−1.85	−1.66	−2.94	**−11.6***
Nainital	0.79	1.95	−0.79	−7.35	−1.99	**−5.57***	−0.97	−3.66	−1.9	−24.07
Ranikhet	−0.59	−0.74	−0.9	−2.09	−0.26	0	0.22	0.28	−0.78	−2.14

*Bold indicates statistical significance at 95% confidence level as per with Mann–Kendall test (+ for increasing and − for decreasing)

parametric approach has been detailed in Table 2. It indicates that the annual rainfall at all the stations is showing decreasing trend with a maximum decrease (−24.07 mm/year) at Nainital with a minimum decrease (−0.262 mm/year) at Mukhim. However, these decreasing trends are significant at 95% confidence level only at stations Hawal Bagh and Mussoorie.

Seasonal analysis of rainfall trends shows that pre-monsoon rainfall shows negative trends for Almora, Hawal Bagh, Mukteshwar, and Ranikhet whereas stations Dehradun, Mukhim, Mussoorie, and Nainital shows increasing trend. Table 2 also indicates that none of these increasing or decreasing trends are statistically significant. Similar analysis of monsoon rainfall shows decreasing trend at all stations except Almora and Mukteshwar. However, none of these increasing/decreasing trends are significant except at Mussoorie (−12.9 mm/year) which shows a statistically significant decreasing trend. In post-monsoon rainfall, all station except Mukhim and Mussoorie are showing decreasing trend. These trends are significant at Mukhim (0.725 mm/year) and Nainital (−5.57 mm/year). The winter rainfall analysis shows decreasing trends at Almora, HawalBagh, Mukhim, Mussoorie and Nainital whereas stations Dehradun, Mukteshwar, and Ranikhet show increasing trends. Winter rainfall trends at two stations namely Almora (−3.53 mm/year) and Hawal Bagh (−3.37 mm/year) are statistically significant.

Conclusions

The mountainous basin is highly sensitive to climate change. Any change in rainfall and temperature highly influences stream flow downstream. The detection of trends and magnitude of variations due to climatic changes in hydro-climatic data, particularly temperature, precipitation and stream flow, is essential for the assessment of impacts of climate variability and change on the water resources of a region. The present study is based on the analysis of trends in rainfall data using parametric (linear regression) and nonparametric (Mann–Kendall test and Sen's estimator of slope) methods on seasonal and annual scales for the Central Himalaya.

The analysis shows that all the stations indicate a decreasing trend in the annual rainfall. These decreasing trends are found statistically significant at the 95% confidence level at two stations HawalBagh and Mussoorie. The seasonal rainfall analysis shows a mix of increasing and decreasing trends. These trends, however, are not significant for pre-monsoon rainfall. Monsoon rainfall is found to be significant at Mussoorie whereas post-monsoon rainfall is significant at two stations. The winter rainfall trends have significant decreasing trends at Almora and HawalBagh. More such analysis is required to examine the trend in other climatological variables in other Himalayan basins. Also, there is need to understand the behavior of this basin to climate change and its future impact to plan and manage the water resources.

References

Bayazit M, Onoz B (2007) to prewritten or not to prewritten in trends analysis? Hydrol Sci J 52 (4):611–624

Basistha A, Arya DS, Goel NK (2009) Analysis of historical changes in rainfall in the Indian Himalayas. Int J Climatol 29:555–572

Bhutiyani MR, Kala VS, Power NJ (2007) Long-term trends in maximum, minimum and mean annual air temperatures across the northwest Himalaya during the twentieth century. Clim Change 85(1–2):159–177

Burn DH, Cunderlick JM, Pietroniero A (2004) Hydrological trends and variability in the Liard river basins. Hydrol Sci J 49(1):53–67

Choudhury BU, Das A, Ngachan SV, Slong A, Bordoloi LJ, Chouwdhury P (2012) Trend analysis of long term weather variables in mid altitude Meghalaya, North-East India. J Agric Phys 12 (1):12–22. ISSN 0973–032X

Chattopadhyay S, Jhajharia D, Chatopadhyay G (2011) Univariate modeling of monthly maximum temperature time series over North East India: neural network versus Yule-walker equation based approach. Meteorolog Appl 18:70–82. doi:10.1002/met.2011

Dimri AP, Ganju A (2007) Wintertime seasonal scale simulation over western Himalaya using RegCM3. Pure Appl Geophys 164(8–9):1733–1746

Dinpashoh Y, Jhajharia D, Fakheri-Fard A, Singh VP, Kahya E (2011) Trends in reference evapotranspiration over Iran. J Hydrol 399:422–433

Douglas EM, Vogel RM, Knoll CN (2000) trends in flood and low flows in the United States: impact of spatial correlation. J Hydrol 240:90–150

Fu G, Chen S, Liu C, Shepard D (2004) Hydro-climatic trends of the Yellow river basin for the last 50 years. Clim Change 65:149–178

Helsel DR, Hirsch RM (1992) Statistical methods in water resources. Elsevier, Amsterdam, p 522

Hirsch RM, Slack JR, Slack RA (1982) Techniques of trend analysis for monthly water quality data. Water Resour Res 18(1):107–121

Hirsch RM, Helsel DR, Cohn TA Gilroy EJ (1993) Statistical treatment of hydrologic data. In: Maidment DR (ed) Handbook of hydrology. McGraw-Hill, New York, USA, pp 17.1–17.52

Jhajharia D, Shrivastava SK, Sarkar S (2009) Temporal characteristics of pan evaporation trend under the humid conditions of Northeast India. Agric. Forest Meteorol. 149:763–770

Jhajharia D, Singh VP (2011) Trends in temperature, diurnal temperature range and sunshine duration in Northern India. Int J Climatol 31(9):1353–1367

Jhajharia D, Dinpashoh Y, Kahya E, Singh VP, Fakheri-Farid A (2011) Trends in reference evapotranspiration in the humid region of North-east India. Hydrol Process 26(3):421–435

Kumar V, Singh P, Jain SK (2005) Rainfall trend over Himachal Pradesh, Western Himalaya, India. In: Proceedings of conference on development of hydro power projects—a prospective challenge, Shimla

Kumar V, Jain SK (2010) Trends in seasonal and annual rainfall and rainy days in Kashmir valley in the last century. QuatInt 212(1):64–69

Lettenmaier DP, Wood EF, Wallis JR (1994) Hydro-climatological trends in the continental United States, 1948–88. J Climate 7:586–607

Partal T, Kahya E (2006) Trend analysis in Turkish precipitation data. Hydrol Process 20:2011–2026

Rai RK, Upadhyay A, Ojha CSP (2010) Temporal variability of climatic parameters of Yamuna river basin: spatial analysis of persistence, trend and periodicity. Open Hydrol J 4:184–210

Salas JD (1993) Analysis and modeling of hydrologic time series. In: Maidment DR (ed) Handbook of hydrology. McGraw-Hill, New York, pp 19.1–19.72

Sharma KP, Moore B, Vorosamarty CJ (2000) Anthropogenic, climatic and hydrologic trends in the kosi basin, Himalaya. Clim Change 47:141–165

Sen PK (1968) Estimates of the regression coefficient based on Kendall's tau. J Am Statist Assoc 63:1379–1389

Singh P, Kumarm V, Thomas T, Arora M (2008a) Change in rainfall and relative humidity in different river basins in the northwest and central India. Hydrol Process 22:2982–2992

Singh P, Kumar V, Thomas T, Arora M (2008b) Basin-wide assessment of temperature trends in the northwest and central India. Hydrol Sci J 53(2):421–433

Tebakari T, Yoshitani J, Suvanpiomol C (2005) Time-space trend analysis in pan evaporation kingdom of Thailand. J Hydrol Eng 10(3):205–215

Yu YS, Zou S, Whittemore D (1993) Non-Parametric trend analysis of water quality data of rivers in Kansas. J Hydrol 150:61–80

Yue S, Hashino M (2003) Temperature trends in Japan: 1900-1990. Theoret Appl Climatol 75: 15–27

Estimation of Changes in Annual Peak Flows in Netravathi River Basin, Karnataka, India

Arega Mulu, T.M. Fasnamol and G.S. Dwarakish

Abstract Flood is one of the most chronic and hazardous natural calamities all over the world. Information about the magnitude of peak streamflow and flood frequency with specified recurrence intervals is necessary to safely and economically design structures that are in, or near streams. The present study is carried out with a view to understand the effects of changing historical peak flows and flood-frequency computations for streams in Netravathi river basin at Bantwal gauging station, Karnataka, India. To accomplish this objective, 39-year records of annual stream flows were tested for historical flood-frequency analyses. The magnitude of annual peak-flow changes over time was computed using the Sen's slope and Mann–Kendall Test. Peak flows with 50, 20, and 1% exceedance probability (2, 5 and 100-year recurrence interval, respectively) were computed for a stream gauge Bantwal in Netravathi river basin and found out 3933, 5238 and 8804 m^3/s, respectively. Sen's slope from showed the decreasing trend of peak flow with the magnitude of −38.355 corresponding to Mann–Kendall Test.

Keywords Trend analysis · Annual daily peak stream flow · Flood frequency AEPs · Sen's slope

Introduction

Flood is one of the most chronic and hazardous natural calamities all over the world. Extreme rainfall events and the resulting floods can take thousands of lives and cause billions of dollars in damage. So, the flood plain management and designs for flood control works, reservoirs, bridges, and other investigations need to

A. Mulu (✉) · T.M. Fasnamol · G.S. Dwarakish
Department of Applied Mechanics and Hydraulics, NITK, Surathkal
Mangalore, India
e-mail: muluarega21@gmail.com

T.M. Fasnamol
e-mail: fasnamol789@gmail.com

V.P. Singh et al. (eds.), *Climate Change Impacts*, Water Science and Technology Library 82, https://doi.org/10.1007/978-981-10-5714-4_16

reflect the likelihood or probability of such events. So the trend analysis of stream flows and frequency studies are equally important to hydrologists.

Trends are important indicators of the temporal variability of phenomena (discharges). By analysing the time sequence of the discharges, we can assess the magnitude and significance of the temporal variability. Hydrologic systems are sometimes impacted by extreme events, such as severe storms, floods, and droughts. The magnitude of an extreme event flood is inversely related to its frequency of occurrence, very severe events occurring less frequently than more moderate events. The objective of frequency analysis of hydrologic data is to relate the magnitude of extreme events to their frequency of occurrence through the use of probability distributions. The hydrologic data analysed are assumed to be independent and identically distributed, and the hydrologic system producing them is considered to be stochastic, space independent and time independent. The hydrologic data employed should be carefully selected so that the assumptions of independence and identical distribution are satisfied. In practice, this is achieved by selecting the annual maximum of the variable being analysed with the expectation that successive observations of this variable from year to year will be independent. Knowledge of the magnitude and probable frequency of recurrence of floods is necessary to the proper design and location of structures such as dams, bridges, culverts, levees, highways, waterworks, sewage disposal plants, and industrial buildings. Bantwal stream gauge for 39 years of recorded been tested for historical flood frequency to provide some insight into future flood frequency.

The main objective of present study is carried out with a view to understand the effects of changing historical peak flows on flood-frequency computations for streams in Netravathi river basin, Karnataka, India and there is also an attempt to study both Mann–Kendall and Sen's Slope statistics for the detection of the trend of peak flows in Nethravathi river basin.

Study Area

The research was conducted at Netravathi river basin. It has an area of about 3,312 km^2. It also has geographical coordinates of 12° 29′ 11″–13° 11′ 11″N latitudes and 74° 49′ 08″ to 75° 47′ 53″E longitude with large elevation difference varying from 0 to 1400 m above MSL (Babar and Ramesh 2013). The Netravathi River, one of the major west flowing rivers in Karnataka, originates in the South of Samse village, at an altitude of approximately 1200 m from mean sea level in the Western Ghats of Karnataka. The river flows westward for about 103 km with a drainage area of 3657 km^2 (Shobitha 2012) and empties into the Arabian Sea, after joining Gurpur River at Mangalore city. The river joined with Mundaja Neriya, Shishla Uppar, Kumaradhara and Beltangady nallas from either side. Average annual rainfall in the region is about 3930 mm. June–September is Southwest monsoon season which contributes 90% of annual rainfall, but the remaining 10%

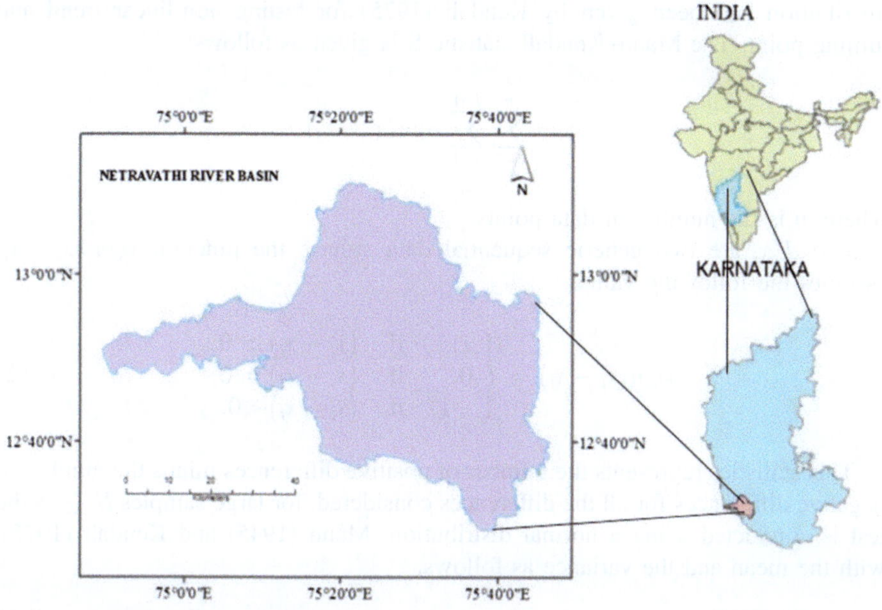

Fig. 1 Location map of the study area

of rain fall is during pre and post monsoon. This river is the major source of water for Bantwal, Mangalore city, Industries, Hydropower production and agricultural activities in the basin (Fig. 1).

Data Used and Methodology

Thirty-nine years of stream flow data of Netravathi river basin at Bantwal stream gauge station from 1971 to 2009 were used for analysis of this study. The data were acquired from central water commission (CWC), India.

Trend of Historical Changes of Annual Peak Flows

In the present study, trend analysis has been done by using non-parametric Mann-Kendall test. This is a statistical method which is being used for studying the spatial variation and temporal trends of hydro-climatic series. A non-parametric test is taken into consideration over the parametric one since it can avoid the problem revived by data skew (Smith 2000). Mann–Kendall test had been formulated by Mann (1945) as non-parametric test for trend detection and the test statistic distribution had been given by Kendall (1975) for testing non-linear trend and turning point. The Mann–Kendall statistic S is given as follows:

$$S = \sum_{i-2}^{n} \sum_{j-1}^{i-1} \operatorname{sign}(x_i - x_j) \tag{1}$$

where, n is the number of data points.

x_i *and* x_j are two generic sequential data values, the function *sign* $(x_i - x_j)$ assumes the following values

$$\operatorname{sign}(x_i - x_j) = \begin{cases} 1, & \text{if} \quad (x_i - x_j) > 0, \\ 0, & \text{if} \quad (x_i - x_j) = 0, \\ -1, & \text{if} \quad (x_i - x_j) < 0. \end{cases} \tag{2}$$

This statistics represents the number of positive differences minus the number of negative differences for all the differences considered, for large samples $N \geq 8$ the test is conducted using a normal distribution, Mann (1945) and Kendall (1975), with the mean and the variance as follows:

$$E[S] = 0$$
$$\mathrm{VAR}(S) = \tfrac{1}{18}\left[n(n-1)(2n+5) - \sum_{p-1}^{g} t_p(t_p - 1)(2t_p + 5)\right] \tag{3}$$

where n is the length of the times series, t is the extent of any given tie (zero difference between compared values) and $\sum t_p$ denotes the summation over all ties. The standardised test statistic Z is given by the following:

$$z = \begin{cases} \frac{S-1}{\sqrt{\mathrm{Var}(S)}} & \text{if} \quad S > 0 \\ 0 & \text{if} \quad S = 0 \\ \frac{S+1}{\sqrt{\mathrm{Var}(S)}} & \text{if} \quad S < 0 \end{cases} \tag{4}$$

The test statistics Z is used as a measure of significance of trend. A positive Z indicates an increasing trend in the time series, while a negative Z indicates a decreasing trend.

Sen's Slope Estimator Test

The magnitude of annual peak-flow changes over time was computed using the Sen's slope (also known as the Kendall–Theil robust line). This slope is computed as the median of all possible pairwise slopes in the temporal data set (Helsel and Hirsch 2002). That is the slope (*Ti*) of all data pairs is computed as shown in below.

$$Ti = \frac{x_j - x_k}{j - k} \tag{5}$$

where, x_j and x_k are considered as data values at time j and k ($j > k$) correspondingly.

The median of these N values of Ti is represented as Sen's estimator of slope and is computed as

$$\text{Qmed} = \begin{cases} T(N+1)/2, & \text{if } N \text{ appears odd} \\ [TN/2 + T(N+2)/2]/2, & \text{if } N \text{ appears even} \end{cases} \tag{6}$$

The magnitude of trend can be analysed by the value of Sen's slope and its positive value indicates the increasing trend and its negative value indicates the decreasing trend.

Flood-Frequency Analyses for Full Periods of Record

Peak flows with 100, 5 and 2-year recurrence intervals were computed for the Bantwal stream gauge station using the full available annual peak-flow record. Extreme Value Type I probability function was used to find the exceedance probabilities and recurrence interval and corresponding streamflow (Chow et al. 1988). Extreme values are selected maximum or minimum values of sets of data. Distributions of extreme values selected from sets of samples of any probability distribution have been shown by Fisher and Tippett (1928) to converge to one of three forms of extreme value distributions. The properties of extreme value type I (EVI) probability distribution function were developed by Gumbel (1941).

EVI probability distribution function for the recurrence interval T is

$$F(x) = \exp\left[-\exp\left(-\frac{x-u}{\alpha}\right)\right] \quad -\infty \le x \le \infty \tag{7}$$

The parameters are estimated as

$$\alpha = \frac{\sqrt{6}s}{\pi} \tag{8}$$

$$u = \bar{x} - 0.5772\alpha \tag{9}$$

where, $\bar{x}$ and s are the mean and standard deviation, respectively. The parameter u is the mode of the distribution (point of maximum probability density). A reduced variate y_T can be defined as

$$y_T = -\ln\left[\ln\left(\frac{T}{T-1}\right)\right] \tag{10}$$

$$x_T = u + \alpha y_T \tag{11}$$

where, x_T is the maximum discharge with the frequency of T years.

Results and Discussions

Trend analysis of Nethravathi river basin has been done in the present study with 39 years of streamflow data from 1971 to 2009. Mann–Kendall and Sen's Slope Estimator has been used for the determination of the trend. The annual daily peak stream flows over 39 years is 9832 m^3/s and minimum annual daily peak stream flow is 1781 m^3/s (Fig. 2). Average peak flow for 39 years is 4175 m^3/s. Sen's slope from Fig. 2 shows the decreasing trend of peak flow with the magnitude of -38.355 corresponding to Mann–Kendall Test.

Flood-Frequency Analysis for Observed Peak Stream Flows

Flood-Frequency Analysis for Observed daily maximum peak flows is done using 50 20 and 1% annual exceedance probabilities (AEPs) (equivalent to 2, 20 and 100 years recurrence interval, respectively). Based up on selected AEPs observed peak flows are 3933, 5238 and 8804 m^3/s for 50, 20 and 1% AEPs as shown in Table 1.

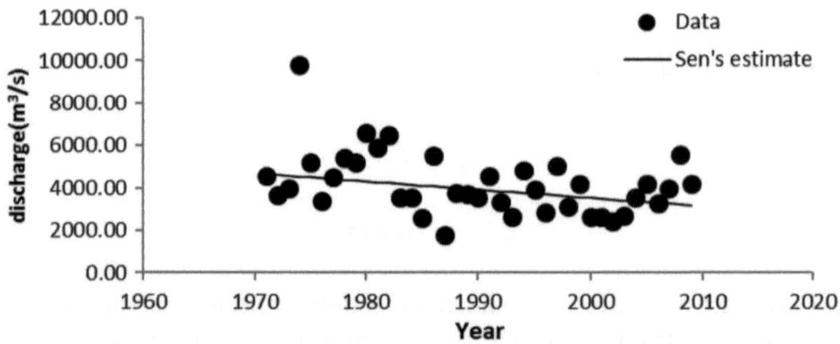

Fig. 2 Annual peak discharge of 39 years

Table 1 Observed peak flows with 50, 20 and 1% annual exceedance probabilities (AEPs)

Stream flow gauging station	Years of record	Annual exceedance probability (%)	Observed stream flow (m^3/s)
Bantwal	39	50	3933
Bantwal	39	20	5238
Bantwal	39	1	8804

Conclusion

The annual daily peak stream flow over 39 years (1971–2009) for Netravathi River Basin at Bantawal gauging station is 9832 m^3/s. Trend analysis of Netravathi River Basin at this gauging station shows negative value of Sen's slope which shows decreasing trend of annual daily peak stream flow corresponding to Mann–Kendall test. Therefore; in Netravathi River Basin at Bantawal gauging station annual daily peak stream flow and flood frequency with 1%, AEPs will decrease in the future.

References

Babar SF, Ramesh H (2013) Analysis of south west monsoon rainfall trend using statistical techniques over Netravathi basin. IJATCE 2:130–136

Chow VT, Maidment DR, Mays LW (1988) Applied hydrology. Editions McGraw-Hill, New York, 572 pp

Fisher RA, Tippett LHC (1928) Limiting forms of the frequency distribution of the largest or smallest member of a sample. In: Mathematical proceedings of the Cambridge philosophical Society vol 24, No 02. Cambridge University Press, UK. pp 180–190

Gumbel EJ (1941) The return period of flood flows. Ann Math Stat 12(2):163–190

Helsel DR, Hirsch RM (2002) Statistical methods in water resources—hydrologic analysis and interpretation: techniques of water-resources investigations of the U.S. geological survey, Chap. A3, book 4, 510 p

Kendall MG (1975) Rank correlation methods, 4th edn. Charles Griffin, London, UK

Mann HB (1945) Nonparametric tests against trend. Econometrica. J Econometric Soc 245–259

Shobitha MP (2012) Estimation of hydropower potential in Netravathi river basin using RS and GIS. M. Tech thesis. NITK Surathkal

Smith LC (2000) Trends in Russian Arctic river-ice formation and breakup, 1917 to 1994. Phys Geogr 21(1):46–56

Potential Impacts of Climate Change on Water Resources in Semi-Arid Region of Chittorgarh, India

Ajit Pratap Singh, Parnika Shrivastava and A.K. Vidyarthi

Abstract Climate change is considered as a substantial anthropogenic global environment threat. This paper presents a framework for assessing climate change impacts on water resources of the Chittorgarh district, Rajasthan (India). Various vulnerability indicators have been considered as the assessment criteria which deal mainly with vulnerability aspects of water availability, climatic conditions, the current status of agriculture/irrigation area, sensitivity, water governance and coping capacity. These factors are finally integrated to evaluate potential impacts of climate change on water resources in the selected region. An outranking method is applied to obtain a temporal ranking of the alternatives, i.e. year-wise performance of the region. To increase the reliability of results, verification is performed using two multi-criteria approaches, namely analytical hierarchy process and social choice methods by taking into consideration of above assessment criteria. A future trend of vulnerability aspects has also been derived using various prediction techniques.

Keywords Climate change · Semi-arid region · Multi-criteria decision methods

---

A.P. Singh (✉)
Civil Engineering Department, Birla Institute of Technology & Science, Pilani 333031, India
e-mail: aps@pilani.bits-pilani.ac.in

P. Shrivastava
Hindustan ECOSOFTT Pvt. Ltd., Rajendra Place, New Delhi 110008, India
e-mail: parnika28@gmail.com

A.K. Vidyarthi
Central Pollution Control Board, New Delhi 110032, India

A.P. Singh
Civil Engineering Department, BITS Pilani-Dubai Campus, DIAC 345055, UAE

V.P. Singh et al. (eds.), *Climate Change Impacts*, Water Science and Technology Library 82, https://doi.org/10.1007/978-981-10-5714-4_17

Introduction

Climate change is defined as the long-term changes in various climatic parameters such as rainfall, temperature, evapotranspiration, etc. Inter-governmental panel on climate change (IPCC) defines climate change as a change in the state of climate that can be identified by changes in the mean and/or the variability of its properties and that persists for an extended period, typically for a decade or longer. Climate change was first addressed in 1992 by United Nations framework convention on climate change (UNFCC). Since then, the importance and significance of the uncertainty in climate change is being realized, and now climate change is recognized as a threat to environment globally. On an average, every decade there is a temperature rise of about 1 °C globally. It adversely effects food production and security, biodiversity of forests and other natural ecosystems, sustained water supply, human health and settlements (Ravindranath 2011). Climate change can be caused because of natural factors such as earth's axial and orbital changes, plate movements, asteroid collision, chemical weathering, etc., and anthropogenic factors such as industrialization, use of fossil fuels, urbanization, excessive agriculture and live-stocks, changing land use pattern, etc. According to IPCC, concentration of greenhouse gases such as carbon dioxide (CO_2), methane (CH_4) and nitrous oxide (N_2O) during the natural and anthropogenic actions contribute significantly towards exponential climate change over the past few decades.

Water is a key natural resource, an essential human need and a valuable national asset. Its management deals with solving problems related to many challenges (Jain 2012; Singh et al. 2007). Though India is gifted with a large number of rivers and about 400×10^{10} m^3 of water is available in Indian territory, there is significant variations in annual rainfall both spatially and temporally. The highest annual rainfall in Cherrapunji is 11,500 mm, whereas the minimum annual rainfall is 215 mm at Jaisalamer. There exist recurring monsoon failures or floods during monsoon periods. Water demand is ever increasing which leads various sociological problems. The problems of water resources becomes even more worsening due to climatic change, which is essentially dependent on two most critical parameters, viz. temperature and precipitation. The rise in temperature leads to global warming phenomena which further disturbs the hydrological cycle. With a rise in temperature, water from the atmosphere will reach earth's surface more as rainfall and less as snow. Also, increasing temperature indicates melting of snow caps and lead to excessive run-off. Rising temperature also increases evapotranspiration from crops and vegetation, leading to higher water demand for agricultural and other beneficial uses of human kind. Industrial sectors have also become a major concern in recent years as they are polluting different components of our environment. Various studies have demonstrated how industrial sectors dealing with steel production, refinery, coal mining, thermal power plant and the chemical industries have been impacting the climate (Vidyarthi et al. 2015).

Rajasthan, the largest state of India (area-wise) falls within the areas of great climate sensitivity. Water sustainability is a prominent concern with increasing population in the state mainly due to the extreme climatic conditions, irregular and low rainfall and reduced per capita water availability (Singh 2008). Chittorgarh, the 14th largest district of Rajasthan (population-wise), is one such climate change effected district located in the Mewar (southern Rajasthan) region which has the total population of about 15 lakhs (2011 Census).

The main objective of the study is to formulate an integrated district profile for assessing the vulnerability of climate change over the temporal scale. To cater this goal, various criteria are being integrated encompassing water, agriculture, climatic, sensitivity and governance conditions. Further, outranking and multi-criteria approaches are applied to the problem to obtain year-wise performance of water resources by deriving ranking on the basis of selected vulnerability criteria. This paper also demonstrates the future climatic condition and its vulnerability using various prediction techniques. The next section highlights the entire study framework including study area, data collection and methodology used with appropriate illustrations.

Methodology

IPCC defines vulnerability as the degree to which a system is susceptible to or unable to cope with adverse effects of climate change, including climate variability and extremes. The paper applies the above principle to the study area using various vulnerability indices mainly focusing on water and agricultural sector.

Study Area

In this study, Chittorgarh district in southern Rajasthan, as shown in Fig. 1, is selected for climate change assessment. The total area of the district is around 7.50 lakhs ha with a sown area of 3.13 lakhs ha, i.e. 41.74% of agricultural land. Nearly, 50% of the area falls in the Banas catchment and average rainfall of the district is around 70.65 mm. There are totally 11 tehsils (sub-divisions) with urban population of 12.59 lakhs and 2.85 lakhs, adding to a total district population of 15.44 lakhs (Statistics: Chittorgarh 2014).

Study Framework

The main goal of this study is to assess the climate change impact on the water resources and study its vulnerability. To cater this goal, five major assessment criteria/indicators are being employed namely water vulnerability index,

Fig. 1 Location of Chittorgarh district in Rajasthan, India (*Source* http://www.india.gov.in/maps)

Fig. 2 Various vulnerability assessment criteria and sub-criteria

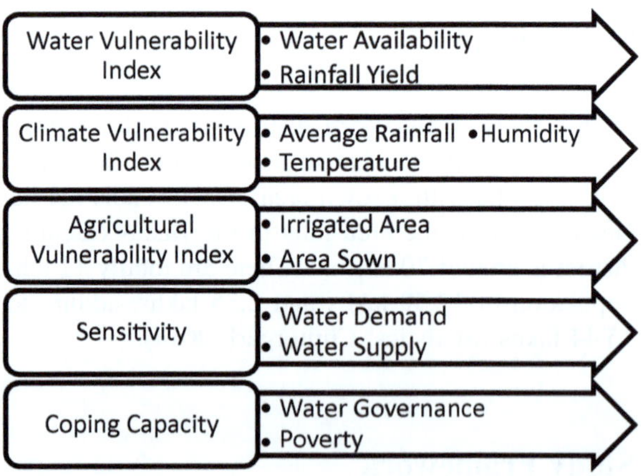

agricultural vulnerability index, climate vulnerability index, sensitivity and coping capacity of the district. Further these criteria are being classified into various sub-criteria, as shown in Fig. 2, elaborating them in a better manner.

Table 1 Input values of sub-criteria for temporal scale

Sub-criteria	1996	1997	1998	1999	2000	2001	2002
Water availability (MCM)	2282	1338	1432	2025	1524	1548	1962
Rainfall yield (MCM)	6454	3289	2768	5460	3069	4152	2551
Average rainfall (mm)	1189	606	510	1006	580	765	470
Temperature (°C)	22.4	23.9	25.5	24.3	26.5	27.8	29.3
Humidity (%)	65	62	51	56	51	51	58
Area irrigated (ha)	231,186	219,473	162,336	184,494	108,267	165,220	72,793
Area sown (ha)	616,778	634,281	607,575	592,622	475,848	556,446	460,021
Water demand (LPCD)	89	92	99	87	98	95	103
Water supply (LPCD)	82	85	80	80	85	83	80
Water governance	Medium	Medium	Good	Medium	Medium	Medium	Good
Poverty, monthly per capita (Rs.)	378.98	404.23	444.35	465.92	512.85	548.98	559.63

Further, a temporal scale is considered from 1996 to 2002 to assess values of the sub-criteria assessment indicators. Lastly, year-wise performance with appropriate ranking has been evaluated using multi-criteria approaches.

Data Collection

The values of various sub-criteria for each of the year is being taken from annual reports of various organizations such as Central Ground Water Board (CGWB), Irrigation Department, Public Health Engineering Department (PHED), Centre for Pollution Control Board (CPCB) Ministry of Water Resources (MoWR), etc. Also, due to limited data available, few parameters' data have been interpolated. Table 1 underlines the input data collected from various agencies.

Data Analysis Using PROMETHEE

For ranking of the temporal scale, based on the various sub-criteria, Preference Ranking Organization Method for Enrichment Evaluation (PROMETHEE) technique has been applied (Singh and Shrivastava 2014). An academic version of Visual PROMETHEE software has been used for complete analysis of the study. The initial stages of the analysis were to input the definition of criteria and alternatives. Further, using statistical calculations and predefined help option, preference function for each alternative has been designed This process enables to estimate the preference function automatically on the basis of numerical data and statistical calculations, by the inbuilt preference function assistance. Finally, positive, negative and net flows have been computed using PROMETHEE-GAIA as shown in Table 2.

Table 2 Positive, negative and net flows of various alternatives (years)

Rank	Year	φ	φ^{+}	φ^{-}
1	**1996**	**0.3661**	**0.5187**	**0.1525**
2	1999	0.1805	0.3544	0.1739
3	2001	0.0065	0.2257	0.2192
4	1997	−0.0699	0.2305	0.3004
5	2000	−0.0850	0.2026	0.2876
6	2002	−0.1379	0.2058	0.3436
7	**1998**	**−0.2603**	**0.1018**	**0.3621**

Results and Discussion

Table 2 gives the ranking of year-wise performance of water resources in the region on the basis of net flow (φ) value. A higher φ value represents a better climate change profile and hence, low vulnerability whereas for lower φ values, higher vulnerability is encountered with a fluctuating climate change profile. Another important result from the above computation is the PROMETHEE Rainbow. It is a dis-aggregated view of the net flow. The actions are displayed from left to right according to the final ranking, as shown in Fig. 3. It helps in visualizing the characteristic profiles of the actions along with weights of criteria. For each action, a multi-coloured bar is drawn. Each slice within the bar corresponds to the contribution of one criterion in the computation of the multi-criteria net flow. Its height is equal to the uni-criterion net flow value multiplied by the weight of the criterion. This way, the multi-criteria net flow value is the sum of all the slices (positive one minus negative one). Larger positive slices (most important good) features of the

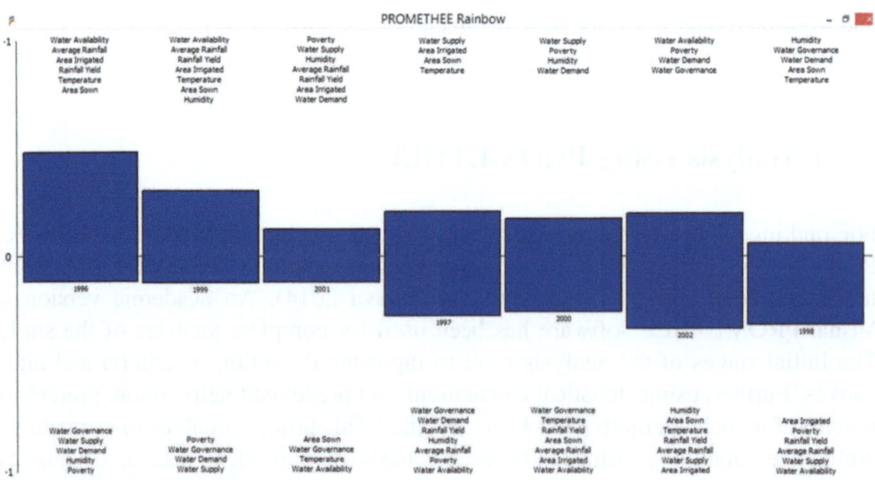

Fig. 3 PROMETHEE rainbow based on net flows

action are on the top and larger negative slices (most important weakness) are on the bottom of the slice.

Hence, according to PROMETHEE technique, the ranking order of vulnerability of climate change has been obtained as 1996 < 1999 < 2001 < 1997 < 2000 < 2002 < 1998 with the lowest and highest vulnerabilities in the years 1996 and 1998, respectively. For verification of these results, two multi-criteria approaches have been used as discussed in the next section.

Verification of Results

The results obtained from data analysis using PROMETHEE approach are verified using two different multi-criteria approaches as discussed below.

Using Analytical Hierarchy Process

In this approach, Expert Choice software has been used to apply concepts of analytical hierarchy process (AHP) in assessing vulnerability of climate change over the temporal scale during 1996–2002. This involves construction of pair-wise decision matrix among the criteria and respective sub-criteria, as shown in Table 3.

Table 3 Criteria and sub-criteria weight allocation using AHP

Criteria	Sub-criteria	Weight
Water vulnerability index		0.350
	Water availability	0.833
	Rainfall yield	0.167
Climate vulnerability index		0.165
	Average rainfall	0.714
	Temperature	0.147
	Humidity	0.138
Agricultural vulnerability index		0.188
	Area irrigated	0.800
	Area sown	0.200
Sensitivity		0.248
	Water demand	0.143
	Water supply	0.857
Coping capacity		0.049
	Water governance	0.167
	Poverty	0.833

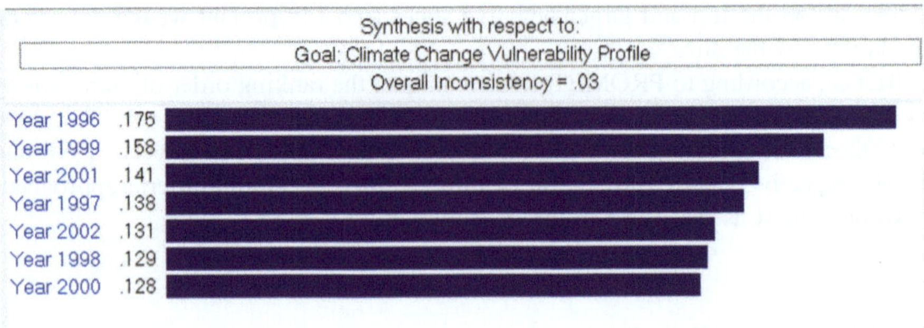

Fig. 4 Ranking of various alternatives (years) using Expert Choice software

The alternatives' decision matrixes are entered in a proportional ratio of their actual numerical values. Finally, as shown in Fig. 4, the ranking order of year-wise performance has been evaluated using criteria weights and alternatives' comparison. The overall inconsistency obtained is 3%, which is within limits of 10%, hence the analysis is consistent and correct.

Hence, according to AHP the ranking order of vulnerability of climate change is 1996 < 1999 < 2001 < 1997 < 2002 < 1998 < 2000 with the lowest and highest vulnerabilities in the years 1996 and 2000, respectively.

Using Social Choice Method

In Social Choice Method, ranking of the alternatives is done strictly by the decision maker based on his/her perception about the overall scenario. Hare system (successive deletion) method is used in the study wherein the deletion of less attractive alternatives take place until the most preferred alternative is found. Table 4 enlists three decision makers' ranking for the temporal scale based on average rainfall, rainfall yield and total irrigated area, respectively.

The successive deletion method is applied, i.e. deleting the less important alternatives till the highest important alternative is encountered. Hence, according to social choice method (Hare system) the ranking of vulnerability in climate change is (lowest vulnerability) 1996 < 1999 < 2000 < 1997 < 2002 < 2001 < 1998 (highest vulnerability).

Table 4 Decision makers evaluation for temporal scale scenarios

Temporal scale	Year 1996	Year 1997	Year 1998	Year 1999	Year 2000	Year 2001	Year 2002
Decision maker 1	1	2	7	4	5	6	3
Decision maker 2	2	4	6	1	5	3	7
Decision maker 3	6	2	5	3	1	7	4

Hence, using the above two multi-criteria approaches ranking obtained from PROMETHEE is compared. It concludes that though the highest vulnerability year (alternative) is varying for each technique, but the year having lowest vulnerability remains the same. Further, an integrated approach can be applied so as to obtain more reliable results. This concludes the climate change vulnerability assessment using previous years' data. Next section attempts to predict the climate change profile in future using prediction techniques.

Climate Change Impact and Its Prediction

According to Narain et al. (2005), the predicted per capita water availability in Rajasthan yearly was 840 m^3 in 2001 which is expected to remain 439 m^3 by the year 2050, beside the nation-wide average of 1,140 m^3 in 2050. Due to the global warming and subsequent rise in temperature, the entire season cycle has been disturbed. The high temperature and changes in precipitation lead to severe droughts in Mewar region in late 1990s. The agriculture is the main occupation in the state is also being severely disturbed due to irregular spells and higher temperature through recent years. Chittorgarh comprising of the majority of rural population is facing various difficulties, decreasing the coping capacity of the area subsequently. Climate change vulnerability profile includes another element of prediction of effects of climate change on water resources in future. This section deals in depth with prediction of various criteria (assessment indicators') values so as to generate a trend, completing the climate change assessment profile. The emission of greenhouse gases is considered as an elementary reason for climate change, according to IPCC. It is estimated that by 2020, 2050 and 2080 in South Asia as a result of future greenhouse emissions, probably annual change in temperature will be 1.36, 2.69 and 3.84 °C, respectively, whereas precipitation changes will be around 2.9, 6.8 and 11.0%, respectively (Lal 2004). The present domestic water demand is around 100 LPCD, which is going to increase about 5% in near future. Hence, with this predicted data, vulnerability profile corresponding to these years have been calculated, as shown in Table 5.

With these projected values, once again the analysis was done using PROMETHEE, which ranked these alternatives (temporal scale) as (lowest

Table 5 Projected numerical values of sub-criteria for 2020, 2050 and 2080

Sub-criteria	Year 2020	Year 2050	Year 2080
Average rainfall (mm)	787.185	817.02	849.15
Temperature (°C)	31.86	33.19	35.34
Water demand (LPCD)	135	145	155
Water governance	Good	Good	Good
Poverty; monthly per capita (Rs.)	758.93	988.65	1252.20

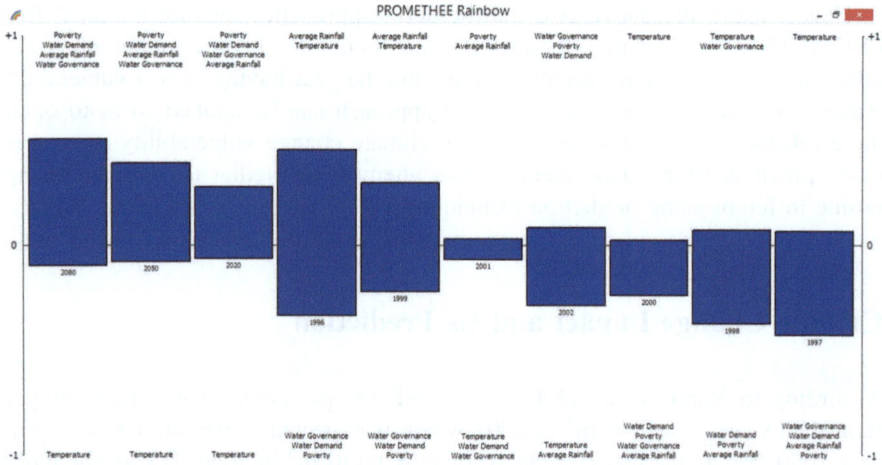

Fig. 5 Predicted PROMETHEE rainbow for future years

vulnerability)
2080 < 2050 < 2020 < 1996 < 1999 < 2001 < 2002 < 2000 < 1998 < 1997 (highest vulnerability), as shown in Fig. 5. The projections done are based on optimistic approach and hence, a better profile is obtained for future years.

This paper deals with the future projections of the climatic conditions and vulnerability index. The approximations used here are actually for the global scale, and hence applying the same to regional level to Chittorgarh leads of various anomalies.

Conclusions

Climate change irregularities are an essential global challenge and effects agricultural, water resources and forest cover. In the study, a climate vulnerability assessment profile was framed to judge the temporal scale (1996–2002) based on various assessment indicators encompassing water, agriculture, climate, water governance and sensitivity. These indicators form the backbone of this paper, and hence their selection, quantification and normalization are very important steps in overall evaluation of climate change impacts.

Evaluation of these indicators on the temporal scale has been primarily performed using PROMETHEE techniques which have generated a preferential ranking among the various years, with 1996 being lowest vulnerable and 1998 highest vulnerable. For verification of the results so obtained, two different multi-criteria approaches were applied though there are many techniques available today (Singh et al. 2015). First, Analytical Hierarchy Process which involved pair-wise comparison of the criteria and their respective sub-criteria. The alternatives were also compared based on their numerical input. Similar to above approach, a ranking order was generated stating 1996 as lowest vulnerability year, whereas 2000 has been found the highest vulnerable year. The third approach applied in the study dealt with Social Choice Method, allocating ranking to temporal scale strictly based upon the opinion of the decision maker. Further, successive deletion of less attractive alternatives until the best alternative is obtained, produced similar results as in the case of PROMETHEE technique.

Lastly, this study also projected the future climatic condition and its vulnerability which is an integral part of vulnerability profile. An optimistic approach is being considered while projecting, leading to a better future. This paper presents an integrated overview of impact of climate change on water resources, agricultural and coping capacity.

References

Jain SK (2012) Sustainable water management in India considering likely climate and other changes. Curr Sci 102(2):177–188

Lal M (2004) Implications of climate change on agricultural productivity and food security in South Asia. Reg Environ Change 11(1):79–94

Narain P, Khan MA, Singh G (2005) Potential for water conservation and harvesting against drought in Rajasthan, India. Working paper 104 drought series: paper 7. International Water Management Institute (IWMI), Colombo, Sri Lanka

Ravindranath N (2011) Climate change vulnerability profiles for North East India. Curr Sci 101 (3):384–394

Singh AP (2008) An integrated fuzzy approach to assess water resources' potential in a watershed. ICFAI J Comput Math 1(1):7–23

Singh AP, Shrivastava P (2014) A comparative study on water quality assessment of a river using AHP and PROMETHEE techniques. In: Proceedings of 19th international conference on hydraulics, water resources, coastal and environmental engineering (hydro 2014 international), 18–20 Dec 2014, Bhopal, pp 880–890

Singh AP, Ghosh SK, Sharma P (2007) Water quality management of a stretch of River Yamuna: an interactive fuzzy multi-objective approach. Int J Water Resour Manage 21:515–532

Singh AP, Srinivas R, Kumar S, Chakrabarti S (2015) Water quality assessment of a river basin under fuzzy multi-criteria framework. Int J Water 9(3):226–247

Statistics: Chittorgarh (2014) Retrieved from home: the official portal of Chittorgarh District (Rajasthan). www.chittorgarh.nic.in. Last accessed on 22 March 2014

Vidyarthi AK, Singh AP, Shrivastava P (2015) Environmental impact assessment of industrial clusters in Central India using fuzzy decision analysis. J Ind Pollut Control 31(2):213–221 alternatives were also compared based on their monetary input. Similar to above approach, a ranking order was prepared among [illegible] where alternatives [illegible] that has been assigned the highest [illegible]. The same approach applied in [illegible] Social [illegible] Method [illegible] ranking [illegible] gal [illegible] based upon the opinion of the [illegible] Further, [illegible] [illegible] of less [illegible] were [illegible] protected [illegible] as in the case of PRI and TDH in [illegible].

Lastly, this study also projected the future climate [illegible] which is an integral part of vulnerability profile. [illegible] approach is [illegible] interventions in view of [illegible] and coping capacity.

References

[illegible]

Ravindranath N [illegible] Climate change [illegible] forests [illegible] [illegible]

Singh AP (2008) [illegible]

[illegible]

Water Availability Under Changing Climate Scenario in Ur River Basin

T. Thomas, S. Goyal, V.C. Goyal and R.V. Kale

Abstract The Bundelkhand region in Central India is facing several environmental issues since the last decade including recurrent droughts, dominant land use changes due to many influencing factors including over exploitation of the natural resources and its degradation, climatic variability and decreased agricultural productivity. The agriculture of the region is mostly rain-fed which has now become a non-lucrative livelihood option for the local population due to the vagaries of the climate and its variability. The threat of climate change which now seems to be real is likely to aggravate the already precarious scenario, which therefore calls for a detailed investigation into the impacts of climate change on the water resources of the region. The Ur river basin has been selected as a pilot basin in Bundelkhand for the development of a decision support system (DSS) integrating climate change, hydrology and livelihood. An attempt has been made to study the impact of climate change by forcing hypothetical climate scenarios on a conceptual water balance model setup for the watershed. The analysis reveals that a 10% reduction in precipitation results in more than 40% reduction in surface runoff whereas a 1 °C increase in temperature results in 6% reduction in surface runoff. A 1 °C rise in

T. Thomas (✉)
National Institute of Hydrology, Central India Hydrology Regional Centre, Bhopal 462042, Madhya Pradesh, India
e-mail: thomas_nih@yahoo.com

S. Goyal
MP Resource Atlas Division, M.P. Council of Science & Technology, Bhopal 462003, Madhya Pradesh, India
e-mail: sgoyalbsg@gmail.com

V.C. Goyal · R.V. Kale
National Institute of Hydrology, Jalvigyan Bhavan, Roorkee 247667 Uttaranchal, India
e-mail: vcgoyal@yahoo.com

R.V. Kale
e-mail: ravikale2610@gmail.com

V.P. Singh et al. (eds.), *Climate Change Impacts*, Water Science and Technology Library 82, https://doi.org/10.1007/978-981-10-5714-4_18

temperature coupled with a 10% reduction in rainfall leads to a further 50% reduction in surface runoff whereas a 2 °C rise in temperature coupled with a 10% reduction in rainfall leads to reduction in surface runoff by 59%. This analysis is being used in the development of a DSS for making effective policy recommendations to assist the decision makers and stakeholders in selecting appropriate water management practices on a sustainable basis.

Keywords Water balance · Supply-demand · Bundelkhand · Water resources management · Climate change

Introduction

The freshwater availability in adequate quantity and quality is crucial for the sustenance of life and economic development of any region. Since water is critical to all forms of life in a watershed including domestic, livestock, agricultural, forest and industrial requirements, any shortages in the availability of water supply poses the greatest threat to the watershed health and productivity. Therefore, the water resources planning and development needs to be carried in an appropriate manner so that the precious resource can be used beneficially on a sustainable basis. This calls for accurate estimation of the supply-demand scenario prior to introducing planning interventions. In semi-arid watersheds particularly located in the rural areas, lack of basic data on the water resources generally leads to erratic planning and unsustainable practices thereby causing depletion of the available water resources. Therefore it is imperative to understand the water balance of an area. However the inter-relationships between the various components of the water balance viz., rainfall, evapotranspiration, groundwater recharge, groundwater draft, surface and groundwater storages are very complex and any intervention in any of the component of the water balance, is ought to have an impact on the other components of the water balance.

The intergovernmental panel on climate change (IPCC) reported that the ecosystems and natural resources shall be affected if the if the projected doubling of atmospheric carbon-di-oxide occurs within the next century (Houghton et al. 1990). It is believed that the climate change will enhance the hydrologic cycle thereby causing changes in the rainfall pattern and its distribution leading to variation in the water storages and fluxes at the land surface, soil moisture storage, groundwater, reservoirs, snowpack, runoff and evapotranspiration. Dickinson (1986) stated that as a consequence of climate change, the terrestrial biosphere will be affected due to the changes in the regional energy balance. This will alter the regional water balance due to seasonal shifts in water balance due to changes in precipitation and other climatic conditions (Eagleson 1986). It is also predicted that the changes in soil moisture and evapotranspiration are likely to have large impacts on water and forest resources (Neilson et al. 1992). Changes in the regional water cycle will influence feedbacks between vegetation and climate Rind (1984).

The monthly water balance models have been found useful for water resources assessment and management on a regional scale by identifying hydrologic consequences of changes in temperature, precipitation, and other climate variables (Gleick 1986; Schaake and Liu 1989; Mimikou et al. 1991; Arnell 1992; Xu and Halldin 1997; Xu and Singh 1998). There are many factors to be considered while selecting a model (Gleick 1986). The purpose of study and data availability is the dominant factors responsible for choice of a particular model (Ng and Marsalek 1992; Xu 1999). Marks et al. (1993) evaluated the potential effects of climate change on runoff and soil moisture in the Columbia river basin using $2xCO_2$ scenario data from the geophysical fluid dynamics laboratory (GFDL) general circulation model (GCM). Calvo (1986) evaluated the Thornthwaite's water balance technique in predicting stream runoff in Costa Rica. Jiang et al. (2007) studied the hydrological impacts of climate change simulated by six hydrological models in the Dongjiang basin, South China. Xiong and Guo (1999) developed a two-parameter monthly water balance model to simulate the runoff of seventy sub-catchments in the Dongjiang, Ganjiang and Hanjiang basins in the south of China. They suggested that this model can be efficiently incorporated in the water resources planning program and the climate impact studies to simulate monthly runoff conditions in the humid and semi-humid regions.

Study Area

The Ur river basin, a tributary of the River Dhasan located in Tikamgarh district of Madhya Pradesh has been selected for carrying out the assessment of water availability under various alternate climate scenarios. The study area represents the typical topography and geology of the Bundelkhand region and is one of the most vulnerable areas in respect of climate change and drought related indicators. Ur river basin lies on the Bundelkhand plateau and extends between latitudes 24° 35′ 00″N and 25° 05′ 00″N and between 78° 50′ 00″E and 79° 10′ 00″E longitudes with a total geographical area of 990.37 km^2. The basin is bounded by Chhattarpur district in the east, Lalitpur district in the west, Jhansi district in the north and Sagar district in the south. The basin is elongated with length of 119 km and an average width of 80 km. The location map of the study area is given in Fig. 1.

The topography of the basin is undulating and comprises of high hills along the ridge line with the elevation varying between 200 and 400 m above mean sea level. The elevation gradually decreases from the southern part of the basin towards the north. The River Ur also flows in a north-easterly direction till its confluence with River Dhasan. Agriculture is the dominant land use (58.6%) followed by scrub land

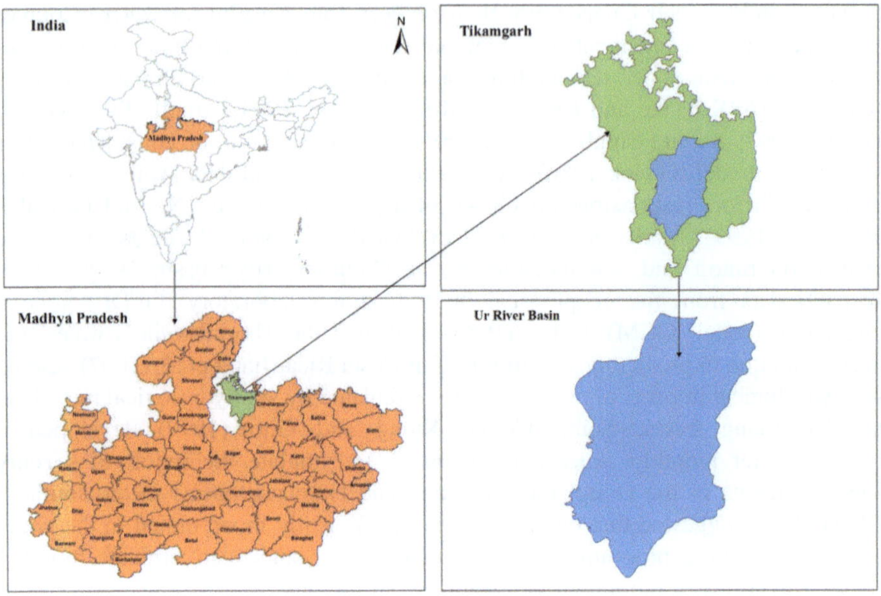

Fig. 1 Index map of the Ur basin

(13.3%). Other land use classes include settlements (2.0%), dense forests (4.5%), water bodies (3.5%), fallow lands (7.0%) and barren (11.1%). The forests are located towards the western portion of the basin, whereas the scrubs are located mostly towards the south-western, western and north-western parts. The agricultural area is well distributed possibly because of a large number of tanks spread all over the basin. The soil in the Ur basin comprises of three dominant soil types. The major portion of the basin is covered by sandy loam soil (68.1%) followed by sandy clay loam (28.5%) and silty clay loam (3.4%) of the total basin area. The land use map and the soil map of the study area are given in Fig. 2a, b. The daily rainfall data of Tikamgarh district comprising of the various blocks located in and around the basin viz., Tikamgarh, Jatara, Baldevgarh and Palera have been obtained from Superintendent of Land Records, Tikamgarh and the daily climatic data including maximum and minimum temperature, relative humidity, wind speed, solar radiation have been obtained from India Meteorological Department (IMD), Pune. The climate of the study area is semi-arid with four distinct seasons.

The winter season extends from December to February followed by the summer season from March to mid-June; rainy season from mid-June to September and the post-monsoon season from October to November. The relative humidity is high during the monsoon season being generally above 70% whereas in summer season the relative humidity is less than 20%.

(a) (b)

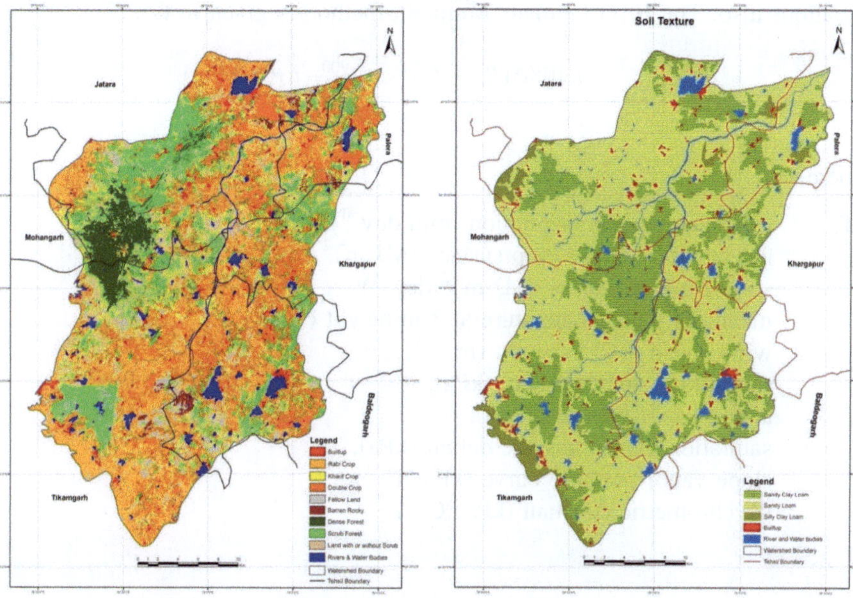

Fig. 2 **a** Land use map **b** soil map

Methodology

Computation of Aerial Average Rainfall

The average areal rainfall has been estimated using the Thiessen polygon method, wherein the representative weights of each of the four influencing rain gauge stations has been derived from the Thiessen Polygon prepared for the basin and the aerial average rainfall for the basin has been computed using Eq. (1),

$$P_{\text{seas}} = \frac{A_1P_1 + A_2P_2 + A_3P_3 + A_4P_4}{A_1 + A_2 + A_3 + A_4} \tag{1}$$

Potential Evapotranspiration

The potential evapotranspiration has been computed by the Penman–Monteith method, which is the sole standard method of determining evapotranspiration as suggested by FAO and has a strong likelihood of correctly predicting ET_o in a wide range of locations and climates and has provision for application in data-short situations also. The FAO Penman–Monteith method is given in Eq. (2),

$$\mathrm{ET_o} = \frac{0.408\Delta(R_\mathrm{n} - G) + \gamma\frac{900}{T+273}u_2(e_\mathrm{s} - e_\mathrm{a})}{\Delta + \gamma(1 + 0.34u_2)} \tag{2}$$

where,

$\mathrm{ET_o}$ reference evapotranspiration (mm day^{-1});
R_n net radiation at the crop surface (MJ m^{-2} day^{-1});
G soil heat flux density (MJ m^{-2} day^{-1});
T mean daily air temperature at 2 m height (°C);
u_2 wind speed at 2 m height (m s^{-1});
e_s saturation vapor pressure (kPa);
e_a actual vapor pressure (kPa);
$e_\mathrm{s} - e_\mathrm{a}$ saturation vapor pressure deficit (kPa);
Δ slope vapor pressure curve (kPa $°\mathrm{C}^{-1}$);
γ psychrometric constant (kPa $°\mathrm{C}^{-1}$).

Soil Conservation Service Curve Number (SCS-CN) Model

The direct surface water runoff has been estimated by the soil conservation service curve number (SCS-CN) model at the outlet of the watershed. The SCS-CN model is based on the single parameter curve number (CN), which depends on the land use, land cover, soil type and the antecedent moisture conditions prevailing in the watershed. The composite curve number (CCN) for the watershed is estimated as 78 using hydrologic soil group and land use for the AMC-II condition (average). The AMC changes to dry or wet conditions depending on the 5-day antecedent rainfall. The direct surface runoff has been estimated using the SCS-CN model given in Eqs. (3)–(5).

$$Q = \frac{(P - I_\mathrm{a})^2}{(P - I_\mathrm{a} + S)} \quad \text{for } P > I_\mathrm{a} \tag{3}$$

$$Q = 0, \quad \text{for } P \leq I_\mathrm{a} \tag{4}$$

$$S = \frac{25{,}400}{\mathrm{CN}} - 254 \tag{5}$$

where,

Q direct surface runoff (mm);
S potential retention (mm);

CN curve number;
I_a initial abstraction = 0.2S for general soils; 0.3S for AMC-I and black soils; and = 0.1S for AMC-III.

Monthly Water Balance Model

The understanding of the catchment response in respect to the changing climate and weather pattern is important in the identification and evaluation of the expected changes in the hydrological components including the evapotranspiration, runoff and recharge to ground water which play a significant role in the water availability and demand scenario within a watershed. The expected changes in the present as well the future scenario can only be quantified based on a simple and complete water balance model, incorporating all the important components of the hydrological cycle which may get affected due to the possible changes in the climate due to the natural and anthropogenic climate forcing. To assess the impacts of the impending climate change on the water resources in the watershed, a water balance model needs to be initially setup and run based on the normal data of rainfall and evapotranspiration based on the average values of the available long-term data. Subsequently, the changes in the climate generally represented by the changes in the rainfall or/and the changes in the temperature can be forced on the model to simulate the catchment responses under alternate climate change scenarios.

The Thornthwaite and Mather (1957) water balance model (TMWB) which is a simple but an effective modeling tool has been employed to analyze the impacts of climate change in the Ur river basin. The TMWB which is a simple model and has already been established as a tool for estimating the hydrological effects of climate change has been chosen as it provides reliable estimation of surface runoff on a monthly time scale using minimal climatic data. As a modification, the potential evapotranspiration (PET) has been estimated by the Penman–Monteith method. The procedure followed in the setup of the model includes:

1. Computation of the average monthly precipitation data (P) using daily station rainfall data by Thiessen Polygon method.
2. Computation of PET on monthly basis using climatic data by Penman–Monteith method.
3. Estimation of the overall water availability scenario with water excess as (+) and water deficit as (−) for each month using Eq. (6),

$$\text{Surplus or Deficit} = P - \text{PET} \tag{6}$$

4. Computation of accumulated potential water losses (APWL) starting from the month in which the $P < \text{PET}$ to account for gross potential deficit during each month using Eq. (7),

$$\text{APWL} = \sum_{0}^{t} \text{Deficit} \quad (7)$$

5. Computation of monthly soil moisture storage. The release of soil moisture is assumed to be an exponential function given by Eq. (8),

$$\text{SM}_t = \text{AWC} * \exp\frac{\text{Ac}(P - \text{PET})}{\text{AWC}} \text{ limited to a maximum of AWC} \quad (8)$$

where, SM_t = actual storage of soil moisture AWC = available water content, i.e. storage capacity of soil moisture zone $\text{Ac}(P - \text{PET})$ = accumulated values of $(P - \text{PET})$.

6. Computation of change in storage (ΔSM_t) for each month as given by Eq. (9),

$$\Delta\text{SM}_t = \text{SM}_t - \text{SM}_{t-1} \quad (9)$$

when the storage remains at capacity level, i.e. AWC, the change in soil moisture, $\text{SM}_t = 0$, but when soil moisture reaches values of less than its capacity, then ΔSM_t is calculated as the difference in soil moisture between soil moisture of present month and soil moisture of the previous month. A negative change in soil moisture (ΔSM_t) implies extraction of water from the soil moisture storage for evapotranspiration, whereas a positive change implies infiltration of water into the soil leading to addition in the soil moisture storage.

7. Computation of actual evapotranspiration is based on the rainfall, PET and ΔSM_t.

$$\begin{aligned}&\{\text{AET}_t = P_t, \quad \text{for } P_t < \text{PET}_t \text{ and } \text{AET}_t = \text{PET}_t, \quad \text{for } P_t > \text{PET}_t\} \text{ for } +\text{ve}\Delta\text{SM}_t\\ &\{\text{AET}_t = P + \Delta\text{SM}_t\} \quad \text{for } -\text{ve}\ \Delta\text{SM}_t\end{aligned} \quad (10)$$

8. The soil moisture deficit starts getting reduced once the precipitation starts getting stored in the soil moisture storage with the onset of the monsoon season. The soil moisture eventually attains field capacity and thereafter the further precipitation excess escapes by gravitational drainage.

9. Thereafter, the computation of the net deficit or net surplus which is based on the change in soil moisture being negative and vice versa is computed as given by Eqs. (11) and (12),

$$\text{Net Deficit}_t = \text{PET}_t - \text{AET}_t \text{ or} \quad (11)$$

$$\text{Net Surplus}_t = P_t - \Delta\text{SM}_t - \text{AET}_t \quad (12)$$

10. The total average runoff (TAVRO) for the first time step during which $P > \text{PET}$ is considered equal to the net surplus estimated in the above step. The total average runoff comprises of surface runoff and detention component. It is assumed that 50% of the TAVRO flows down the stream as surface runoff (SRO) and the balance is detained in the watershed as detention storage (DETN).
11. However for the subsequent months of the analysis, the total average runoff is computed as given by Eq. (13),

$$\text{TAVRO}_t = \text{Net Surplus}_t + \text{DETN}_{t-1} \tag{13}$$

12. Therefore, about 50% of the surplus water that is available for runoff in any month actually runs off as SRO. The rest of the surplus is detained in the subsoil, ground water, small lakes and canals and is available for runoff during the subsequent month.

Assessment of Climate Change Impacts

The impact of climate change on the water resources systems in the Ur river watershed has to be understood for developing a Decision Support System linking the climate change aspects also. The impact of the climate change can be understood by the scenario analysis based on the climatic data. The climate data pertaining to various scenarios are available for many general circulation models (GCM) simulated based on the historical data and is able to give the future climatic data. However owing to the computational constraints, the GCM simulations are available at coarse resolutions and cannot be directly applied for basin scale studies for hydrological application. Under such circumstances, the coarse resolution GCM data needs to be downscaled to a finer resolution so that it can be applied for hydrological applications.

In this study, an effort has been made to study the impacts of climate change using hypothetical scenarios of decrease in precipitation and increase in temperature, both on a standalone basis as well as on a combined basis and thereby analyzing the impact on the water availability under each scenario. The scenarios considered include decrease in normal precipitation by 10, 20 and 30%; increase in temperature by 1, 2, 3 °C and combinations of both precipitation decrease and temperature increase scenarios. The precipitation decrease as well as the temperature increase has been considered, as both will lead to reduced water availability in the basin. The model has initially been setup with the normal rainfall and PET data, and subsequently the inputs have been varied based on these scenarios. The increase in temperature is accounted by the increased potential evapotranspiration, which has been estimated separately for each degree rise in temperature. The comparison of the surface runoff availability with the under normal conditions and alternate climate scenarios helps to understand the impacts of possible climate change. The use of the downscaled climate data based on various scenarios from the GCM will definitely give a better idea of the complex mechanisms and their interaction leading to future water availability, even though these analyses still involve considerable uncertainties.

Results and Discussion

The long-term daily average rainfall at each of the four stations have been obtained by taking the average daily rainfall values during 1999–00 to 2009–10 and the monthly average rainfall at these stations have been computed thereafter and the mean areal rainfall computed using the Thiessen Polygon Method. The area of influencing raingauge stations is given in Table 1. It can be observed that the rain gauges at Jatara and Tikamgarh have maximum influence followed by Baldevgarh and the rain gauge at Palera has minimal influence on the rainfall distribution in the basin.

The PET has been estimated by the Penman–Monteith method, considering the normal climatic data available at Tikamgarh (Table 2). The average daily evapotranspiration at Tikamgarh is 4.30 mm/day.

The Thornthwaite water balance model has been set up for the Ur river basin. The available water capacity (AWC) has been considered to be 150 mm as the soil mostly comprises of sandy clay loam. The water balance computations are given in Table 3. The computations for the accumulated potential water loss (APWL) starts from the month of October when $P <$ PET and continues up to June. No APWL is observed in the computations during July, August and September as the $P >$ PET during this period and situation is of surplus water. The soil moisture (SM) is at its full capacity only during August and September after which it starts reducing up to June. The actual evapotranspiration from the basin is at the potential rate only during July, August and September, when $P >$ PET. Considerable surface runoff is observed during August to December after which minimal flows are sustained in the river. The total runoff observed in the basin is 162.1 mm for the normal rainfall of 844.3 mm. The runoff coefficient based on the normal climatic data is 0.19 which seems to reasonable, but cannot be validated as the catchment is ungauged. The graph showing the temporal variation of the important water balance components

Table 1 Thiessen weights of influencing rain gauge stations in Ur river basin

S. No.	Raingauge station	Influencing area (km^2)	Thiessen weight
1	Tikamgarh	313.37	0.32
2	Baldevgarh	275.63	0.28
3	Jatara	324.66	0.33
4	Palera	76.70	0.07

Table 2 Climate and evapotranspiration at Tikamgarh

Month	Min temp (°C)	Max temp (°C)	Humidity (%)	Wind (km/day)	Sun (h)	Rad (MJ/m^2/day)	ET_o (mm/day)
January	7.3	24.7	74	94	9	16.4	2.49
February	9.1	27.2	70	95	10	19.8	3.21
March	13.9	33.4	61	99	9.1	21.1	4.30
April	20	38.9	42	96	9.8	24	5.64
May	24.7	42	40	117	10.2	25.3	6.82
June	26.4	39.8	57	183	7.7	21.6	6.69
July	24.3	33.4	79	229	6.1	19.1	4.90
August	23.5	31.5	85	198	5.3	17.4	3.99
September	22.7	32.5	80	115	6.6	18.1	4.07
October	17.7	33.2	66	73	8.6	18.6	3.93
November	11.4	29.8	68	70	9.5	17.4	3.13
December	7.6	25.8	69	78	9.0	15.6	2.47

Table 3 Water balance computations for Ur river basin

Water balance components	Jun	Jul	Aug	Sep	Oct	Nov
Precipitation (*P*)	73.8	263.8	282.0	163.9	29.8	2.8
PET (mm)	200.7	151.9	123.7	122.1	121.8	93.9
Surplus/deficit (*P* − PET)	−126.9	111.9	158.3	41.8	−92.0	−91.1
Accumulated potential water loss (APWL)	−039.4	0.0	0.0	0.0	−92.0	−183.2
Soil moisture (SM)	0.1	112.1	150.0	150.0	81.2	44.2
Change in soil moisture (ΔSM)	−0.2	111.9	37.9	0.0	−68.8	−37.0
Actual ET (AET)	74.0	151.9	123.7	122.1	98.6	39.7
Net deficit	126.7	0.0	0.0	0.0	23.2	54.2
Net surplus	0.0	0.0	120.4	41.8	0.0	0.0
Total average runoff (TAVRO)	0.2	0.0	120.4	102.0	51.0	25.5
Surface runoff (SRO)	0.1	0.0	60.2	51.0	25.5	12.8
Detention	0.1	0.0	60.2	51.0	25.5	12.8
Water balance components	Dec	Jan	Feb	Mar	Apr	May
Precipitation (*P*)	4.4	10.7	6.0	3.7	0.9	2.4
PET (mm)	76.6	77.2	89.9	133.3	169.2	211.4
Surplus/deficit (*P* − PET)	−72.2	−66.5	−83.8	−129.6	−168.3	−209.0
Accumulated potential water loss (APWL)	−255.3	−321.8	−405.7	−535.2	−703.5	−912.5
Soil moisture (SM)	27.3	17.6	10.0	4.2	1.4	0.3
Change in soil moisture (ΔSM)	−16.9	−9.8	−7.5	−5.8	−2.9	−1.0
Actual ET (AET)	21.3	20.5	13.5	9.6	3.8	3.4
Net deficit	55.3	56.7	76.3	123.7	165.4	208.0
Net surplus	0.0	0.0	0.0	0.0	0.0	0.0
Total average runoff (TAVRO)	12.8	6.4	3.2	1.6	0.8	0.4
Surface runoff (SRO)	6.4	3.2	1.6	0.8	0.4	0.2
Detention	6.4	3.2	1.6	0.8	0.4	0.2

All units are in mm

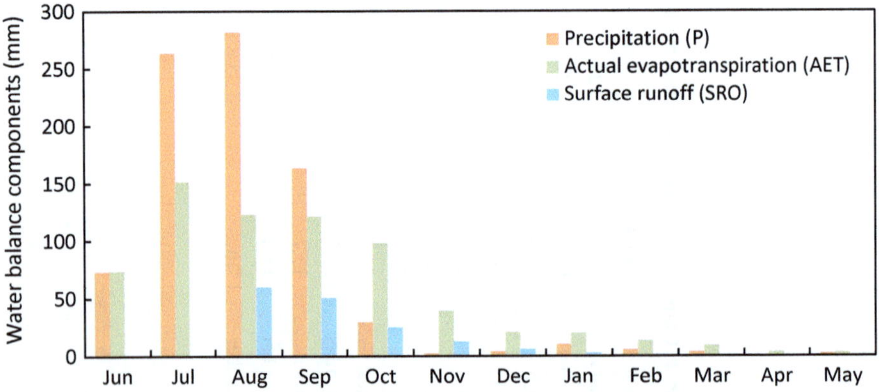

Fig. 3 Temporal variation in important hydrological components

Table 4 Comparison of seasonal surface runoff (MCM)

Year	TMWB model	SCS-CN model
1999–00	468.75	300.55
2000–01	60.97	44.65
2001–02	77.87	20.53
2002–03	170.04	108.52
2003–04	297.60	205.90
2004–05	67.64	76.65
2005–06	39.59	63.24
2006–07	20.91	22.01
2007–08	0.00	0.04
2008–09	356.92	333.02
2009–10	71.93	76.55

based on the normal climatic data is given in Fig. 3. The computation of the monthly water balance helps to identify and quantify the important hydrological components during various months of a water year. The analysis reveals that about 20% of the precipitation is converted into surface runoff, whereas the remaining water gets stored in ground water aquifers, lakes, detention storages including initial abstraction and evapotranspiration losses.

In order to study the range of variation of individual water balance components, the water balance model has been subsequently run based on the observed rainfall and climatic data during the period between 1999–00 and 2009–10. The comparison of the surface runoff generated by this model as well as that obtained by the SCS-CN method during the monsoon season have been compared and is given in Table 3. It has been observed that the seasonal surface runoff generated by TMWB model is higher than that generated by the SCS-CN model on most occasions as SCS-CN produces the direct runoff only instead of the total surface runoff which includes base flow. Also the SCS-CN model computed the runoff on a daily basis, and therefore the rainfall pattern and its distribution also gets reflected in the catchment response whereas the TMWB model computes the surface runoff on a monthly basis based in the water available in the soil moisture storage (Table 4).

However, it can be observed that both models reproduce the surface runoff with reasonable degree of accuracy, even though the individual model responses cannot be validated due to lack of observed stream flow data as the basin is ungauged. The comparison of the seasonal runoff generated by both models is given in Fig. 4.

Assessment of Climate Change Impacts

Since the TMWB model is able to produce the surface runoff satisfactorily, it can therefore be readily applied for analyzing the impacts of climate changing by forcing various scenarios and studying the runoff pattern emerging from the catchment under changed climatic conditions. The scenarios considered include

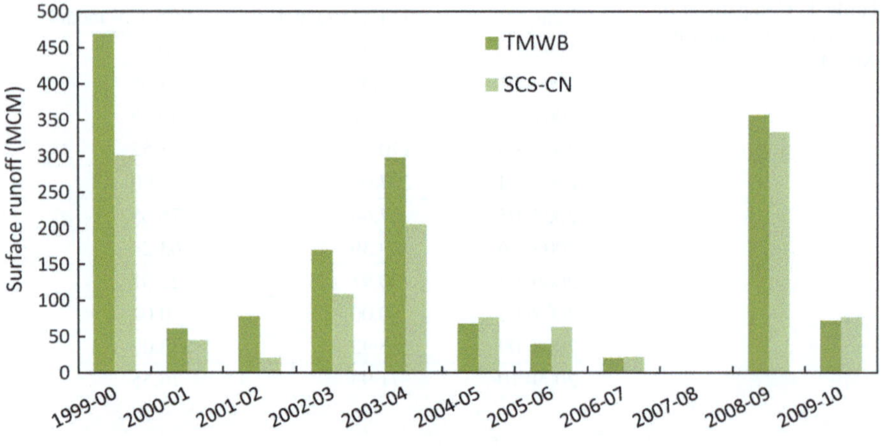

Fig. 4 Comparison of seasonal surface runoff by SCS-CN and TMWB models

decrease in normal precipitation by 10, 20 and 30%; increase in temperature by 1, 2, 3 °C; and combinations of both precipitation decrease and temperature increase scenarios. The TMWB model has been run separately for each of these scenarios and the change in the surface water availability compared amongst various scenarios. The comparison of the surface water availability under various alternate climate scenarios is given in Fig. 5. It has been observed that there is substantial variation in the catchment response in the form of generated surface runoff for each of the climate scenario.

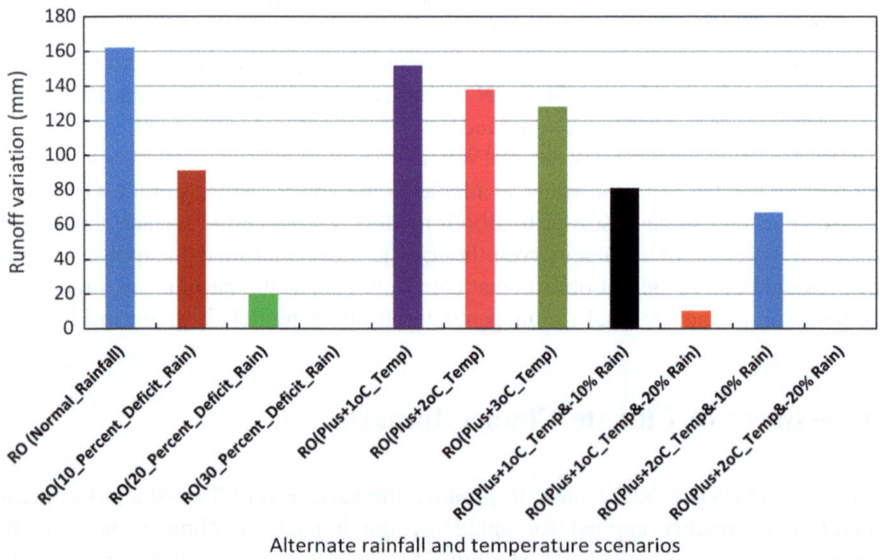

Fig. 5 Comparison of surface runoff under various climate scenarios

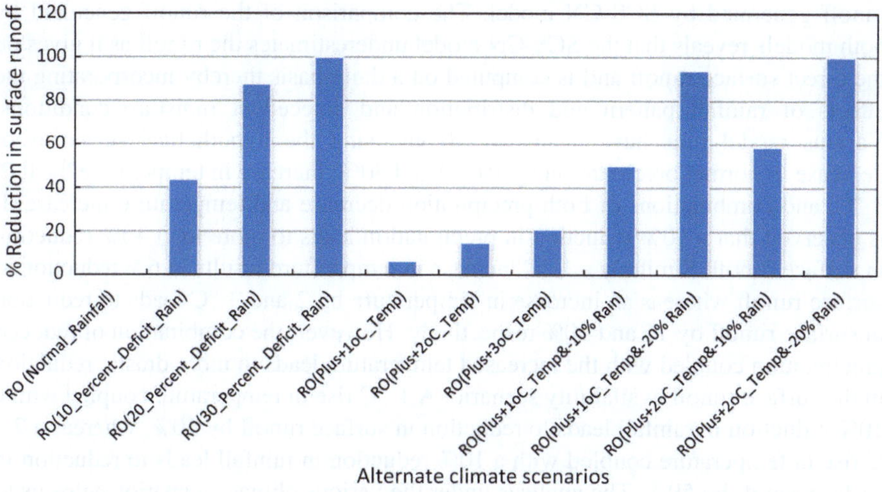

Fig. 6 Percentage reduction in surface runoff under various climate scenarios

The change in surface runoff under various alternate climate scenarios can be better understood by critically analyzing Fig. 6 gives the percentage reduction in runoff. It is observed from the analysis that a 10% reduction in precipitation leads to more than 40% reduction in surface runoff whereas a 20% reduction in precipitation leads to more than 80% reduction in surface runoff. Similarly for various temperature scenarios, a 1 °C increase in temperature which leads to increase in PET ultimately results in 6% reduction in surface runoff whereas an increase in temperature by 2 and 3 °C leads to reduction in surface runoff by 15 and 20%, respectively.

However, the combination of reduced precipitation coupled with the increased temperature leads to more drastic reduction in the surface runoff availability scenario. A 1 °C rise in temperature coupled with a 10% reduction in rainfall leads to reduction in surface runoff by 50% whereas a 2 °C rise in temperature coupled with a 10% reduction in rainfall leads to reduction in surface runoff by 59%. The analysis under the various climate scenarios helps to understand the change in the hydrologic regime due to the changing climate scenario which has become more pronounced in recent times and is expected to continue so with greater impacts in the future too.

Conclusions

The range of variation of individual water balance components has been studied using the TMWB model and simulation performed during 1999–00 to 2009–10. The surface runoff generated by the TMWB model has been compared with the runoff generated by SCS-CN model. The comparison of the runoff generated by both models reveals that the SCS-CN model underestimates the runoff as it gives on the direct surface runoff and is computed on a daily basis thereby incorporating the effects of rainfall pattern and distribution and antecedent moisture conditions. Various model runs have been carried out using the hypothetical scenarios of decrease in normal precipitation by 10, 20 and 30%; increase in temperature by 1, 2, 3 °C, and combinations of both precipitation decrease and temperature increase. It is observed that a 10% reduction in precipitation leads to more than 40% reduction in surface runoff. Similarly a 1 °C increase in temperature results in 6% reduction in surface runoff whereas an increase in temperature by 2 and 3 °C leads to reduction in surface runoff by 15 and 20% respectively. However, the combination of reduced precipitation coupled with the increased temperature leads to more drastic reduction in the surface runoff availability scenario. A 1 °C rise in temperature coupled with a 10% reduction in rainfall leads to reduction in surface runoff by 50%, whereas a 2 ° C rise in temperature coupled with a 10% reduction in rainfall leads to reduction in surface runoff by 59%. The analysis under the various climate scenarios helps us to understand the change in the hydrologic regime due to the changing climate scenario which has become more pronounced in recent times and is expected to continue so with greater impacts in the future too. However, the use of downscaled GCM datasets of precipitation and temperature for various future emission scenarios give better representation of the climate change impacts as the relationship between the temperature rise and precipitation is very complex and therefore the use of hypothetical scenarios can only throw an insight into the approximate changes in the water balance components.

References

Arnell NW (1992) Factors controlling the effects of climate change on river flow regimes in humid temperature environment. J Hydrol 132:321–342

Calvo JC (1986) An evaluation of Thornthwaite's water balance technique in predicting stream runoff in Costa Rica, Hydrol Sci J 31(1):51–60

Dickinson RE (1986) How will climate change? The climate system and modelling of future climate. In: Bolln B, Doos BR, Jdger J, Warrick RA (eds) The greenhouse effect, Climatic change ecosystem scope 29, Wiley, Chichester pp 207–270

Eagleson PS (1986) The emergence of global-scale hydrology. Water Resour Res 22(9):6s–14s

Gleick PH (1986) Methods for evaluating the regional hydrologic impacts of global climatic change. J Hydrol 88:97–116

Houghton JT, Jenkins GJ, Ephraums JJ (eds) (1990) Climate change: the IPCC scientific assessment. Cambridge University Press, Cambridge

Jiang T, Chen YD, Xu C, Chen X, Singh VP (2007) Comparison of hydrological impacts of climate change simulated by six hydrological models in the Dongxiang basin, South China. J Hydrol 336(3–4):316–333

Marks D, King GA, Dolph J (1993) Implications of climate change for the water balance of the Columbia river basin, USA. Clim Res 2:203–213

Mimikou MA, Hadjisavva PS, Kouvopoulos YS (1991) Regional effects of climate change on water resources systems. In: van de Ven FHM, Gutknecht D, Loucks DP, Salewicz KA (eds) Hydrology for the water management of large river basins. IAHS Publ. No. 201, pp 173–182

Neilson RP, King GA, Koerper G (1992) Toward a rule based biome model. Landscape Ecol pp 27–43

Ng HY, Marsalek J (1992) Sensitivity of streamflow simulation to changes in climatic inputs. Nord Hydrol 23:257–272

Rind D (1984) The influence of vegetation on the hydrologic cycle in a global climate model. In: Hansen JD, Takahashi T (eds) Climate processes and climate sensitivity. Geophysical Monograph 5, American Geophysical Union, Washington, DC, USA.

Schaake JC, Liu LZ (1989) Development and application of simple water balance models to understand the relationship between climate and water resources. In: Kavvas ML (ed) New directions for surface water modelling (Proceedings of the Baltimore Symposium, May 1989). IAHS Publication, No. 181, pp 345–352

Thornthwaite CW, Mather JR (1957) Instructions and tables for computing potential evapotranspiration and the water balance. Laboratory of Climatology, Publications in Climatology, vol 10, No 3, Centerton, N. J., pp 185–311

Xiong L, Guo S (1999) A two-parameter monthly water balance model and its application. J Hydrol 216:111–123

Xu C-Y (1999) From GCMs to river flow: a review of downscaling methods and hydrologic modeling approaches. Prog in Phys Geogr 23(2):229–249

Xu C-Y, Halldin S (1997) The effect of climate change on river flow and snow cover in the NOPEX area simulated by a simple water balance model. Nordic Hydrol 28:273–282

Xu C-Y, Singh VP (1998) A review on monthly water balance models for water resources investigation and climatic impact assessment. Water Resour Manage 12:31–50

Water Sustainability Assessment Under Climatic Uncertainty—A Case Study of Chhattisgarh (India)

Surendra Kumar Chandniha and M.L. Kansal

Abstract Water sustainability and vulnerability in terms of economic, social, and environmental terms is a holistic concept that affects the society at large. Assessing water sustainability is a decision-making problem that involves multiple stakeholders and perspectives. Precipitation is one of the key parameters in water resources sustainability. In the present case study, techniques based on Standardized Precipitation Index (SPI) has been advocated for assessment of drought and non-drought conditions at regional level. These conditions are subcategorized into drought, near normal, and wet conditions, whereas near normal and wet conditions are considered as non-drought events. On the basis of long-term rainfall time series, the performance parameters such as reliability, resilience, vulnerability, and relative vulnerability are assessed and utilized for quantification of district wise water sustainability index (WSI). Finally, an overall WSI is estimated at the regional level using the distance-based approach on the basis of most pessimistic, most optimistic, and neutral approach. The proposed methodology is illustrated with the help of a real time case study for the state of Chhattisgarh in India. Further, in order to study the impact of climate change on drought conditions, two possible scenarios, i.e., HadCM3 A2 and B2 are considered to generate the future downscaled rainfall series. On the basis of the generated rainfall series and by using the proposed methodology, WSI for all the districts of the Chhattisgarh state has been assessed.

Keywords Sustainability · SPI · Reliability · Resilience · Vulnerability Chhattisgarh

S.K. Chandniha (✉) · M.L. Kansal
Department of Water Resources Development and Management, Indian Institute of Technology Roorkee, Roorkee 247667, Uttarakhand, India
e-mail: chandniha.surendra@gmail.com

M.L. Kansal
e-mail: mitthan@gmail.com

V.P. Singh et al. (eds.), *Climate Change Impacts*, Water Science and Technology Library 82, https://doi.org/10.1007/978-981-10-5714-4_19

Introduction

Water security is a holistic concept that protects the society against related threats, risks, or vulnerability. In other terms, water sustainability affects the society in terms of economic, environmental, and social aspects. Assessing water sustainability is a decision-making problem that involves the hydrological uncertainties such as rainfall and the perspective of the decision maker and their index of optimism. Further, droughts are generally classified into four categories, i.e., meteorological, hydrological, agricultural, and socioeconomic. Meteorological drought is associated with inadequacy of normal precipitation. Hydrological drought is related to the inadequacy of water resources to meet the demands. Agricultural drought is related to soil moisture or wilting point of the site specific cropping system. Socioeconomic drought is based on ensuing economic consequences for the region (Maity et al. 2012). Also, drought is termed as climatic entity and its occurrence is a natural phenomenon. Recent years are witness of global drought scenarios which have been identified by various researchers (Sridhar et al. 2008). Drought is best characterized by multiple climatological and hydrological aspect and is also a good indicator of climate change (Mishra and Singh 2010). The term meteorological drought is used to identify those conditions which indicate that the precipitation amount has decreased with respect to long-term average values (Khalili et al. 2011).

Standardized precipitation index (SPI) is one of the well-known parameters to assess the long-term behavior of precipitation in terms of drought or non-drought conditions. Drought condition represents the below normal availability of water. Various researchers have carried out drought analysis using SPI. Various indices suggested by different researchers are as follows: Munger's Index (Munger 1916), Precipitation Effectiveness Index (Thornthwaite 1931), Blumenstock's Index (Blumenstock 1942), antecedent precipitation index (API) (McQuigg 1954; Waggoner and O'Connell 1956), palmer drought severity index (PDSI) (Palmer 1965; Alley 1984; Dai et al. 2004), Rainfall Anomaly Index (RAI) (Van Rooy 1965; Moron 1994), Drought Area Index (Bhalme and Mooley 1980; Bhalme et al. 1983), Standardized Precipitation Index (SPI) (McKee et al. 1993; Guttman 1998, 1999a, b), Effective precipitation (Byun and Wilhite 1999), and Normalized antecedent precipitation index (Heggen 2001). This study focusses on meteorological drought which is based on SPI.

Suitable strategies are required for smooth implementation of various water sector activities and sustainable development in future. Commonly accepted definition of sustainable development is reported in Brundtland Commission's report (Brundtland et al. 1987), which says that sustainable development is that development which meets the needs of current generations without compromising the ability of future generations to meet their own needs. There are three major indicators of sustainable development, which are economic, environment, and social (Pahl-Wostl 2002) which further depends on various sub indicators (Loucks and Gladwell 1999). Sustainability assessment covers the review process of planning, implementation and benefits accrued from water resources development and management in qualitative and quantitative terms. Various researchers (Raskin et al. 1996; Loucks and Gladwell 1999; Salameh 2000; Lawrence et al. 2002; Sullivan 2002) have carried out sustainability analysis by focusing on these three indicators.

From the adaptation and mitigation point of view, water sustainability index (WSI) is one of the best indicators of representing the sustainable development of a system. Therefore, in this study, a distance-based WSI is assessed by using the various performance parameters of SPI such as reliability, resilience, and vulnerability (Hashimoto et al. 1982). The proposed methodology is demonstrated through a case study of Chhattisgarh state in India. The WSI has been quantified for each district and thereafter, it is assessed for the entire state using the distance-based approach as per historical and future projected time series of rainfall.

Terminology Used in the Assessment of WSI

Standardized Precipitation Index (SPI)

The SPI, which is a probability index and shows a better presentation of abnormal wetness or dryness, was proposed by (McKee et al. 1993) to quantify the precipitation deficit or excess on multiple time scales (monthly, 3-monthly, 6-monthly, yearly and 2-yearly). In order to estimate SPI, long-term precipitation data is required. Multiple long-term time series data is fitted into a probability distribution (generally a two parametric gamma distribution). The fitted distribution is evaluated using goodness-of-fit (GoF) tests and then transformed into a normal distribution where the mean SPI is zero. Zero mean reflect the normal condition while positive and negative values indicate the non-drought and drought conditions. The parameters of gamma distribution depend on the observed precipitation and their spatial variation. The newly fitted series obtained from the gamma distribution is expressed as a function of precipitation amount, mean, and the standard deviation of the fitted series (McKee et al. 1993; Guttman 1999a, b; Komuscu 1999; Lana et al. 2001; Wu et al. 2007). Finally, SPI is calculated as

$$\mathrm{SPI} = \frac{x_i - \bar{x}}{\sigma} \tag{1}$$

where x_i individual observed precipitation; $\bar{x}$ and σ are the mean and standard deviation of the fitted series.

Generally, in a long-time series, the value of SPI varies from −3 to +3. Negative and positive scales reflect the dryness/wetness of a year. In this study, SPI has been categorized into seven subcategories, namely, extremely wet (EW), severely wet (SW), moderately wet (MW), near normal (NNC), moderately dry (MD), severely dry (SD), and extremely dry (ED) and its corresponding ranges are shown in Table 1.

Table 1 SPI range and corresponding wetness/dryness category

SPI range	Category of wet/dry conditions	Drought/non-drought
2.00 and above	Extremely wet (EW)	Non-drought events
1.50 to 1.99	Severely wet (SW)	
1.00 to 1.49	Moderately wet (MW)	
−0.99 to 0.99	Near normal (NN)	Near normal
−1.00 to −1.49	Moderately dry (MD)	Drought events
−1.50 to −1.99	Severely dry (SD)	
−2.00 and less	Extremely dry (ED)	

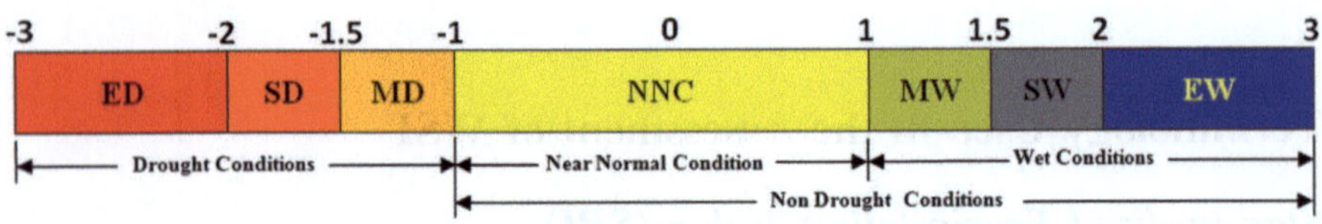

Fig. 1 SPI based drought and non-drought conditions

These classes are grouped under two major conditions such as Drought and Non-drought (Normal and Wet) as shown in Fig. 1.

Probability Density Function Used in SPI

The main purpose of fitting any distribution in different time scales is to identify the general behavior or pattern of the time series. Three distributions, namely, gamma, Weibull, and log-normal have been considered for the purpose. Weibull distribution is identified as the heavy-tailed distribution in precipitation fitting (Vicente-Serrano et al. 2011; Zhu et al. 2011; Yusof et al. 2014). Gamma distribution is found to be best fitted for estimation of SPI (McKee et al. 1993, 1995; Hayes et al. 1999; Shiau 2006). The probability density function (PDF) and cumulative distribution function (CDF) of the gamma distribution are as follows:

$$f(x;\alpha,\beta) = x^{(\alpha-1)}\frac{e^{\frac{x}{\beta}}}{\beta^{\alpha}\Gamma(\alpha)} \quad \text{for } x > 0 \tag{2}$$

$$f(x;\alpha,\beta) = \int_0^x f(u;\alpha,\beta)\,\mathrm{d}u = \frac{\gamma\left(\alpha,\frac{x}{\beta}\right)}{\Gamma(\alpha)} \tag{3}$$

for $x > 0$ and $\alpha, \beta > 0$ where, $\Gamma(\alpha) = \int_0^{\infty} x^{\alpha-1}\mathrm{e}^{-x}\,\mathrm{d}x$ and $\gamma(s,x) = \int_0^x t^{s-1}\mathrm{e}^{-t}\,\mathrm{d}t$ is the lower incomplete gamma function, α is the shape parameter, and β is the scale parameter.

Thus, for carrying out SPI analysis, alpha (shape) and beta (scale) parameters in the gamma probability density function are estimated for each station on yearly basis. As suggested by Stacy and Mihram (1965), the maximum likelihood concept is used to estimate α and β as mentioned below:

$$\hat{\alpha} = \frac{1}{4A}\left(1 + \sqrt{1 + \frac{4A}{3}}\right) \quad (4)$$

$$\hat{\beta} = \frac{\bar{x}}{\hat{\alpha}} \quad (5)$$

where

$$A = \ln(\bar{x}) - \frac{\sum \ln(x)}{n}$$

n = number of precipitation observations.

Goodness-of-Fit Tests

Goodness-of-fit (GOF) tests are used to test the fitness of a distribution to a set of observations and measures. In order to test the compatibility of the proposed distribution with the observed data, a statistical goodness-of-fit (GOF) test is carried out. GOF typically summarizes the discrepancy between observed and expected values from the model in question. The best-fitted distribution is then chosen based on the minimum error. These errors may be estimated by any of the following test statistics: Kolmogorov–Smirnov (KS), Anderson–Darling (AD), or Chi-square.

The KS test statistics is defined as

$$D = \max_{1 \leq i \leq N}\left(\left|F(x_i) - \frac{i-1}{N}\right|, \left|\frac{i}{N} - F(x_i)\right|\right) \quad (6)$$

where x_i is the increasing ordered data, F is the theoretical cumulative distribution, and N is the sample size.

Anderson–Darling (AD) is used to compare the model on the basis of observed CDF with an expected CDF for a particular distribution. The AD test statistics is defined as

$$A^2 = -n - \frac{1}{n}\sum_{i-1}^{n}(2i-1)[\ln F(X_i) + \ln(1 - F(X_{n+1=i}))] \quad (7)$$

where F is the CDF of the specified distribution and X_i are the ordered data.

Chi-square test statistics is defined as

$$\chi^2 = \sum_{i=1}^{k} \frac{(O_i - E_i)^2}{E_i} \tag{8}$$

where O_i is the observed frequency for bin i, and E_i is the expected frequency for bin i calculated by

$$E_i = F(x_2) + F(x_1)$$

where F is the CDF of probability distribution being tested, and x_1, x_2 are the limits for bin i.

Although there is no optimal choice for the number of bins (k), there are several formulas which can be used to calculate this number based on the sample size (N). For example, EasyFit uses

$$k = 1 + \log_2 N$$

Various statistical software is available in the market which can suggest the best-fitted distribution for any time series data. In this study EasyFit-5.4 (professional) is used for estimating the GoF for testing the type of best-fitted distribution of the rainfall series.

Reliability–Resilience–Vulnerability for SPI Time Series

The concept of Reliability, Resilience, and Vulnerability in water resources systems was incorporated by Hashimoto et al. (1982). In this study, the same concept is used for SPI time series. As mentioned in Table 1 and Fig. 2, the variation of SPI is termed as satisfactory (non-drought) and unsatisfactory (drought) conditions. The fluctuation of SPI depends upon precipitation at a particular location. Let X_1, X_2, X_3, …, X_N be the time series of SPI having data series length (N) and SPI at any time is represented by X_T. If $X_T \geq NN$ condition, it is considered as satisfactory conditions (S), here Z_T is 1 and if $X_T \leq NN$ condition, it is considered as unsatisfactory conditions or failure condition (U), here Z_T is 0. Index Z_T signifies a satisfactory or unsatisfactory state of a system (Bradley et al. 2003).

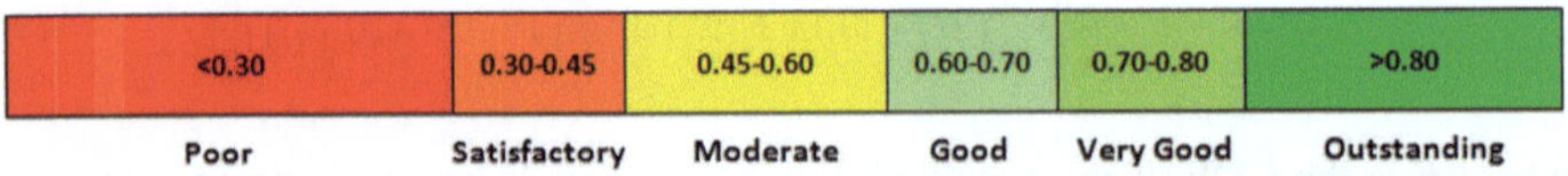

Fig. 2 Qualitative classification of WSI

$$Z_T = \begin{cases} 1, & \text{if} \quad X_T \in S \\ 0, & \text{if} \quad X_T \in U \end{cases} \tag{9}$$

An index, W_T, is defined to capture the transition between S and U state (Hashimoto et al. 1982) $W_T = \begin{cases} 1, & \text{if} \quad X_T \in U \quad \text{and} \quad X_{T+1} \in S \\ 0, & \text{otherwise.} \end{cases}$

Reliability

Reliability is defined by the probability that a system is in a satisfactory state (Hashimoto et al. 1982). Mathematically, reliability α is stated as

$$\alpha = P(X_T \ \in \ S) \tag{10}$$

where S = the satisfactory stage as started before. From the time series, α can be computed as

$$\alpha = \underset{n \to \infty}{Lt} \frac{1}{n} \sum_{t=1}^{n} Z_T$$

where $Z_T = 1$, if $X_T \ \in \ S$ and $Z_T \ = 0$, if $X_T \ \in \ U$ (a unsatisfactory set).

The reliable or satisfactory conditions are identified as the SPI values varying from −1 to +3. SPI values below −1 are considered as unsatisfactory condition. Therefore, reliability can be estimated as a ratio of total number of satisfactory events to the total number of events in the time series, i.e.,

$$\text{Reliability } (S_{\text{Rel}}) = \frac{\text{Number of satisfactory values (wet condition)}}{\text{Total number of values (lenght of time series)}}. \tag{11}$$

Resilience

Resilience is a measure that indicates how quickly the system can return to a satisfactory stage (Near normal condition) after it has fallen below the satisfactory threshold (Hashimoto et al. 1982; Sandoval-Solis et al. 2010). This can be defined as the ratio of the probability of transition from the unsatisfactory (drought condition) to the satisfactory stage (near normal condition) and the probability of failure, i.e.,

$$\gamma = \frac{P(X_T \in U, X_{T+1} \in S)}{P(X_T \in U)} \tag{12}$$

where S and U are as defined earlier. The numerator, probability of transition from the unsatisfactory to the satisfactory stage and is denoted by ρ. In the long run, the number of times the system transforms from the satisfactory to the unsatisfactory stage and from the unsatisfactory to the satisfactory stage will be the same. Therefore, it can be eventually expressed as $\rho = P(X_T \in U, X_{T+1} \in S) = P(X_T \in S, X_{T+1} \in U)$. From the time series, ρ can be calculated as

$$\alpha = \underset{n \to \infty}{Lt} \frac{1}{n} \sum_{t=1}^{n} W_T \tag{13}$$

where W_T = event of transformation from the satisfactory to the unsatisfactory stage (or vice versa) and $W_T = 1, \text{if } X_T \in S \text{ and } X_{T+1} \in U \text{ and } W_T = 0$ otherwise. The denominator of Eq. 11 can be expressed as $P(X_T \in U) = 1 - P(X_T \in S)$. Again $P(X_T \in S)$ is expressed as reliability α as explained before. Thus, Eq. (12) can be expressed as

$$\gamma = \frac{\rho}{1 - \alpha} \tag{14}$$

A number of time steps of satisfactory values (non-drought condition) following an unsatisfactory value (drought condition) are counted to the total number time steps of unsatisfactory values. The ratio of these two numbers of time steps is termed as resilience of the system.

$$\text{Relsilience } (S_{\text{Res}}) = \frac{\text{Number of time steps of satisfactory value follows an unsatisfactory value}}{\text{Number of unsatisfactory values}} \tag{15}$$

Vulnerability

Vulnerability is a measure of severity of a failure event, once it has occurred (Hashimoto et al. 1982; Sandoval-Solis et al. 2010). It is defined as

$$\upsilon = \sum_{j \in U} s_j e_j \tag{16}$$

where s_j = the numerical indicator of severity for an observation x_j, which belongs to the unsatisfactory state; e_j = the probability of that x_j, corresponding to s_j which is the most unsatisfactory and severe outcome that occurs from the set of unsatisfactory states. In the context of SPI, vulnerability is a probability weighted average of the deficit conditions (with respect to the *NN* condition) of failure events. Thus, the events below the *NN* condition is the severity indicator and vulnerability is measured in terms of the Expected Extent and Expected Duration of the SPI vulnerability. Mathematically,

$$\text{Vulnerability } (S_{\text{Vul}}) = \text{EE}_{\text{unsatisfactory values}} \times \text{ED}_{\text{unsatisfactory values}} \tag{17}$$

$$\text{EE}_{\text{unsatisfactory values}} = \frac{\text{Cumulative extent of failure}}{\text{Number of individual failure events}} \tag{18}$$

$$\text{ED}_{\text{unsatisfactory values}} = \frac{\text{Total number of failure events in a time series}}{\text{Number of continuous series of failure events}} \tag{19}$$

where

$\text{EE}_{\text{unsatisfactory values}}$ = Expected extent of unsatisfactory values
$\text{ED}_{\text{unsatisfactory values}}$ = Expected duration of unsatisfactory values.

Relative Vulnerability

Relative vulnerability is a dimensionless vulnerability measure and is defined as

$$\text{Relative Vulnerability} = \frac{\upsilon}{\sum_{T \in J_n} X_T} \tag{20}$$

where υ is calculated using Eq. (14), J_n refers to the unsatisfactory period that υ represents and X_T is Maximum Vulnerability among all alternatives at time step *T*. Alternatively, it may be defined as the ratio of vulnerability and maximum extent of failure event in entire time series.

$$\text{Relative Vulnerability } (S_{\text{Relative Vul}}) = \frac{\text{Vulnerability } (S_{\text{Vul}})}{\text{Maximum extent of failure event in a entire time series}} \tag{21}$$

Water Sustainability

Since reliability, resilience, and vulnerability are the criteria of water sustainability, these can be combined to assess the water sustainability. A multiplicative index is proposed by various researchers (Hashimoto et al. 1982; Sandoval-Solis et al. 2010; Maity et al. 2012) for aggregating these three criteria as shown in Eq. 22.

$$\text{Water Sustainability Index} = \text{Reliability} \times \text{Resiliency} \times (1 - \text{Relative Vulnerability}) \quad (22)$$

In order to represent the water sustainability in qualitative terms, it is proposed to categorize it subcategories such as poor, satisfactory, moderate, good, very good, and outstanding as shown in Fig. 2.

Distance-Based Multicriteria Decision Making

A Multicriteria Decision-Making (MCDM) problem means either a multi attribute or a multi objective decision problem or both. Water sustainability assessment is also a MCDM problem in the sense that it is based on reliability, resilience, and vulnerability of seasonal or annual rainfall. There are a number of possible solution types in MCDM theory which depend on the type of problem and the required solution. Some of these are: Value and utility theory, distance-based techniques—goal and compromise programming, outranking techniques, sequential solution method, the $\in$ constraint method, weighting method, and the surrogate worth trade-off (SWT) analysis technique. In this study, a distance-based goal programming approach is used for assessing the water sustainability on the basis of three criteria, i.e., reliability, resilience, and vulnerability of SPI. Goal programming allows the decision maker (DM) to specify a target for each objective criterion. A preferred solution is then defined as the one that minimizes the sum of the deviations from the prescribed set of target values. The basic principle is based on the measures of the distance between a feasible pay-off vector and the goal point. In case of uncertainty, the decision making is carried using following criteria:

1. Laplace;
2. Mini-max;
3. Savage Regret;
4. Hurwicz.

In case of uncertainties, these criteria differ in the degree of conservatism of the decision maker.

Hurwicz criteria reflect a range of decision-making attitudes from the most optimistic to the most pessimistic. If α varies 0–1 and $V(a_i, S_j)$ represent gain then selected action on the chosen action must be associated with

$$\underset{ai}{\text{Max}} \left\{ \alpha \underset{Sj}{\text{Max}} V(a_i, S_j) + (1 - \alpha) \underset{S_j}{\text{Min}} V(a_i, S_j) \right\}$$

If $V(a_i, S_j)$ is loss

$$\underset{ai}{\text{Min}}\left\{\alpha \underset{Sj}{\text{Min}} V(a_i - S_j) + (1-\alpha)\underset{S_j}{\text{Max}} V(a_i, S_j)\right\}$$

If $\alpha = 0$ represent the conservative approach (i.e., max-min or mini-max criteria).
If $\alpha = 1$ represent the optimistic approach (best of the best).
If $\alpha = 0.5$ represents the neutral approach (Laplace approach).

Proposed Methodology for Water Sustainability Assessment

This study uses the concept of SPI as a measure of water sustainability at district level. Once the SPI is calculated for each of the district, it estimates the reliability, resilience and vulnerability of SPI for these districts. Thereafter, the overall WSI for the entire state is estimated by following the distance-based approach as mentioned above. The proposed methodology is shown in the form of a flow chart in Fig. 3.

The steps of proposed methodology are:

1. Arrange the observed rainfall time series and determine the statistics of the series.
2. Since two-parameter gamma distribution works well for the SPI, the gamma distribution is fitted in the observed rainfall series and the parameters of the distribution are estimated along with the GoF. One can use the software like EasyFit for the purpose.
3. On the basis of the two parameters of the fitted gamma distribution, prepare the calculated time series of the rainfall and estimate the mean and standard deviation of the fitted series.
4. Calculate the SPI as suggested in Eq. 1 and prepare the series of SPI.
5. Using the classification shown in Table 1, identify the years as dry or wet.
6. Use the series as mentioned in step 5 to determine the reliability, resilience, vulnerability, and relative vulnerability of the series.
7. Estimate the district wise water sustainability using Eq. 22.
8. Using the district wise WSI, the state level WSI is estimated on the basis of distance from the ideal point depending on the index of optimism (such as maximum of maximum (pessimistic), minimum of maximum (optimistic) and average of maximum (neutral)) approach applied on the basis of SPI values of all the districts.

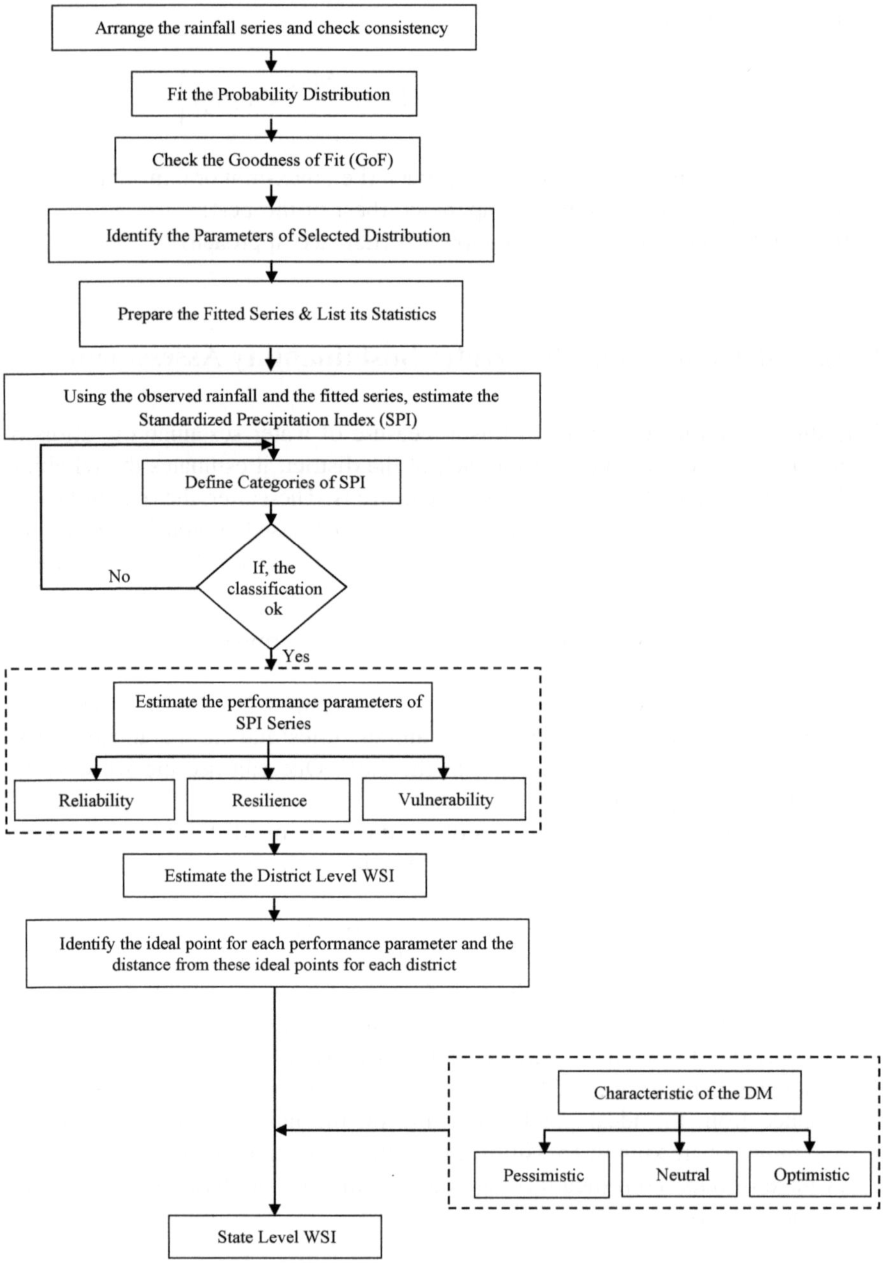

Fig. 3 Proposed methodology for estimating WSI

Case Study

An attempt has been made to estimate the water sustainability for the state of Chhattisgarh on the basis of district wise historical and downscaled monthly rainfall under suggested climate change scenarios. In the observed data for each district, two-parameter gamma distribution is fitted and the parameters of the distribution are estimated. On the basis of the estimated parameters of the distribution, fitted time series for each of the district is prepared. Further, GoF is checked on the basis of Kolmogorov–Smirnov (KS), Anderson Darling (AD) and Chi-Squared statistics test. For the historical rainfall series, the district wise gamma distribution statistics are calculated and summarized in Table 2.

The values of SPI for all the districts are shown in Table 3. For illustration purpose, the SPI variation for Koriya district (D10) is shown in Fig. 4.

On the basis of SPI, the WSI for each district is calculated. For example, calculations for the Koriya district are as follows:

Total number of events = 111; Non-drought events = 96;
Drought events = 15; **Reliability** of the system = 96/111 = **0.86**
No. of times satisfactory value (*s*) follow an unsatisfactory value = 10;
Hence, **Resilience** = 10/15 = **0.67**
Cumulative extent of failure = **9.73**;
Expected extent of unsatisfactory values = 9.73/15 = **0.65**
Maximum extent of failure event in an entire time series = **1.73**
Number of continuous series of failure events = **10**
Expected duration of unsatisfactory values = 15/10 = **1.5**
Hence, **vulnerability** = 0.65 × 1.5 = **0.97**, and
Relative Vulnerability = 0.97/1.73 = **0.56**
Therefore, **WSI = 0.86 × 0.67 × (1−0.56) = 0.254, i.e., Poor**.

On the same lines, WSI for all the districts is calculated and is shown in Table 4.

From the Table 4, it may be noted that on the basis of historical rainfall data, the Bilaspur district has the maximum water sustainability, i.e., 0.57. The worst is the condition of Dhamtari district with water sustainability of 0.09. Although, reliability is more in case of Dhamtari district as compared to Bilaspur, but the overall sustainability is worse due to high vulnerability. Graphically, the maps of reliability, resilience, vulnerability, relative vulnerability, and WSI are prepared on the historical record basis and are shown in Fig. 5a–e.

Table 2 Two parametric gamma distribution statistics and GoF values for Chhattisgarh (1901–2011)

Gamma distribution statistics		D1	D2	D3	D4	D5	D6	D7	D8	D9	D10	D11	D12	D13	D14	D15	D16
Shape factor		37.3	13.6	34.1	38.9	42.9	47.9	31.9	35.2	34.6	34.1	32.5	46.9	40	46.2	36.3	29.3
Scale factor		33.8	56.6	32.2	32	28	27.2	44.1	35	28.8	33.4	39.9	27.1	34	27.2	34.7	43.3
Goodness-of-Fit (GOF)																	
Kolmogorov Smirnov	Statistics	0.04	0.06	0.04	0.05	0.06	0.06	0.08	0.05	0.06	0.04	0.11	0.06	0.06	0.04	0.05	0.07
	p value	0.98	0.85	0.98	0.92	0.88	0.73	0.49	0.89	0.85	0.98	0.15	0.78	0.88	0.98	0.9	0.68
Anderson–darling	Statistics	0.24	0.38	0.27	0.31	0.23	0.28	0.59	0.25	0.42	0.34	1.28	0.3	0.36	0.29	0.32	0.54
Chi-squared	Statistics	3.73	4.05	4.03	4.11	3.34	2.97	9.37	3.15	1.68	4.13	5.4	5.9	5.58	1.89	5.74	7.02
	p value	0.71	0.67	0.67	0.66	0.77	0.81	0.15	0.79	0.95	0.66	0.49	0.43	0.05	0.93	0.45	0.32

where *D1* Bastar; *D2* Bilaspur; *D3* Dantewada; *D4* Dhamtari; *D5* Durg; *D6* Janjgir; *D7* Jashpur; *D8* Kanker; *D9* Korba; *D10* Koriya; *D11* Kawardha; *D12* Mahasamund; *D13* Raigarh; *D14* Raipur; *D15* Rajnandgaon; *D16* Surguja district of Chhattisgarh

Table 3 SPI categories and corresponding scores and their distribution of different districts

Category	Number of years followed by different categories during historical time series (1901–2011)																
	SPI ranges	D1	D2	D3	D4	D5	D6	D7	D8	D9	D10	D11	D12	D13	D14	D15	D16
Extremely wet	2.00 and above	2	4	5	4	2	3	1	1	0	2	1	5	2	4	2	0
Severely wet	1.50 to 1.99	8	6	2	2	6	6	7	6	8	4	4	2	7	4	4	7
Moderately wet	1.00 to 1.49	6	7	8	12	12	7	8	14	11	15	9	8	7	8	15	9
Near normal	−0.99 to 0.99	76	76	80	78	74	77	73	73	77	75	85	77	77	76	72	78
Moderately dry	−1.00 to −1.49	11	12	11	5	7	8	15	9	7	7	6	9	8	8	11	10
Severely dry	−1.50 to −1.99	5	5	3	8	8	8	4	5	4	5	1	8	8	10	4	3
Extremely dry	−2.00 and less	3	1	2	2	2	2	3	3	4	3	5	2	2	1	3	4

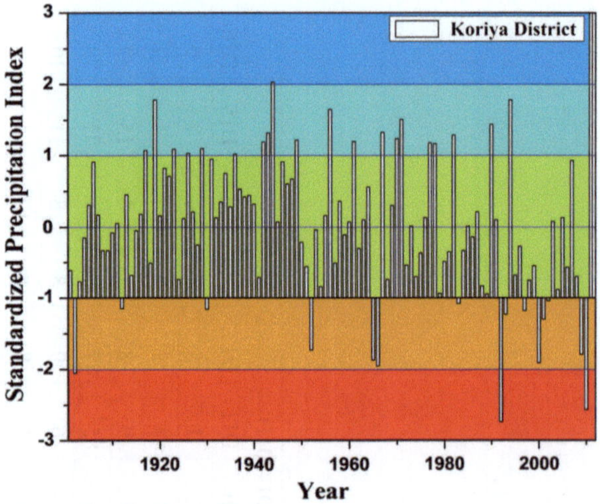

Fig. 4 Graphical representation of SPI for Koriya district in Chhattisgarh

In order to assess the state level WSI, the WSIs for all the districts are used as the input data. For example, from Table 4, one can determine the ideal points for the parameters reliability, resilience, and relative vulnerability. Thereafter, the distance from the ideal points is calculated for each of the district as shown in Table 5. On the basis of the maximum distance from the ideal point and with different levels of optimism, the WSI for the entire state is calculated as shown in Table 5

It may be noticed that if the decision maker (DM) is pessimistic in approach, he will select the maximum of maximum distance from the ideal point as the critical point, and hence the overall WSI. In Table 5, it is maximum for relative vulnerability for district Dhamtari district (i.e., 0.86) and hence the WSI for the state is 0.14. If the DM is optimistic in approach, he will consider the minimum of the maximum distance from the ideal points of the performance criteria. For example, minimum of maximum distance among all the districts is for Dantewada district, i.e., 0.28. Therefore, WSI for the state will be 0.72. If the DM is neutral and unbiased, i.e., neither optimistic nor pessimistic, then one can take the average of maximum distance of all the districts. For example, it is 0.53 in this case. Therefore, the WSI for the state can be taken as 0.47.

WSI Under HadCM3-A2 Climate Change Scenario

The district wise gamma distribution statistics are summarized in Table 6.

The category wise frequency distribution of SPI for all the districts is shown in Table 7.

Using the SPI values for each district, the reliability, resilience and vulnerability are assessed. Thereafter, the WSI for all the districts is calculated. The results are shown in Table 8.

Table 4 Estimated values of WSI parameters for each districts of Chhattisgarh, India

Particular	D1	D2	D3	D4	D5	D6	D7	D8	D9	D10	D11	D12	D13	D14	D15	D16
Total events	111	111	111	111	111	111	111	111	111	111	111	111	111	111	111	111
Drought events	19	18	15	15	17	18	21	15	15	15	12	19	18	19	18	17
Non-drought events	92	93	95	96	94	93	89	94	96	96	99	92	93	92	93	94
No. of times satisfactory value (s) follow an unsatisfactory value	15	18	13	11	13	12	13	9	11	10	7	14	12	14	14	10
Cumulative failure extent	9.22	6.78	7.89	9.96	9.54	10.87	10.56	10.15	11.35	9.73	12.31	10.22	11.1	10.27	10.34	10.94
Max. failure extent in entire series	1.81	1.16	2.15	1.05	1.5	1.52	1.63	2.29	2.02	1.73	2.55	1.21	1.32	1.06	1.87	2.42
No. of continuous failure events	15	18	13	11	13	12	13	9	11	10	7	14	12	14	14	10
Reliability	0.83	0.84	0.86	0.86	0.85	0.84	0.8	0.85	0.86	0.86	0.89	0.83	0.84	0.83	0.84	0.85
Resilience	0.79	1.00	0.87	0.73	0.76	0.67	0.62	0.6	0.73	0.67	0.58	0.74	0.67	0.74	0.78	0.59
Vulnerability	0.61	0.38	0.61	0.91	0.73	0.91	0.81	1.13	1.03	0.97	1.76	0.73	0.93	0.73	0.74	1.09
Relative vulnerability	0.34	0.32	0.28	0.86	0.49	0.6	0.5	0.49	0.51	0.56	0.69	0.6	0.7	0.69	0.39	0.45
Districts wise WSI	0.43	0.57	0.53	0.09	0.33	0.23	0.25	0.26	0.31	0.25	0.16	0.24	0.17	0.19	0.39	0.27

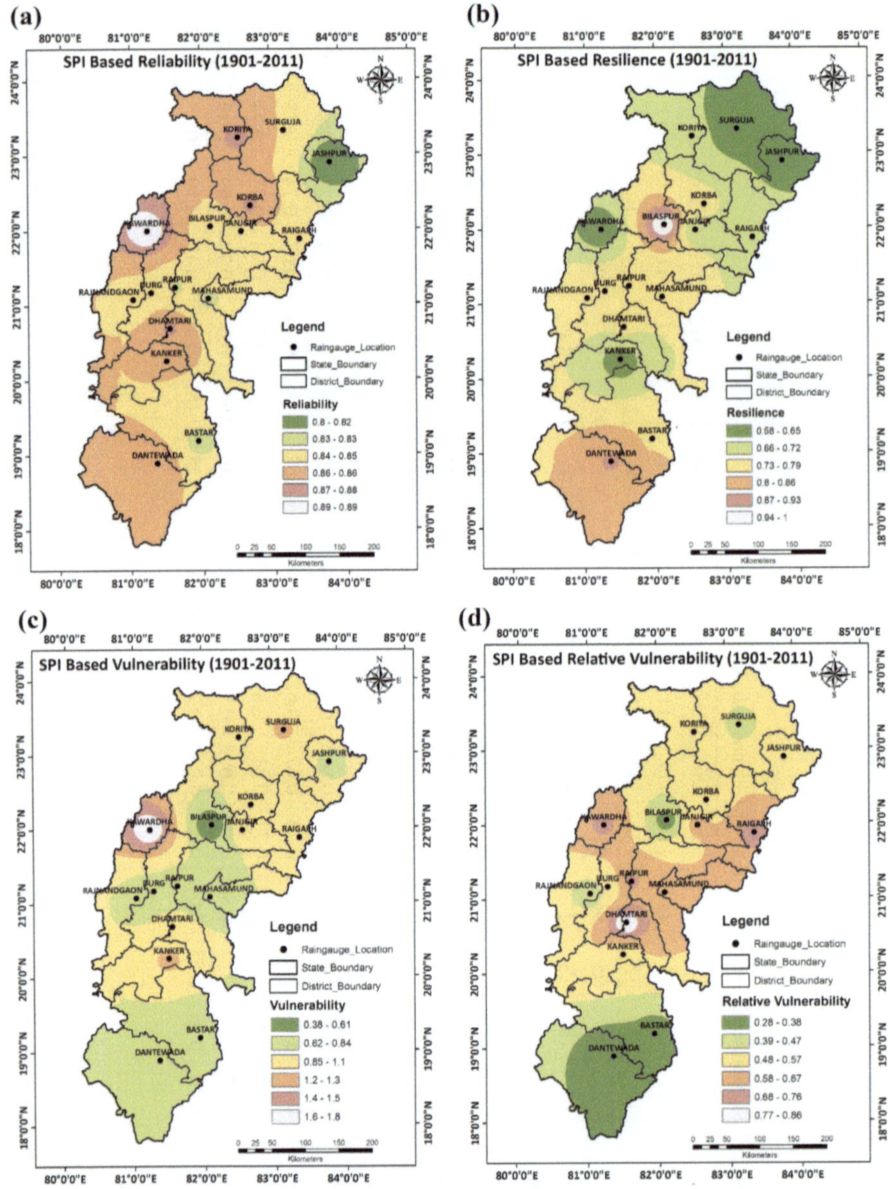

Fig. 5 Spatial map of **a** reliability, **b** resilience, **c** vulnerability, **d** relative vulnerability and **e** water sustainability index

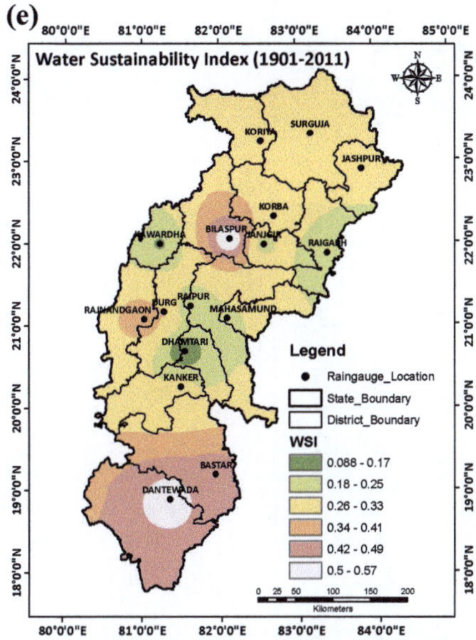

Fig. 5 (continued)

From the Table 8, it may be noted that on the basis of downscaled rainfall data, the Raipur district has the maximum water sustainability, i.e., 0.57. The worst is the condition of Durg district with water sustainability of 0.29. Although, reliability is more in case of Durg district as compared to Raipur, but the overall sustainability is worse due to high vulnerability.

In order to assess the state level WSI, the WSIs for all the districts are used as the input data. For example, from Table 8, one can determine the ideal points for the parameters reliability, resilience, and relative vulnerability. Thereafter the distance from the ideal points is calculated for each of the district as shown in Table 9. On the basis of the maximum distance from the ideal point and with different levels of optimism, the WSI for the entire state is calculated as shown in Table 9.

It may be noticed that if the decision maker (DM) is pessimistic in approach, he will select the maximum of maximum distance from the ideal point as the critical point, and hence the overall WSI. In Table 9, it is maximum for relative vulnerability for district Rajnandgaon district (i.e., 0.54), and hence the WSI for the state is 0.46. If the DM is optimistic in approach, he will consider the minimum of the maximum distance from the ideal points of the performance criteria. For example, minimum of maximum distance among all the districts is for Raipur district, i.e., 0.26. Therefore, WSI for the state will be 0.74. If the DM is neutral and unbiased, i.e., neither optimistic nor pessimistic, then one can take the average of maximum distance of all the districts. For example, it is 0.46 in this case. Therefore, the WSI for the state can be taken as 0.54.

Table 5 Values of overall WSI using distance-based approach (Historical 1901–2011)

Particular	D1	D2	D3	D4	D5	D6	D7	D8	D9	D10	D11	D12	D13	D14	D15	D16
Distance from ideal point (Rel.)	0.17	0.16	0.14	0.14	0.15	0.16	0.20	0.15	0.14	0.14	0.11	0.17	0.16	0.17	0.16	0.15
Distance from ideal point (Res.)	0.21	0.00	0.13	0.27	0.24	0.33	0.38	0.40	0.27	0.33	0.42	0.26	0.33	0.26	0.22	0.41
Distance from ideal point (Rel. vul.)	0.34	0.32	0.28	0.86	0.49	0.60	0.50	0.49	0.51	0.56	0.69	0.60	0.70	0.69	0.39	0.45
Maximum distance in terms of (Rel., Res., and Rel. Vul.)	0.34	0.32	**0.28**	**0.86**	0.49	0.60	0.50	0.49	0.51	0.56	0.69	0.60	0.70	0.69	0.39	0.45

Overall water sustainability index based on distance-based approach

Approaches	Distance	Overall WS (1-distance)
Maximum of maximum (pessimistic approach)	**0.86**	0.14
Minimum of maximum (optimistic approach)	**0.28**	0.72
Average of maximum distance for all districts (neutral approach)	0.53	**0.47**
Overall water sustainability index (WSI)		**0.47**

where *D1* Bastar; *D2* Bilaspur; *D3* Dantewada; *D4* Dhamtari; *D5* Durg; *D6* Janjgir; *D7* Jashpur; *D8* Kanker; *D9* Korba; *D10* Koriya; *D11* Kawardha; *D12* Mahasamund; *D13* Raigarh; *D14* Raipur; *D15* Rajnandgaon; *D16* Surguja District of Chhattisgarh

Table 6 Values of two parametric gamma distribution statistics and GoF for each district of Chhattisgarh during 2011–2099

Gamma distribution statistics		D1	D2	D3	D4	D5	D6	D7	D8	D9	D10	D11	D12	D13	D14	D15	D16
Shape factor		53.1	19.3	30.9	88.7	67.2	70.3	38.8	49.0	70.8	41.6	62.1	74.0	66.3	53.0	48.8	46.2
Scale factor		22.5	43.7	37.4	12.5	16.5	16.4	34.8	23.8	16.1	27.2	19.6	15.3	18.1	20.2	24.6	26.1
Goodness-of-Fit (GOF)																	
Kolmogorov Smirnov	Statistics	0.07	0.07	0.07	0.08	0.05	0.06	0.07	0.06	0.08	0.06	0.06	0.04	0.06	0.05	0.07	0.06
	p value	0.77	0.69	0.68	0.61	0.94	0.89	0.78	0.86	0.55	0.90	0.88	1.00	0.88	0.97	0.81	0.91
Anderson-darling	Statistics	0.49	0.69	0.36	0.35	0.12	0.29	0.29	0.34	0.43	0.35	0.39	0.20	0.33	0.21	0.26	0.24
Chi-squared	Statistics	4.73	7.58	2.16	2.73	0.75	2.46	4.09	2.66	3.18	3.49	3.81	1.38	3.73	4.14	1.62	2.17
	p value	0.58	0.27	0.90	0.84	0.99	0.87	0.66	0.85	0.79	0.74	0.70	0.97	0.71	0.66	0.95	0.90

Table 7 SPI categories and corresponding scores and their distribution of different districts

Category	Number of years followed by different categories during historical time series (2011–2099)																
	SPI ranges	D1	D2	D3	D4	D5	D6	D7	D8	D9	D10	D11	D12	D13	D14	D15	D16
Extremely wet	2.00 and above	2	3	4	1	3	0	2	3	2	3	1	1	0	3	2	2
Severely wet	1.50 to 1.99	3	6	1	5	2	6	6	1	7	6	7	4	8	4	4	6
Moderately wet	1.00 to 1.49	10	6	10	5	10	10	7	7	6	5	8	11	9	8	9	6
Near normal	−0.99 to 0.99	58	62	60	62	60	59	60	64	62	61	58	58	59	59	59	60
Moderately dry	−1.00 to −1.49	9	7	9	9	9	9	8	6	4	7	9	9	5	10	8	9
Severely dry	−1.50 to −1.99	4	4	3	5	3	3	4	6	6	5	4	4	6	3	4	4
Extremely dry	−2.00 and less	3	1	2	2	2	2	2	2	2	2	2	2	2	2	3	2

Table 8 Estimated values of WSI parameters of each districts of Chhattisgarh, India

Particular	D1	D2	D3	D4	D5	D6	D7	D8	D9	D10	D11	D12	D13	D14	D15	D16
Total events	89	89	89	89	89	89	89	89	89	89	89	89	89	89	89	89
Drought events	16	12	14	16	14	14	14	14	12	14	15	15	13	15	15	15
Non-drought events	73	77	75	73	75	75	75	75	77	75	74	74	76	74	74	74
No. of times satisfactory value(s) follow an unsatisfactory value	13	10	11	14	10	10	13	12	9	12	10	11	10	13	11	12
Cumulative failure extent	9.12	5.48	6.67	8.53	7.82	7.82	7.67	8.87	6.94	7.06	7.59	8.26	7.72	7.14	8.87	7.57
Max. failure extent in entire series	1.67	1.11	1.18	1.89	1.51	1.51	1.4	1.45	1.53	1.18	1.57	1.64	1.55	2.13	1.48	1.54
No. of continuous failure events	13	10	11	14	10	10	13	12	9	12	10	11	10	13	11	12
Reliability	0.82	0.87	0.84	0.82	0.84	0.84	0.84	0.84	0.87	0.84	0.83	0.83	0.85	0.83	0.83	0.83
Resilience	0.81	0.91	0.79	0.88	0.71	0.79	0.93	0.86	0.75	0.86	0.71	0.73	0.77	0.93	0.79	0.80
Vulnerability	0.70	0.55	0.61	0.61	0.78	0.69	0.59	0.74	0.77	0.59	0.76	0.75	0.77	0.55	0.81	0.63
Relative vulnerability	0.42	0.49	0.51	0.32	0.52	0.49	0.42	0.51	0.50	0.50	0.48	0.46	0.50	0.26	0.54	0.41
Districts wise WSI	0.39	0.40	0.32	0.49	0.29	0.34	0.45	0.35	0.32	0.36	0.31	0.33	0.33	0.57	0.30	0.39

Table 9 Values of overall WSI using distance-based approach (Historical 1901–2011)

Particular	D1	D2	D3	D4	D5	D6	D7	D8	D9	D10	D11	D12	D13	D14	D15	D16
Distance from ideal point (Rel.)	0.17	0.16	0.14	0.14	0.15	0.16	0.20	0.15	0.14	0.14	0.11	0.17	0.16	0.17	0.16	0.15
Distance from ideal point (Res.)	0.21	0.00	0.13	0.27	0.24	0.33	0.38	0.40	0.27	0.33	0.42	0.26	0.33	0.26	0.22	0.41
Distance from ideal point (Rel. Vul.)	0.42	0.49	0.51	0.32	0.52	0.49	0.42	0.51	0.50	0.50	0.48	0.46	0.50	0.26	0.54	0.41
Maximum distance in terms of (Rel., Res., and Rel. Vul.)	0.42	0.49	0.51	0.32	0.52	0.49	0.42	0.51	0.50	0.50	0.48	0.46	0.50	0.26	0.54	0.41

Overall water sustainability index based on distance-based approach		
Approaches	Distance	Overall WSI (1-distance)
Maximum of maximum (pessimistic approach)	**0.54**	0.46
Minimum of maximum (optimistic approach)	**0.26**	0.74
Average of maximum distance for all districts (neutral approach)	0.46	0.54
Overall water sustainability index (WSI)	**0.54**	

where *D1* Bastar; *D2* Bilaspur; *D3* Dantewada; *D4* Dhamtari; *D5* Durg; *D6* Janjgir; *D7* Jashpur; *D8* Kanker; *D9* Korba; *D10* Koriya; *D11* Kawardha; *D12* Mahasamund; *D13* Raigarh; *D14* Raipur; *D15* Rajnandgaon; *D16* Surguja District of Chhattisgarh

WSI Under HadCM3-B2 Climate Change Scenario

For the projected rainfall series (HadCM3-B2), the district wise gamma distribution statistics are summarized in Table 10.

The category wise distribution of SPI for all the districts is shown in Table 11.

Using the SPI values for each district, the reliability, resilience, and vulnerability are assessed. Thereafter, the WSI for all the districts is calculated. The results are shown in Table 12.

From Table 12, it may be noted that on the basis of downscaled rainfall data, the Dhamtari district has the maximum water sustainability, i.e., 0.60. The worst is the condition of Kanker district with water sustainability of 0.19. Although, reliability is more in case of Dhamtari district as compared to Kanker, but the overall sustainability is worse due to high vulnerability.

In order to assess the state level WSI, the WSIs for all the districts are used as the input data. For example, from Table 12, one can determine the ideal points for the parameters reliability, resilience, and relative vulnerability. Thereafter, the distance from the ideal points is calculated for each of the district as shown in Table 13. On the basis of the maximum distance from the ideal point and with different levels of optimism, the WSI for the entire state is calculated as shown in Table 13.

It may be noticed that if the decision maker (DM) is pessimistic in approach, he will select the maximum of maximum distance from the ideal point as the critical point, and hence the overall WSI. In Table 13, it is maximum for relative vulnerability for district Kanker district (i.e., 0.73), and hence the WSI for the state is 0.27. If the DM is optimistic in approach, he will consider the minimum of the maximum distance from the ideal points of the performance criteria. For example, minimum of maximum distance among all the districts is for Janjgir district, i.e., 0.20. Therefore, WSI for the state will be 0.80. If the DM is neutral and unbiased, i.e., neither optimistic nor pessimistic, then one can take the average of maximum distance of all the districts. For example, it is 0.37 in this case. Therefore, the WSI for the state can be taken as 0.63.

Conclusions

In this paper, a new approach has been suggested for assessment of WSI for the regional level study. The proposed methodology uses the concepts of SPI, reliability, resilience, and vulnerability, and the index of optimism of the decision maker. The approach has been named as the “Distance based approach” for assessment of WSI. The suggested methodology has been applied for assessment of WSI for the state of Chhattisgarh in India. It has been estimated using the historical rainfall data as well as the projected rainfall series under the two possible scenarios under climate change conditions HadCM3 A2 and B2. On the basis of historical rainfall series, the WSI for the Chhattisgarh state was found to be 0.47 with neutral

Table 10 Values of two parametric gamma distribution statistics and GoF for each district of Chhattisgarh during 2011–2099

Gamma distribution statistics		D1	D2	D3	D4	D5	D6	D7	D8	D9	D10	D11	D12	D13	D14	D15	D16
Shape factor		81.5	6.3	51.9	81.5	74.8	66.7	41.1	67.1	63.9	40.7	58.8	75.4	59.3	53.6	62.2	46.4
Scale factor		14.8	21.6	22.3	14.8	14.9	17.5	33.0	17.4	17.9	27.8	20.8	15.2	20.4	20.5	19.3	26.1
Goodness-of-Fit (GOF)																	
Kolmogorov Smirnov	Statistics	0.05	0.09	0.06	0.05	0.06	0.05	0.08	0.05	0.06	0.08	0.07	0.06	0.06	0.06	0.04	0.09
	p value	0.98	0.47	0.86	0.98	0.85	0.97	0.62	0.97	0.91	0.53	0.75	0.90	0.93	0.93	1.00	0.44
Anderson-darling	Statistics	0.18	1.16	0.32	0.18	0.25	0.36	0.46	0.21	0.42	0.65	0.41	0.31	0.43	0.37	0.13	0.72
Chi-squared	Statistics	1.85	15.23	6.77	1.85	8.44	3.00	13.04	0.54	4.14	8.91	0.97	7.08	2.51	3.14	2.15	18.03
	p value	0.93	0.02	0.34	0.93	0.21	0.81	0.04	1.00	0.66	0.18	0.99	0.31	0.87	0.79	0.90	0.01

Table 11 SPI categories and corresponding scores and their distribution of different districts

Category	Number of years followed by different categories during historical time series (2011–2099)																
	SPI ranges	D1	D2	D3	D4	D5	D6	D7	D8	D9	D10	D11	D12	D13	D14	D15	D16
Extremely wet	2.00 and above	3	0	2	2	2	3	2	2	3	2	3	3	3	2	2	2
Severely wet	1.50 to 1.99	3	5	2	2	4	3	6	4	3	2	3	2	3	2	5	5
Moderately wet	1.00 to 1.49	8	4	12	7	6	6	6	9	6	10	5	7	6	7	7	6
Near normal	−0.99 to 0.99	60	62	59	65	61	60	61	60	64	61	66	63	61	65	58	65
Moderately dry	−1.00 to −1.49	7	8	8	7	14	14	8	5	9	7	9	12	13	7	13	6
Severely dry	−1.50 to −1.99	5	7	3	3	0	1	3	9	2	4	1	0	1	4	2	3
Extremely dry	−2.00 and less	3	3	3	3	2	2	3	0	2	3	2	2	2	2	2	2

Table 12 Estimated values of WSI parameters of each districts of Chhattisgarh, India

Particular	D1	D2	D3	D4	D5	D6	D7	D8	D9	D10	D11	D12	D13	D14	D15	D16
Total events	89	89	89	89	89	89	89	89	89	89	89	89	89	89	89	89
Drought events	15	18	14	13	16	17	14	14	13	14	12	14	16	13	17	11
Non-drought events	74	71	75	76	73	72	75	75	76	75	77	75	73	76	72	78
No. of times satisfactory value(s) follow an unsatisfactory value	11	11	12	13	13	14	13	11	11	12	11	12	12	12	15	10
Cumulative failure extent	8.37	9.66	7.88	8.39	6.76	6.25	7.68	7.51	6	7.12	5.93	6.14	6.01	7.78	7.02	7.8
Max. failure extent in entire series	1.28	1.53	1.5	2.18	1.72	2.19	1.54	0.93	2.24	1.31	1.98	1.96	2.18	2.27	1.67	2.24
No. of continuous failure events	11	11	12	13	13	14	13	11	11	12	11	12	12	12	15	10
Reliability	0.83	0.8	0.84	0.85	0.82	0.81	0.84	0.84	0.85	0.84	0.87	0.84	0.82	0.85	0.81	0.88
Resilience	0.79	0.65	0.86	1	0.81	0.82	0.93	0.85	0.85	0.92	0.92	0.86	0.8	0.92	0.88	1
Vulnerability	0.76	0.88	0.66	0.65	0.52	0.45	0.59	0.68	0.55	0.59	0.54	0.51	0.5	0.65	0.47	0.78
Relative vulnerability	0.59	0.57	0.44	0.3	0.3	0.2	0.38	0.73	0.24	0.45	0.27	0.26	0.23	0.29	0.28	0.35
Districts wise WSI	0.26	0.22	0.41	0.6	0.46	0.53	0.48	0.19	0.55	0.43	0.58	0.53	0.51	0.56	0.51	0.57

where *D1* Bastar; *D2* Bilaspur; *D3* Dantewada; *D4* Dhamtari; *D5* Durg; *D6* Janjgir; *D7* Jashpur; *D8* Kanker; *D9* Korba; *D10* Koriya; *D11* Kawardha; *D12* Mahasamund; *D13* Raigarh; *D14* Raipur; *D15* Rajnandgaon; *D16* Surguja District of Chhattisgarh

Table 13 Values of overall WSI using distance-based approach (Historical 2011–2099)

Particular	D1	D2	D3	D4	D5	D6	D7	D8	D9	D10	D11	D12	D13	D14	D15	D16
Distance from ideal point (Rel.)	0.17	0.20	0.16	0.15	0.18	0.19	0.16	0.16	0.15	0.16	0.13	0.16	0.18	0.15	0.19	0.12
Distance from ideal point (Res.)	0.21	0.35	0.14	0.00	0.19	0.18	0.07	0.15	0.15	0.08	0.08	0.14	0.20	0.08	0.12	0.00
Distance from ideal point (Rel. Vul.)	0.59	0.57	0.44	0.30	0.30	0.20	0.38	0.73	0.24	0.45	0.27	0.26	0.23	0.29	0.28	0.35
Maximum distance in terms of (Rel., Res., & Rel. Vul.)	0.59	0.57	0.44	0.30	0.30	0.20	0.38	0.73	0.24	0.45	0.27	0.26	0.23	0.29	0.28	0.35

Overall water sustainability index based on distance-based approach		
Approaches	Distance	Overall WSI (1-distance)
Maximum of maximum (pessimistic approach)	**0.73**	0.27
Minimum of maximum (optimistic approach)	**0.20**	0.80
Average of maximum distance for all districts (neutral approach)	0.37	0.63
Overall water sustainability index (WSI)		**0.63**

where *D1* Bastar; *D2* Bilaspur; *D3* Dantewada; *D4* Dhamtari; *D5* Durg; *D6* Janjgir; *D7* Jashpur; *D8* Kanker; *D9* Korba; *D10* Koriya; *D11* Kawardha; *D12* Mahasamund; *D13* Raigarh; *D14* Raipur; *D15* Rajnandgaon; *D16* Surguja District of Chhattisgarh

index of optimism but can vary from 0.14 to 0.72 from the most pessimistic and most optimistic feeling of the DM. Further, under climate change HadCM3-A2 possible scenario, the WSI index for the Chhattisgarh state is assessed as 0.54, which varies from 0.46 to 0.74 under most pessimistic and most optimistic feeling of the DM. Under HadCM3-B2 scenario, the overall WSI has been assessed as 0.63 which varied from 0.27 to 0.80 under most pessimistic and most optimistic feeling of the DM. The adopted methodology may also help to assess the status of WSI in regional level study using the SPI. It is one of the simplest techniques to access the WSI because its only depends upon precipitation. It may help for water resources planning and mitigation for future perspectives.

References

Alley WM (1984) The Palmer drought severity index: limitations and assumptions. J Clim Appl Meteorol 23(7):1100–1109

Bhalme HN, Mooley DA (1980) Large-scale droughts/floods and monsoon circulation. Mon Weather Rev 108(8):1197–1211

Bhalme H, Mooley D, Jadhav S (1983) Fluctuations in the drought/flood area over India and relationships with the southern oscillation. Mon Weather Rev 111(1):86–94

Blumenstock G (1942) Drought in the United States analyzed by means of the theory of probability. United States Department of Agriculture, Economic Research Service

Bradley AA, Hashino T, Schwartz SS (2003) Distributions-oriented verification of probability forecasts for small data samples. Wea Forecasting 18(5):903–917

Brundtland G, Khalid M, Agnelli S, Al-Athel S, Chidzero B, Fadika L, Okita S (1987) Our common future ('Brundtland report')

Byun H-R, Wilhite DA (1999) Objective quantification of drought severity and duration. J Clim 12 (9):2747–2756

Dai A, Trenberth KE, Qian T (2004) A global dataset of palmer drought severity index for 1870–2002: relationship with soil moisture and effects of surface warming. J Hydrometeorol 5 (6):1117–1130

Guttman NB (1998) Comparing the palmer drought index and the standardized precipitation index1. Wiley Online Library

Guttman NB (1999a) Accepting the standardized precipitation index: a calculation algorithm. Wiley Online Library

Guttman NB (1999b) Accepting the standardized precipitation index: a calculation algorithm. J Am Water Resour Assoc 35(2):311–322

Hashimoto T, Stedinger JR, Loucks DP (1982) Reliability, resiliency, and vulnerability criteria for water resource system performance evaluation. Water Resour Res 18(1):14–20

Hayes MJ, Svoboda MD, Wilhite DA, Vanyarkho OV (1999) Monitoring the 1996 drought using the standardized precipitation index. Bull Am Meteor Soc 80(3):429–438

Heggen RJ (2001) Normalized antecedent precipitation index. J Hydrol Eng 6(5):377–381

Khalili D, Farnoud T, Jamshidi H, Kamgar-Haghighi AA, Zand-Parsa S (2011) Comparability analyses of the SPI and RDI meteorological drought indices in different climatic zones. Water Resour Manage 25(6):1737–1757

Komuscu A (1999) Using the SPI to analyze spatial and temporal patterns of drought in Turkey. Drought Netw News 11(1):7–13

Lawrence P, Meigh J, Sullivan C (2002) The water poverty index: an international comparison. Keele Economics Research Papers. www.keele.ac.uk/depts/ec/web/wpapers/kerp0219.pdf

Lana X, Serra C, Burgueño A (2001) Patterns of monthly rainfall shortage and excess in terms of the standardized precipitation index for Catalonia (NE Spain). Int J Climatol 21(13):1669–1691

Loucks DP, Gladwell JS (1999) Sustainability criteria for water resource systems. Cambridge University Press

Maity R, Sharma A, Nagesh Kumar D, Chanda K (2012) Characterizing drought using the reliability-resilience-vulnerability concept. J Hydrol Eng 18(7):859–869

McKee TB, Doesken NJ, Kleist J (1995) Drought monitoring with multiple time scales. In: Ninth conference on applied climatology. American Meteorological Society, Boston

McKee TB, Doesken NJ, Kleist J (1993) The relationship of drought frequency and duration to time scales. In: Proceedings of the 8th conference on applied climatology, American Meteorological Society Boston, MA, 179–183

McQuigg J (1954) A simple index of drought conditions. Weatherwise 7(3):64–67

Mishra AK, Singh VP (2010) A review of drought concepts. J Hydrol 391(1):202–216

Moron V (1994) Guinean and Sahelian rainfall anomaly indices at annual and monthly scales (1933–1990). Int J Clim 14(3):325–341

Munger TT (1916) Graphic method of representing and comparing drought intensities. Mon Weather Rev 44(11):642–643

Pahl-Wostl C (2002) Towards sustainability in the water sector–The importance of human actors and processes of social learning. Aquat Sci 64(4):394–411

Palmer WC (1965) Meteorological drought, vol 30. US Department of Commerce, Weather Bureau, Washington, DC, USA

Raskin PD, Hansen E, Margolis RM (1996) Water and Ormance evaluation. Water Resour Res 18 (1):14–20

Salameh E (2000) Redefining the water poverty index. Water Int 25(3):469–473

Sandoval-Solis S, McKinney D, Loucks D (2010) Sustainability index for water resources planning and management. J Water Resour Plan Manag 137(5):381–390

Shiau J (2006) Fitting drought duration and severity with two-dimensional copulas. Water Resour Manage 20(5):795–815

Sridhar V, Hubbard KG, You J, Hunt ED (2008) Development of the soil moisture index to quantify agricultural drought and its "user friendliness" in severity-area-duration assessment. J Hydrometeorol 9(4):660–676

Stacy EW, Mihram GA (1965) Parameter estimation for a generalized gamma distribution. Technometrics 7(3):349–358

Sullivan C (2002) Calculating a water poverty index. World Dev 30(7):1195–1210

Thornthwaite CW (1931) The climates of North America: according to a new classification. Geogr Rev 633–655

Van Rooy M (1965) A rainfall anomaly index independent of time and space. Notos 14:43–48

Vicente-Serrano SM, López-Moreno JI, Beguería S, Lorenzo-Lacruz J, Azorin-Molina C, Morán-Tejeda E (2011) Accurate computation of a streamflow drought index. J Hydrol Eng 17 (2):318–332

Waggoner ML, O'Connell TJ (1956) Antecedent precipitation index. Wkly Weather Crop Bull 43:6–7

Wu H, Svoboda MD, Hayes MJ, Wilhite DA, Wen F (2007) Appropriate application of the standardized precipitation index in arid locations and dry seasons. Int J Climatol 27(1):65–79

Yusof F, Hui-Mean F, Suhaila J, Yusop Z, Ching-Yee K (2014) Rainfall characterisation by application of standardised precipitation index (SPI) in Peninsular Malaysia. Theoret Appl Climatol 115(3–4):503–516

Zhu X, Bothe O, Fraedrich K (2011) Summer atmospheric bridging between Europe and East Asia: Influences on drought and wetness on the Tibetan Plateau. Quatern Int 236(1):151–157

Part IV
Low Flow and Drought

Coupling of Tennant Concept with Standardized Precipitation Index (SPI) for the Prediction of Environmental Flow Condition from Rainfall in Upper Narmada Basin

Kumar Amrit, S.K. Mishra and R.P. Pandey

Abstract In this study, an effort has been made to describe the environmental flow condition of a watershed using standardized precipitation index (SPI), a drought index based on the precipitation. Mohegaon, Manot, Hridaynagar, and Sher are the four watersheds of upper Narmada Basin that have been taken for the analysis. The purpose of this study is to derive relationship between environmental flow condition and the corresponding estimates of standardized precipitation index which is used as measure of drought conditions. It is expected that these relationship will be useful in determining environmental flow conditions in ungauged sub-basins using rainfall data only. The study revealed an excellent relationship between SPI and percentage of average annual flow as the value of coefficient of determination are greater than 0.75. The analysis indicates that for each of the four watersheds, the percentage of average annual flow increases with the increase in the value of SPI. It can be concluded that the relationships presented in this paper will be useful for estimating the EF condition for ungauged watersheds.

Keywords Tennant method · Narmada basin · SPI · Environmental flow

Introduction

Water is necessary for life and it is regarded an important natural resource for the well-being of society and existence of life on the earth. The water demand is increasing day by day which exceeds the water supply due to increasing water

---

K. Amrit (✉) · S.K. Mishra
Department of Water Resources Development & Management, IIT Roorkee, Roorkee 247667, India
e-mail: amrit27upadhyay@yahoo.com

R.P. Pandey
National Institute of Hydrology, Roorkee 247667, India

V.P. Singh et al. (eds.), *Climate Change Impacts*, Water Science and Technology Library 82, https://doi.org/10.1007/978-981-10-5714-4_20

demand for growing population and industrialization. For the conservation of natural ecosystem, water has to be preserved in the rivers and it should be clean so that a healthy ecosystem can be maintained. The minimum amount of water required for the survival of rivers is known as environmental flow (EF). For a healthy ecosystem, environmental flows are one of the important factors. The minimum supply of water is maintained in the steams called EF requirement which helps in the sustainability of aquatic lives and other natural ecosystem. The lack of requisite flow will influence the whole ecosystem. So the EF is necessary to carry out the needs of animal, vegetation, and aquatic lives which depend on the river for their sustainability.

The socioeconomic development and climate change has affected the global hydrological cycles, threatening human water security, the health of aquatic environments, and river biodiversity largely during past few decades (Vörösmarty et al. 2010; Jacobsen et al. 2012; Van Vliet et al. 2013). These situations attract the attention towards the assessment of environmental flow requirement (EFR) and water scarcity (Vörösmarty et al. 2010; Kirby et al. 2014). Thus EFR is defined as the quality, quantity, and timing of the water flows required maintaining freshwater and estuarine ecosystems and the human livelihoods and well-being that depend upon these ecosystems (Brisbane Declaration 2007). More than 200 methods are available and being used worldwide to calculate EFR to maintain healthy rivers (Tharme 2003). These methods can be grouped into four categories: hydrological approach, hydraulic rating, habitat simulation, and holistic methods.

Drought is a natural calamity, occurring due to less than average rainfall over a given period of time at given space which consequently leads to stream flow reduction and short-term water deficit. Further, the droughts cause lowering of water levels in lakes, reservoirs, tanks, etc. There are various drought indices were developed which are very useful for drought monitoring based on different parameters. Some of the common drought indices used to assess the meteorological droughts in India, are standardized precipitation index (SPI) (McKee et al. 1993), Effective Drought Index (EDI) (Byun and Wilhite 1999), and percentage departure of annual and seasonal rainfall from corresponding mean are applied for identification of onset, termination and quantification of severity of drought events.

The SPI is used to quantify the dry and wet conditions based on the precipitation, and Tennant method is used to describe the environmental flow condition of a river from severe degradation to flushing flow, i.e., whether the river runs dry or have maximum flow based on the flow data. Since both SPI and Tennant Method are used to describe the dry and wet conditions based on the different parameters so, there might have some possibility to establish a relationship between these two methods.

Despite the fact there exist a large number of EF methods, none of them has the efficacy to predict EF for ungauged watersheds, i.e. using rainfall only. On the other hand, SPI has the efficacy to describe a similar dry or wet situation, but in terms of drought. Thus, there exists a possibility to explore a relationship between these two, for EF prediction from rainfall, useful for ungauged watersheds, which forms the primary objective of this study.

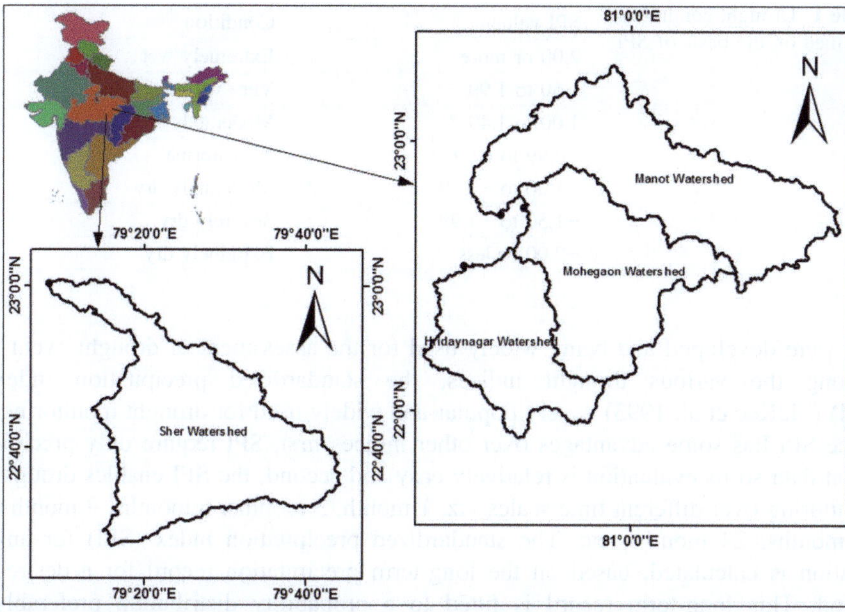

Fig. 1 Index map of study area

Study Area

Narmada is the largest west-flowing river of the Indian peninsula. It is one of the important rivers of India. The Narmada basin extends over an area of 98,796 km^2 and lies between longitudes 72° 32′E to 81° 45′E and latitudes 21° 20′N to 23° 45′ N. Here for the study, four catchments of Upper Narmada basin are selected based on the availability of data viz. Mohegaon, Manot, Hridaynagar, and Sher having the catchment area of 3978 km^2, 4884 km^2, 3370 km^2 and 2901 km^2 respectively. The region has sub tropical and sub humid climate with average annual rainfall ranging from 1000 mm to 1400 mm. The summer is very hot and winter is quite cold. The area comprises of both flat and undulating lands covered with timber, grasses, and cultivated land. Soils vary from black to mixed red soils. The location map of the study catchments are presented in Fig. 1.

Methodology

Standardized Precipitation Index

Drought monitoring is based on identification and quantification of various drought characteristics viz. frequency, duration, and severity. The magnitude of deficit can also be assessed for the past drought events. There are so many methods and indices

Table 1 Drought conditions classified on the basis of SPI

SPI values	Condition
2.00 or more	Extremely wet
1.50 to 1.99	Very wet
1.00 to 1.49	Moderately wet
−0.99 to 0.99	Near normal
−1.00 to −1.49	Moderately dry
−1.50 to −1.99	Severely dry
−2.00 or less	Extremely dry

that were developed and being widely used for the assessment of drought events. Among the various drought indices, the standardized precipitation index (SPI) (McKee et al. 1993) is very popular and widely used for drought monitoring. Since SPI has some advantages over other indices first, SPI require only precipitation data so its evaluation is relatively easy and second, the SPI enables drought monitoring over different time scales viz. 1 month, 3 months, 6 months, 9 months, 12 months, 24 months, etc. The standardized precipitation index (SPI) for any location is calculated, based on the long-term precipitation record for a desired period. This long-term record is fitted to a probability distribution preferably gamma distribution, which is then transformed to a normal distribution so that the mean SPI for the location and desired period is zero (McKee et al. 1993; Edwards and McKee 1997). SPI is normalized index representing the occurrence of an observed rainfall when compared with the average rainfall of a particular location over a long reference period. SPI values represent the deviation of rainfall from long-term mean. SPI values describes from extremely wet to extremely dry condition. Negative SPI values represent the deficiency in rainfall while positive values of SPI show the surplus rainfall. The magnitude of SPI negative values are used to classify the severity of drought event. Higher the negative SPI value, more severe the drought event are likely to occur. Table 1 represents the different conditions classified on the basis of SPI values by McKee et al. (1993).

Tennant Method

This method was developed by Donald Tennant in Montana region of USA Tennant (1975, 1976a, b), also called as Montana approach and developed for the needs of fish. It used 58 cross-section and 38 different flows of 11 streams in Wyoming, Montana, and Nebraska (Mann 2006). He established a relationship between aquatic habitat suitability and flow using subjective assessment of habitat quality and the empirical hydraulic data obtained from cross-channel transects. This method is based on the assumption that to uphold good stream environment some percentage of average flow is required. According to Tennant, for short-term survival the average depth and velocity of flow should be at least 0.3 m and 0.25 m/s

Table 2 Tenant method for EFR assessment

Flow condition	Oct–Mar	Apr–Sep
Flushing flow	200% AAF	200% AAF
Optimum range of flow	60–100% AAF	60–100% AAF
Outstanding	40% AAF	60% AAF
Excellent	30% AAF	50% AAF
Good	20% AAF	40% AAF
Fair or degrading	10% AAF	30% AAF
Poor or minimum	10% AAF	10% AAF
Severe degradation	10% AAF to zero flow	10% AAF to zero flow

respectively and the depth between 0.45 and 0.6 m and velocity ranging from 0.45 to 0.6 m/s found to be optimal for the fish. These conditions were found at 10 and 30% of average annual flow (AAF) respectively in the different streams he studied. The different flow conditions based on percentage of average annual flow for low (October–March) and high (April–September) flow periods (Tennant 1976a, b) are given in Table 2.

In this study, the SPI on 9-month time scale for the month of June was evaluated for every year. In the same manner the average flow of 9 months (October–June) was calculated for every year which is termed as annual flow. The average of annual flow of all the years is calculated and the percentage of annual flow of a particular year with respect to average annual flow is termed as percentage of average annual flow of that year. The analysis has been done to check the relation between the percentage of average annual flow which describes the environment flow condition and SPI, the index based on precipitation.

Results and Discussion

The analysis has been done for the different catchments to explore the relationship between environmental flow condition and SPI. The EF condition is described on the basis of percentage of average annual flow (AAF) (Tennant 1976a, b). The study is concern with non-monsoon season (October–June).

Mohegaon Catchment

The rainfall and runoff data of 1982–1989 were used. The average flow of the 9 months (October–June) for each year was computed to estimate AAF, and %AAF to describe different flow conditions of the catchment. On the other hand, SPI for the same 9 months for each year was determined. A plot between %AAF and SPI is shown in Fig. 2. As seen from the figure, as SPI increases, %AAF also increases to describe a similar condition. The value of R^2 is 0.764, which shows a very good fit. Thus, from the available rainfall data, the EF condition of this catchment can be easily ascertained with the help of SPI.

Hridaynagar Catchment

The rainfall and flow data of 9 years (1981–1989) were used. Following the similar procedure, as above, the derived %AAF is plotted against the corresponding SPI for Hridaynagar catchment in Fig. 2. $R^2 = 0.818$ exhibits an excellent relationship between %AAF and the corresponding SPI. Thus, EF condition for this catchment can be ascertained using SPI.

Manot Catchment

The data of 8 years (1982–1989) were used, and the requisite plot with $R^2 = 0.864$ is shown in Fig. 2, again indicating an excellent fit.

Sher Catchment

Similar to the above, Fig. 2 shows an excellent %AAF-SPI relation with $R^2 = 0.993$ with rainfall-runoff data of 6 years (1978–1983), leading to similar inference as above.

The flow condition corresponding to SPI values for each of the four catchments is presented in Fig. 2.

Fig. 2 Percentage of average annual flow corresponding to different SPI values in four catchments

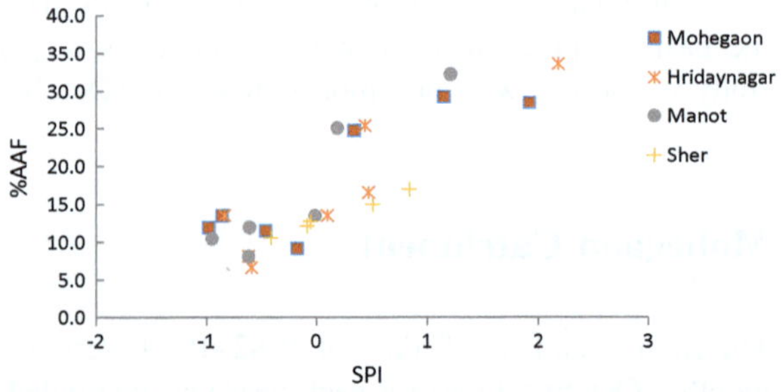

Conclusion

The study has been done over the four catchments viz. Mohegaon, Manot, Hridaynagar and Sher of Narmada basin in Madhya Pradesh. The analysis has been done for different time period in different catchments based the data availability. The rainfall and flow data of these catchments were used in the analysis. The rainfall data were used to calculate the SPI on 9-month time scale for the month of June for these catchments. Average annual flows in the sub-basins were estimated using the flow data. The percentage of average annual flow which describes the environmental flow condition has been estimated for the lean period (October–June) for every year. The plot of %AAF and SPI for all the watersheds over which study has been done shows an increasing trend. The R^2 value for each of the watershed has been found to be greater than 0.75 which shows the best fit. The analysis revealed that in the lean period for a particular watershed the percentage of average annual flow increases with the corresponding increase in the SPI value, i.e., higher the SPI value, environmental flow condition is good.

Acknowledgements The authors wish to thank Indian Institute of Technology Roorkee for providing the needful space and resources during the study. This work was financially supported by the Ministry of Human Resources and Development, New Delhi.

References

Brisbane Declaration (2007) The brisbane declaration: environmental flows are essential for freshwater ecosystem health and human well-being, September 3–6, Brisbane, Australia. Available at http://www.eflownet.org/viewinfo.cfm?linkcategoryid=4&linkid=13&siteid=1&FuseAction=display

Byun HR, Wilhite DA (1999) Objective quantification of drought severity and duration. J Clim 12:2747–2756

Edwards DC, McKee TB (1997) Characteristics of 20th century drought in the United States at multiple scales. Atmospheric Science Paper No. 634, 1–30 May

Jacobsen D, Milner A, Brown L, Dangles O (2012) Biodiversity under threat inglacier-fed river systems. Nat Clim Change 2:361–364

Kirby JM, Connor J, Ahmad MD, Gao L, Mainuddin M, (2014) Climate change and environmental water reallocation in the Murray–Darling Basin: impacts on flows, diversions and economic returns to irrigation. J. Hydrol. http://dx.doi.org/10.1016/j.jhydrol.2014.01.024 (in press)

Mann JL (2006) Instream flow methodologies: an evaluation of the tennant method for higher gradient streams in the national forest system lands in the Western U.S. M.Sc. thesis, Colorado State University, Fort Collins, CO, 143 p

McKee TB, Doesken, NJ, Kleist J, (1993) The relationship of drought frequency and duration to time scales. Paper presented at 8th conference on applied climatology, American Meteorological Society, Anaheim, CA

Tennant DL (1975) Instream flow regimens for fish, wildlife, recreation and related environmental resources. U.S. Fish and Wildlife Service, Federal Building, Billings, MT, 30 p

Tennant DL (1976a) Instream flow regimens for fish, wildlife, recreation and related environmental resources. In: Orsborn JF, Allman CH (eds) Proceedings of the symposium and specialty conference on instream flow needs. American Fisheries Society, Bethesda, pp 359–373

Tennant DL (1976b) Instream flow regimens for fish, wildlife, recreation, and related environmental resources. Fisheries 1(4):6–10

Tharme RE (2003) A global perspective on environmental flow assessment: emerging trends in the development and application of environmental flow methodologies for rivers. River Res Appl 19:397–441

Van Vliet MTH, Franssen WHP, Yearsley JR, Ludwig F, Haddeland I, Letten-maier DP, Kabat P (2013) Global river discharge and water temperature under climate change. Glob Environ Change 23:450–464

Vörösmarty CJ, McIntyre PB, Gessner MO, Dudgeon D, Prusevich A, Green P, Glidden S, Bunn SE, Sullivan CA, Liermann CR, Davies PM (2010) Global threats to human water security and river biodiversity. Nature 467:555–561

Assessment of Drought in Balangir District of Odisha, India Using Drought Indices

A. Sudarsan Rao, Jyotiprakash Padhi and Bitanjaya Das

Abstract Droughts are normally triggered due to lack of rainfall and prolonged droughts have a multiplying effect and mount tremendous stress on natural resources leading to scarcity of water, food, and fodder. Balangir district is a chronically drought-prone area. Therefore this study was carried out to determine drought indices, drought index (DI), moisture indicator (MI), hydrothermal coefficient (HTC), and standardized precipitation index (SPI) to have a preliminary idea to forecast the possibility of occurrence of droughts for the 14 blocks of the district by treating each block as a unit. The drought indices were developed by utilizing historical rainfall data and temperature data during 1961–2007. Marginal droughts return period varies from 2–4 years, 4–16 years for moderate droughts and 24–48 years in case of severe droughts for all the blocks of Balangir district. Deogaon, Loisinga, and Titlagarh are chronically drought-prone areas; therefore these blocks need urgent attention from drought point of view. Balangir, Loisinga, Patnagarh, Puintala, Tentulikhunti, and Titlagarh blocks faced extreme agricultural drought conditions as per 3 months' SPI value. Extreme hydrological drought conditions faced by Belpara, Balangir, Loisinga, Patnagarh, Puintala, and Tentulikhunti blocks according to 12 months' SPI value. Agalpur, Bongomunda, Deogaon, Khaprakhol, Muribahal, Saintala, and Tureikela blocks did not experience any extreme dry events based on 3, 6, and 12 months' SPI value. The analysis of these drought indices led to several useful and practicable inferences for better understanding the drought attributes of the study area. For this reason, this study will help in planning drought preparedness and its mitigation in a realistic and appropriate manner.

Keywords Drought · Drought index · Moisture indicator · Hydrothermal coefficient · SPI index

A. Sudarsan Rao
Department of Water Resources, Government of Odisha, Cuttack, India

J. Padhi (✉) · B. Das
School of Civil Engineering, KIIT University, Bhubaneswar 751024, Odisha, India
e-mail: erjppadhi@gmail.com

B. Das
e-mail: bdasfce@kiit.ac.in

V.P. Singh et al. (eds.), *Climate Change Impacts*, Water Science and Technology Library 82, https://doi.org/10.1007/978-981-10-5714-4_21

Introduction

The climate of a region is determined by long-term average, frequency, and extremes of several weather variables, notably precipitation, and temperature. In a large semiarid country such as India, precipitation is precious and varies both in space and time (Patel et al. 2007). Thus, any departure in precipitation patterns seldom leads to widespread natural disasters such as drought and floods affecting natural habitats, ecosystems and, importantly, agricultural and economic sectors. Drought, in particular, is considered by many to be the most complex but least understood of all natural hazards, affecting more people than any other hazard (Hagman 1984). Drought develops in a region and gradually spreads to the adjoining areas over a period of time and hence can be considered as a creeping phenomenon. It occurs in all agro-climatic zones although the magnitude and impacts differ in different zones. Drought differs from most other natural hazards in many ways, especially in the sense that its onset and termination is difficult to predict (McKee et al. 1993).

The Indian sub-continent experiences tropical monsoon climate and about 70–80% of rainfall occurs in the summer monsoon. The Indian sub-continent is very much prone to natural calamities, i.e., droughts and floods every year in some part of the country or other either individually or collectively during the monsoon (June to September) period. Twenty-one large scale droughts have occurred in the past century in India in the years 1891, 1896, 1899, 1905, 1911, 1915, 1918, 1920, 1941, 1951, 1965, 1966, 1972, 1974, 1979, 1982, 1986, 1987, 1988, 1999, 2000 (Rao 1997; Narain et al. 2000). Sinha Ray and Shewale (2000) have indicated that El Nino brings about adverse climatic changes in different part of the country and is also partly responsible for drought. The State of Odisha experiences floods, cyclones, and droughts almost on a regular basis either separately or collectively. According to final report on Community Disaster Resilience Fund 2009, in between 1955 and 2008, Odisha has experienced 28 years of flood, 19 years of drought and 7 years of cyclone along with the Super Cyclone in 1999. In between 1990 and 2008, Odisha has experienced 12 years of flood, 5 years of drought, one Super Cyclone and many depressions and cyclones.

A study on the drought-prone areas and chronically drought-affected areas of India by Appa Rao (1991) indicated that most of the drought-prone areas identified above are in either arid or semiarid regions where droughts occur more frequently. Therefore, arid and semiarid regions with little or no development of irrigation facilities are by far the worst affected due to droughts. Drought results insignificant impacts, regardless of level of developments although the character of these impacts will differ profoundly (Subbiah 1993; Benson and Clay 1998, 2000; Wilhite 2000; Wilhite and Vanyarkho 2000). Prolonged droughts have a multiplying effect and mount tremendous stress on natural resources leading to scarcity of water, food, and fodder (Rao 1997; Narain et al. 2000). Droughts being a climatic anomaly, hence, even if a methodology to forecast droughts is possible, droughts are inevitable.

Therefore, there remains a need for implementation good management practices to ensure better crop and to keep the agriculture sector sustainable.

The characteristics of droughts such as intensity, duration, etc., vary from each other from place to place. Deficiency in rainfall or severity it causes due to the deficiency of rainfall is referred to as intensity. The intensity of drought is measured by the deviation of a climatic variable from normal. This intensity together with period affected determines the impacts of drought. Drought indices, in general, enable the detection of the onset of drought events and enable their severity to be measured, thereby allowing an examination of the spatial and temporal characteristics of drought, and comparisons between different regions to be made (Alley 1984). There are more than 150 drought indices that exist and many more new indices came into account in the last decades; and not only many drought indices are developed every year across the globe but also sincere attempts also have been made reviewing the drought indices and the different climatic parameters such as precipitation, soil moisture, vegetation moisture, land surface temperature, humidity, land cover change, etc., which plays direct or indirect role in development of drought indices (Wang and Qu 2009; Mishra and Singh 2010; Zhang et al. 2010; Zargar et al. 2011) and improving the existing ones. The most commonly used drought indices include the Palmer drought severity index (PDSI) and the moisture anomaly index (Z-index) (Palmer 1965), the standardized precipitation index (SPI) (McKee et al. 1993, 1995), aridity index (Gore and Sinha Ray 2002) and Percent Normal, Deciles (Gibbs and Maher 1967). The various indices that are in widespread use nowadays are standardized precipitation index, drought index (Rathore 2005), moisture indicators (Budyko 1958; Hounam et al. 1975) and hydrothermal coefficients (Selyaninov 1930). However, selection strongly depends upon the requirement like availability of resources/data, field of application of interest, specific boundary conditions, and according to the necessity of spatial or temporal resolution.

Pandey et al. (2008) have used the geographical information system based spatial and time series information modeling (SPATSIM) and daily water resources assessment modeling (DWRAM) software for drought analysis on monthly and daily basis respectively and its spatial distribution in both dry and wet years. SPATSIM utilizes standardized precipitation index (SPI), effective drought index (EDI), deciles index and departure from long-term mean and median; and DWRAM employs only EDI. They have stated that the analysis of data from the Kalahandi and Nuapada districts of Odisha (India) revealed that droughts in this region occurred with a frequency of once in every 3–4 years, droughts occurred in the year when the ratio of annual rainfall to potential evapo-transpiration was less than 0.6 and SPI, EDI, and annual deviation from the mean showed a similar trend of drought severity. Sahoo (2015) analyzed the various drought indices such as standard precipitation index, reclamation drought index (RDI), normalized difference vegetation index (NDVI), and streamflow drought index (SDI) for KBK districts of Odisha by taking the entire district as a unit.

Though it is not possible to avoid a drought, it is quite feasible to be better prepared for coping with it. Therefore it is necessary to understand the rainfall pattern and use various drought indices in different ways to appraise quantitatively the severity (Pandey et al. 2008). Very few studies have been conducted to assess the drought condition at the block level using various drought indices by utilizing the historical rainfall and temperature data. In this study, an attempt has been made for identifying various drought indices in different blocks of Balangir district to assess the drought conditions in that area which will help the people to remain in well preparedness and to take precautionary measures to meet any adverse impacts in such drought conditions.

Description of Study Area

This study was conducted in the Balangir district of the state Odisha, India with a geographical area of 65,755 km^2 (Fig. 1). The Balangir district is encompassed between latitude of 20° 09′N and 21° 05′N and Longitude 82° 41′E and 83° 42′E. The district comprises of 14 blocks (Belpara, Balangir, Deogaon, Khaprakhol, Muribahal, Patnagarh, Puintala, Saintala, Titlagarh, Tentulikhunti, Bongomunda, Tureikela, Loisinga, and Agalpur). The climate in the district is generally cold in winter and hot in summer with temperature ranging from 13 to 42 °C and normal annual rainfall is 1041 mm. Drought index (DI), moisture indicator (MI), hydrothermal coefficient (HTC), and standard precipitation index (SPI) were estimated for the 14 blocks of the district by treating each block as a unit. These climatic indices were developed by utilizing the rainfall and temperature data for the past 47 years (1961–2007) for all the blocks whereas 38 years (1970–2007) of data used for the Balangir district as a whole. The consistency of the observed rainfall data is checked by double mass analysis and the rainfall data is found to be consistent.

Droughts are classified into three categories: meteorological (it is a situation where there is more than 25% decrease in precipitation from normal over an area), hydrological (when meteorological drought prolonged results in depletion of surface and groundwater indicates hydrological drought), and agricultural drought (it occurs when the soil moisture and rainfall are inadequate during the growing season to support healthy crop growth to maturity) as per National commission on Agriculture in India. According to India Meteorological Department (IMD), if the deficiency (%) of rainfall from normal is $\leq 25\%$, it is known as marginal drought, moderate if deficiency lies between 26 and 50% and if the deficiency is more than 50%, then drought is said to be severe. Hydrothermal coefficient (HTC), a non-dimensional indicator popularly known as G.T. Selyaninov's hydrothermal coefficient has been widely used (Selyaninov 1930) in Russia to indicate the aridity condition of an area. HTC is defined as

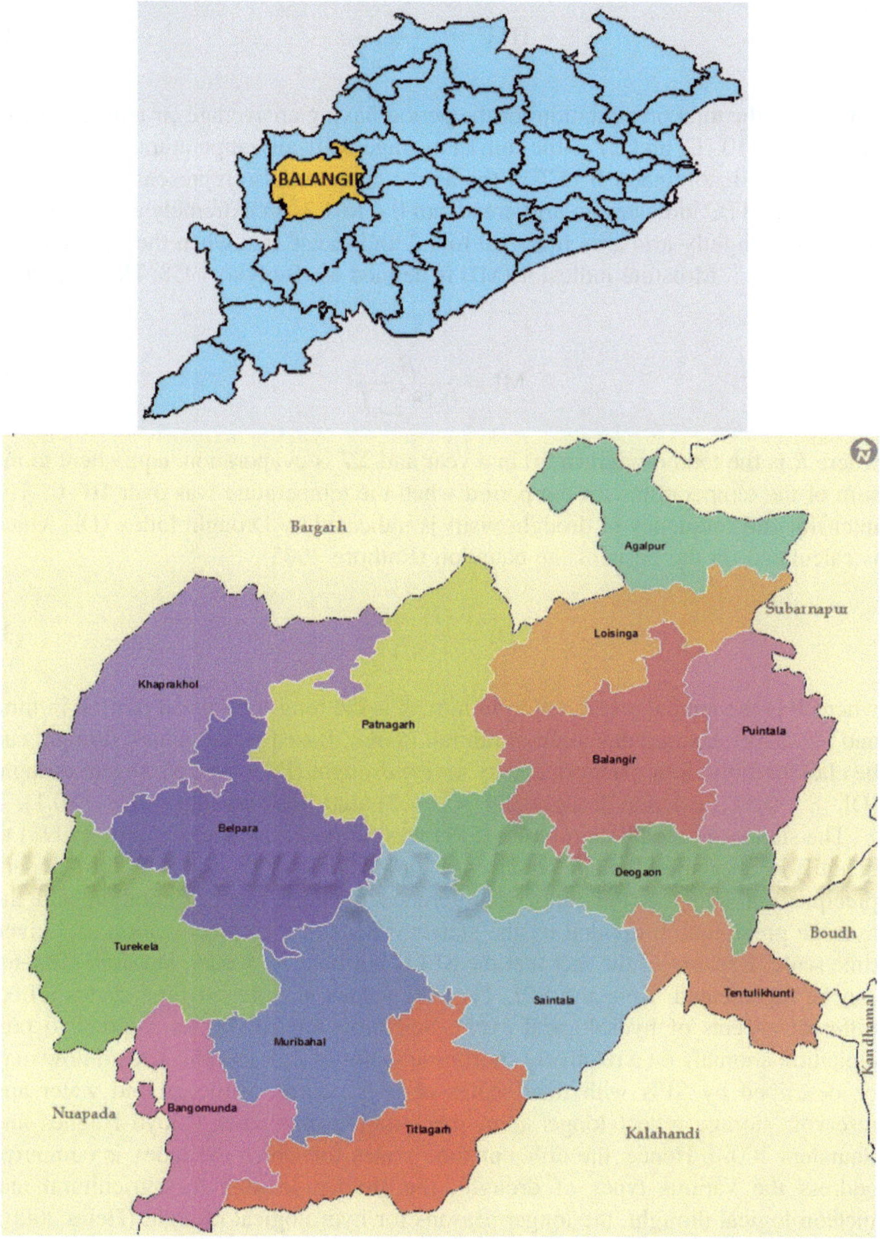

Fig. 1 The study area of different blocks of Balangir district of Odisha, India

$$\mathrm{HTC} = \frac{R}{0.1 \sum T}, \tag{1}$$

where R is the total precipitation for the period having an average air temperature of greater than 10 °C; and ΣT is the sum of average daily air temperatures for the same period. The denominator $0.1\Sigma T$ in the above equation also represents evaporation quite well. HTC indicates an arid area from 0.4 to 1.3, an extremely arid area from 0.4 to 0.7, slightly arid area from 1.0 to 1.3 and moist area when the coefficient is more than 1.3. Moisture indicator (MI) is defined as (Budyko 1958; Hounam et al. 1975),

$$\mathrm{MI} = \frac{R}{0.18 \sum T}, \tag{2}$$

where R is the total rainfall (mm) in a year and ΣT is evaporation, equivalent to the sum of the temperatures for the period when the temperature was over 10° C. The intensity and frequency of drought visits is indicated by Drought Index (DI) which is calculated on the basis of the equation (Rathore 2005),

$$\mathrm{DI} = \frac{(P - X)}{\mathrm{SD}}, \tag{3}$$

where P is the annual precipitation in mm, X is the long-term mean rainfall in mm, and SD is the standard deviation of rainfall in mm. Based on DI values, drought can be classified into four categories: very severe drought (DI ≤ -0.8), severe drought (DI ≥ -0.5), moderate drought (DI ≤ -0.2), and light drought (DI ≤ -0.1).

The standard precipitation index (SPI) was proposed by McKee et al. (1993) to quantify precipitation deficits or surpluses on a variety of time scales. The standard precipitation index is the ratio of difference between the precipitation and the average precipitation divided by the standard deviation of precipitation at a given time scale. Because of the fact that the SPI is normalized, wetter and drier climates can be represented using the SPI. The SPI values at different time scales reflect different aspects of hydrological cycle. Soil moisture conditions respond to precipitation anomaly on a relatively short time scales of 2–3 months, stream flow may be described by SPIs with time scales of 2–6 months, while ground water and reservoir storage reflect longer term precipitation anomalies (Lloyd-Hughes and Saunders 2002). Hence, the different time scales for which the index is computed address the various types of drought; the shorter seasons for agricultural and meteorological drought, the longer seasons for hydrological drought (Heim 2000). Many studies have also demonstrated that a short-term (3-month) SPI provides an indication of the seasonal anomaly in precipitation and thereby considered as an agricultural drought indicator (McKee et al. 1993; Hayes et al. 1999; Ji and Peters 2003).

According to McKee et al. (1993), there are seven classifications based on SPI values; extremely dry (SPI ≤ -2.0), severely dry ($-1.5 > \mathrm{SPI} > -1.99$), moderately dry ($-1.0 > \text{SPI} > -1.49$), near normal ($-0.99 < \text{SPI} < 0.99$), moderately wet ($1.0 < \text{SPI} < 1.49$), very wet ($1.5 < \text{SPI} < 1.99$), and extremely wet ($\text{SPI} \geq 2.0$). In this study, SPI calculated for 3, 6, and 12 months' time scale. SPI for September was calculated for 3 months' time scale and SPI December for 6 months' time scale for all the years. SPI September value was estimated because it covers the rainfall for July, August, and September, which contributes maximum rainfall during monsoon season.

Results and Discussion

Maximum, minimum, and average annual rainfall of Balangir district was 1765, 614, and 1132 mm, respectively. Highest monsoon rainfall (i.e., 1657 mm) occurred in the year 2001 whereas minimum amount of rainfall (i.e., 562 mm) observed in 1996. Average monsoonal rainfall observed in the district was 1041 mm. Similarly maximum amount of non-monsoon rainfall took place in 1990 and in the year 1989, least amount of rainfall occurred. The district experienced 16 marginal and 5 moderate drought events out of 38 years (1970–2007) and no severe drought event was observed. Marginal droughts return period varies from 2–4 years, 4–16 years for moderate droughts and 24–48 years in case of severe droughts for all the blocks of Balangir district. Table 1 shows the number of marginal, moderate and severe droughts occurred over 47 years for the fourteen blocks of the Balangir district and also shows the probability of occurrence of those drought events. It can be observed from Table 1 that Belpara, Saintala, and Tureikela blocks did not face any severe drought events over this period of time. Deogaon and Titlagarh experienced highest number of marginal drought events (i.e., 23) whereas Muribahal faced maximum number of moderate drought events (i.e., 11). Deogaon, Saintala, and Titlagarh experienced 28 drought events (i.e., marginal, moderate and severe) whereas Belpara faced least number of all drought events (i.e., 21). If the drought occurs in an area with a probability of $0.2 \leq P \leq 0.4$, the area is classified as drought-prone area and if the probability of occurrence is greater than 0.4, the area can be termed as chronically drought-prone area (Subramanya 2013). As per the above definition, Deogaon, Loisinga, and Titlagarh are chronically drought-prone areas. Within the same district, drought events occurred differently among various blocks due to spatial and temporal distribution in the amount of rainfall.

Table 2 shows the value of HTC, DI and MI for all the blocks of Balangir district and also for the district as a whole for marginal droughts. It can be observed from Table 2 that, DI value varies from 0.00 to −0.09 that means all the blocks experienced light drought which also matched the condition of normal drought. MI value was lowest for Tureikela block (i.e., 0.55) and maximum for Agalpur block (i.e., 0.70). Tureikela block categorized as arid area (as HTC value was 0.99) whereas the remaining blocks comes under slightly arid area, as the HTC values varied from 1.01 to 1.26.

Table 1 Probability and nos. of marginal, moderate and severe droughts in different blocks of Balangir District

Block	Drought events (nos.)			Probability of occurence		
	Marginal	Moderate	Severe	Marginal	Moderate	Severe
Balangir	17	4	2	0.35	0.08	0.04
Belpara	12	9	0	0.25	0.19	0.00
Deogaon	23	4	1	0.48	0.08	0.02
Khaprakhol	17	6	1	0.35	0.13	0.02
Patnagarh	19	4	2	0.40	0.08	0.04
Saintala	19	9	0	0.40	0.19	0.00
Puintala	14	7	2	0.29	0.15	0.04
Muribahal	12	11	1	0.25	0.23	0.02
Agalpur	16	6	1	0.33	0.13	0.02
Loisinga	21	3	1	0.44	0.06	0.02
Titlagarh	23	4	1	0.48	0.08	0.02
Bongomunda	15	8	1	0.31	0.17	0.02
Tureikela	19	7	0	0.40	0.15	0.00
Tentulikhunti	18	6	2	0.38	0.13	0.04

Table 2 Drought index (DI), moisture indicator (MI) and hydrothermal coefficients (HTC) for marginal droughts in different blocks of Balangir district

Block	Marginal drought		
	Drought index	Moisture indicator	Hydrothermal coeff.
Belpara	−0.06	0.68	1.22
Bolangir	−0.04	0.56	1.01
Deogaon	−0.05	0.61	1.09
Khaprakhol	−0.04	0.58	1.04
Muribahal	−0.02	0.55	0.99
Patnagarh	0.00	0.59	1.07
Puintala	0.00	0.66	1.20
Saintala	−0.03	0.60	1.08
Titlagarh	−0.02	0.60	1.08
Tentulikhunti	−0.03	0.61	1.09
Bongomunda	−0.06	0.65	1.18
Tureikela	−0.09	0.55	0.99
Loisinga	−0.04	0.59	1.07
Agalpur	−0.04	0.70	1.26
District as a whole	−0.08	0.59	1.07

Drought index (DI), moisture indicator (MI), and hydrothermal coefficients (HTC) for moderate droughts in different blocks of Balangir district are presented in Table 3. DI values were comparatively lower during moderate drought (i.e., −0.82

Table 3 Drought index (DI), moisture indicator (MI) and hydrothermal coefficients (HTC) for moderate droughts in different blocks of Balangir district

Block	Moderate drought		
	Drought index	Moisture indicator	Hydrothermal coeff.
Belpara	−1.25	0.51	0.92
Balangir	−1.05	0.42	0.76
Deogaon	−1.10	0.46	0.83
Khaprakhol	−1.18	0.42	0.76
Muribahal	−1.07	0.40	0.73
Patnagarh	−1.02	0.45	0.81
Puintala	−1.04	0.49	0.87
Saintala	−0.82	0.44	0.80
Titlagarh	−1.12	0.44	0.80
Tentulikhunti	−1.46	0.42	0.75
Bongomunda	−1.00	0.48	0.86
Tureikela	−0.95	0.44	0.79
Loisinga	−1.07	0.45	0.81
Agalpur	−1.03	0.48	0.87
District as a whole	−1.17	0.44	0.79

to −1.46) in comparison to marginal drought condition (0.00 to −0.09). Similar trend is also observed for moisture indicator values for all the blocks. The maximum MI value was observed for Belapara block (i.e., 0.51) and minimum for Muribahal block (i.e., 0.40). All the blocks categorized under arid area as the HTC value varied from 0.73 to 0.92. DI, MI and HTC for severe droughts in different blocks of Balangir district are presented in Table 4.

Table 4 Drought index (DI), moisture indicator (MI) and hydrothermal coefficients (HTC) for severe droughts in different blocks of Balangir district

Block	Severe drought		
	Drought index	Moisture indicator	Hydrothermal coeff.
Balangir	−2.18	0.35	0.62
Deogaon	−2.37	0.26	0.46
Khaprakhol	−2.32	0.28	0.51
Muribahal	−1.81	0.30	0.54
Patnagarh	−1.95	0.28	0.51
Puintala	−1.91	0.25	0.45
Titlagarh	−2.17	0.30	0.54
Tentulikhunti	−2.03	0.31	0.55
Bongomunda	−1.85	0.31	0.55
Loisinga	−2.37	0.31	0.55
Agalpur	−2.03	0.34	0.61

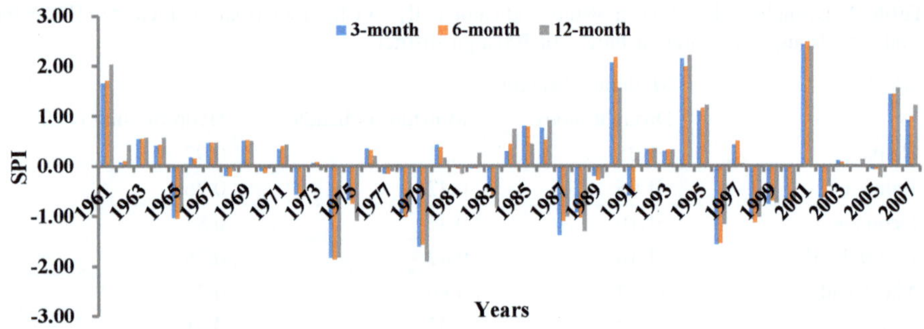

Fig. 2 SPI for 3, 6, and 12-month time scales during the period 1961–2007 in Agalpur

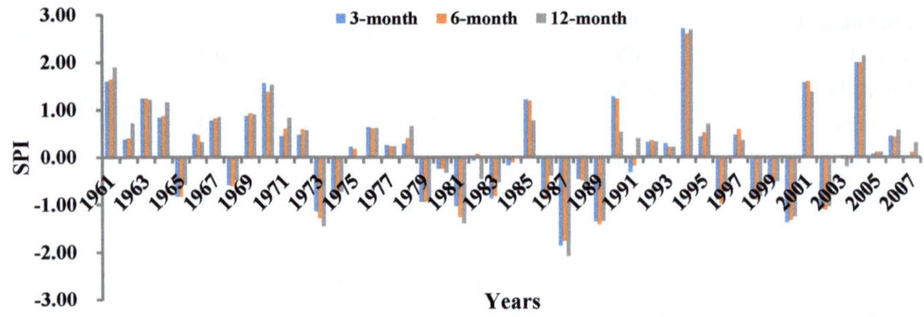

Fig. 3 SPI for 3, 6, and 12-month time scales during the period 1961–2007 in Belpara

The maximum MI value was observed for Balangir block (i.e., 0.35) and minimum for Puintala block (i.e., 0.25). 11 blocks categorized under extremely arid area as the HTC value varied from 0.4 to 0.7, which also justified the severe drought condition. Drought index varied from −1.81 to −2.37 for the blocks experiencing. Based on these DI values, these blocks faced very severe drought conditions.

It can be observed from Fig. 2 that, the block Agalpur experienced severe dry conditions in the years 1974, 1979, and 1996 according to 3, 6, and 12 months' SPI scales. During these years, the block faced agricultural as well as hydrological drought. Extreme wet condition was detected in 2001. In 1987, moderately dry condition was observed as per 3, 6, and 12 months' SPI values. For this reason, the block experienced agricultural as well as hydrological drought but the severity was less as compared during the years 1974, 1979, and 1996. Most of the times, the block experienced near normal condition. In the years 1973, 1989, and 2000 moderately dry condition was observed in Belapara block (Fig. 3). As per 6 and 12 months' time SPI, the block faced moderately dry condition whereas near normal condition as per 3-months' time of SPI in 1981. In the year 1987, severely dry condition for 3 and 6 months' time scale (i.e., agricultural drought) whereas

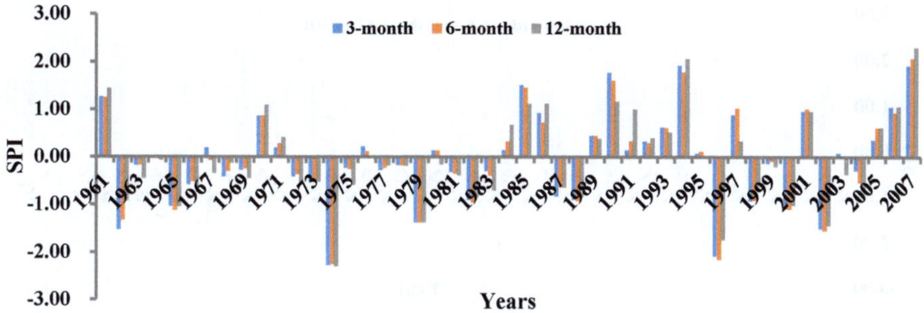

Fig. 4 SPI for 3, 6, and 12-month time scales during the period 1961–2007 in Balangir

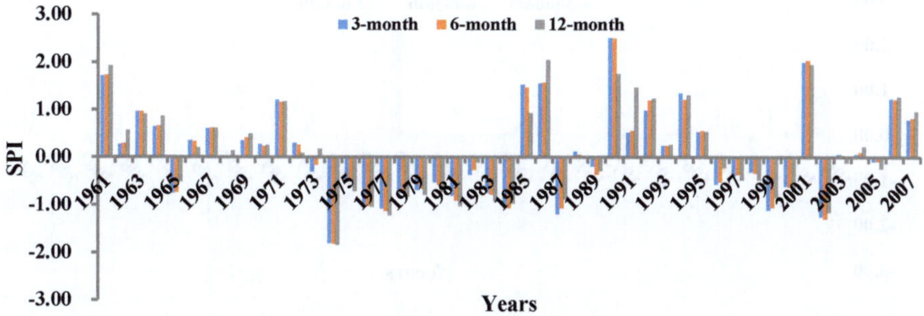

Fig. 5 SPI for 3, 6, and 12-month time scales during the period 1961–2007 in Bongomunda

extremely dry condition for 12 months' SPI (hydrological drought). Extremely wet condition was experienced by the block in 1994.

Balangir block experienced moderately dry condition during the years 1965, 1979, and 2000 whereas extremely dry situation observed in 1974 (Fig. 4). In 1996, the severity of agricultural drought was extreme as per 3 and 6 months' SPI values whereas severely dry for 12 months' SPI value. Severely dry, moderately dry and near normal condition was experienced for 3, 6, and 12 months' SPI value respectively in 1963. Therefore the severity of drought decreased as the SPI time scale increased from 3 months to 12 months period which implies the availability of soil moisture over 12 months period was highest whereas lowest over 3 months' time scale. During 2002 severe agricultural drought condition was experienced by the block as per 3 and 6 months' SPI value and as per 12 months SPI value moderate hydrological drought condition was observed. Extremely wet condition was detected as per 6 and 12 months' SPI whereas severely wet condition experienced according to 3 months' SPI value in 2007.

It can be observed from Fig. 5 that, Bongomunda block experienced moderate agricultural and hydrological drought situations during 1976, 1977, and 2002 according to 3, 6, and 12 months' time scales. During the years 1984, 1987, and 1999, moderately dry condition was detected as per 3 and 6 months' time scales

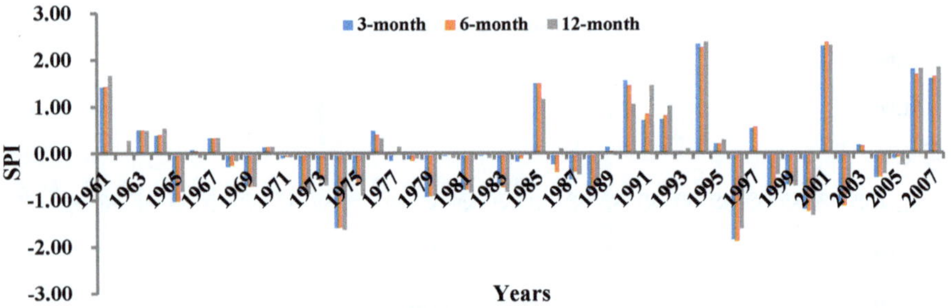

Fig. 6 SPI for 3, 6, and 12-month time scales during the period 1961–2007 in Deogaon

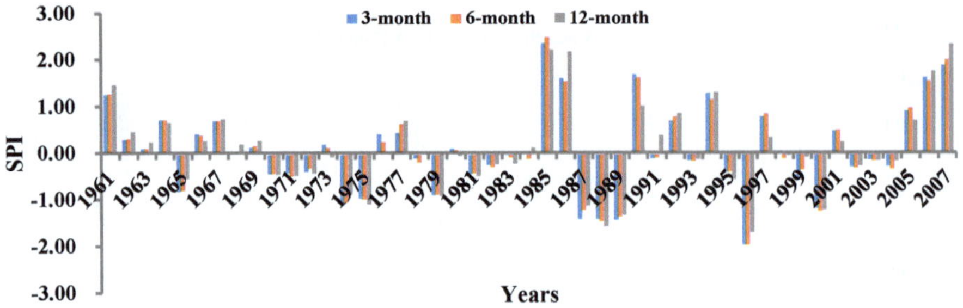

Fig. 7 SPI for 3, 6, and 12-month time scales during the period 1961–2007 in Khaprakhol

SPI value whereas near normal situation for 12-month SPI value. Severe agricultural and hydrological drought was detected in 1974 and very wet situation observed in 2001. During 1990, extremely wet condition was experienced as per 3 and 6 months' time scales whereas very wet situation was detected based on 12-month time scale SPI value. Severely dry and moderately dry condition was experienced by Deogaon block in the year 1974 and 2000 respectively (Fig. 6). Similarly, very wet condition was observed during 2006 and 2007 whereas extremely wet situation in 1994 and 2001. In 2002, the block faced moderate agricultural drought condition according to 3 and 6 months' SPI value whereas near normal situation for 12-month SPI value.

It can be observed from Fig. 7 that, Khaprakhol block experienced moderately dry conditions during the years 1987, 1988, 1989, and 2000. Severely dry condition observed in 1996 within the block. Very wet conditions were detected in 2006 and 2007 whereas extremely wet situation in 1985. Loisinga block experienced moderately dry situation in the years 1965, 1996, and 1999 (Fig. 8). Severely dry condition was observed in 1974 and extremely dry situation was detected in 1979. In the year 1972, moderately dry situation observed as per 12 months' time scale SPI whereas near normal condition as per 3 and 6 months' time scales SPI. Extremely wet condition was observed in 1994. SPI value for the 3, 6, and 12 months' time scales are presented in Fig. 9 for Muribahal block. Severely dry

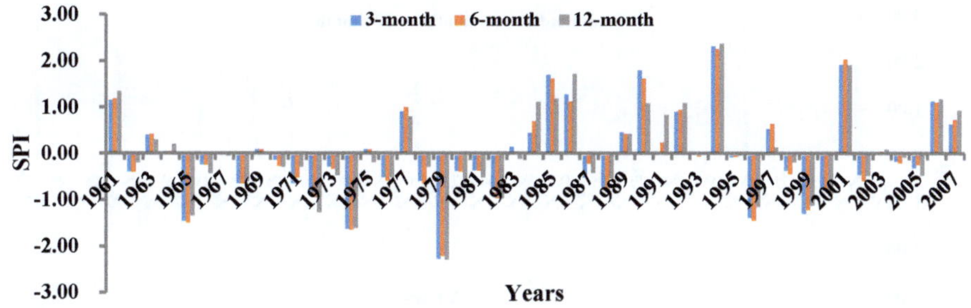

Fig. 8 SPI for 3, 6, and 12-month time scales during the period 1961–2007 in Loisinga

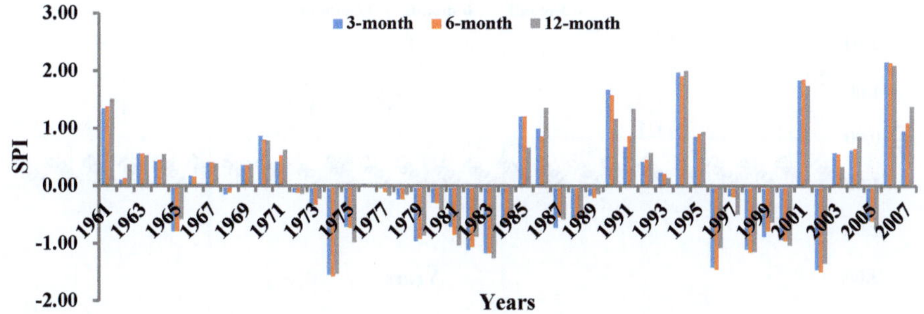

Fig. 9 SPI for 3, 6, and 12-month time scales during the period 1961–2007 in Muribahal

condition was observed in 1974. In 1981, moderately dry situation as per 12 months' time scale whereas near normal condition occurred according to 3 and 6 months' time scale SPI. During 1982, moderately dry condition was experienced as per 3 and 6 months' time scale and near normal according to 12 months' time scale SPI. Moderately dry situation was experienced by the block during the years 1983, 1996, and 1998.

Severely dry condition was detected in 2002 and during 2006, extremely wet condition observed in the block. Moderately dry condition as per 3 and 6 months' time scale and near normal according to 12 months' time scale SPI value in Patnagarh block (Fig. 10). Near normal situation observed during the years 1983 and 1988 as per 3 and 6 months' time scales whereas moderately dry according to 12 months' time scale SPI. Extremely dry situation was experienced by the block during 1987. Moderately dry situation was observed in the years 1989 and 1996 whereas severely dry condition faced by the block in 2000. Extremely wet situation occurred in 2001.

According to 3 and 6 months' time scales, moderately dry situation was observed whereas near normal condition as per 12 months' time scales SPI during 1965 in Puintala block (Fig. 11).

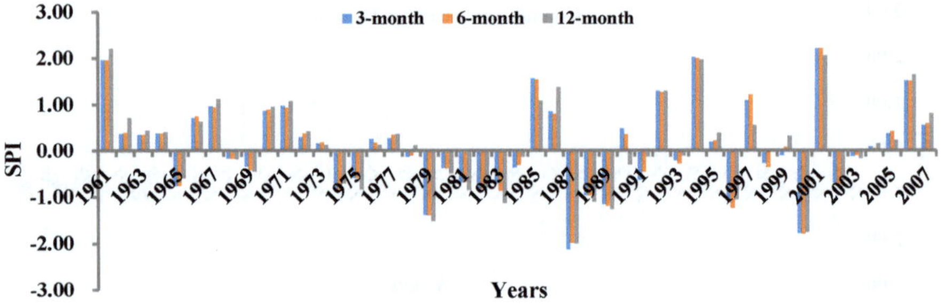

Fig. 10 SPI for 3, 6, and 12-month time scales during the period 1961–2007 in Patnagarh

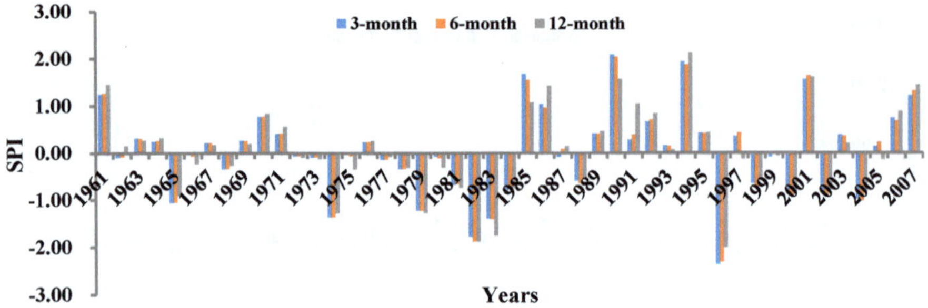

Fig. 11 SPI for 3, 6, and 12-month time scales during the period 1961–2007 in Puintala

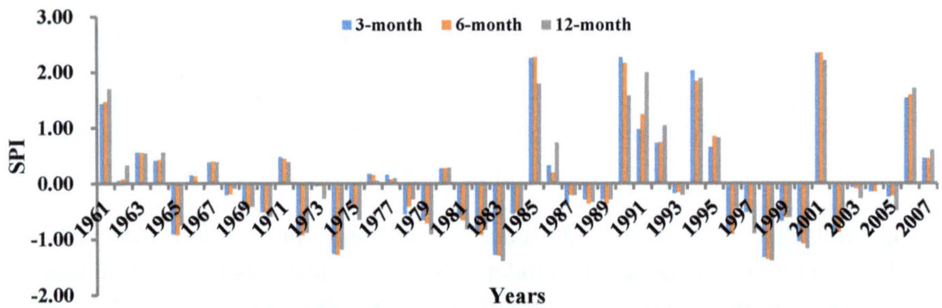

Fig. 12 SPI for 3, 6, and 12-month time scales during the period 1961–2007 in Saintala

Moderately dry condition was experienced by the block in the years 1974 and 1979 and severely dry condition in 1982. In 1983, moderately dry situation as per 3 and 6 months' SPI whereas severely dry according to 12 months' SPI. Extremely dry condition occurred in 1996. According to 6 months' SPI, moderately dry condition and near normal as per 3 and 12 months' SPI value in 2004. Very wet situation experienced by the block in 2001. It can be observed from Fig. 12, moderately dry condition occurred in Saintala block in the years 1974, 1983, 1998,

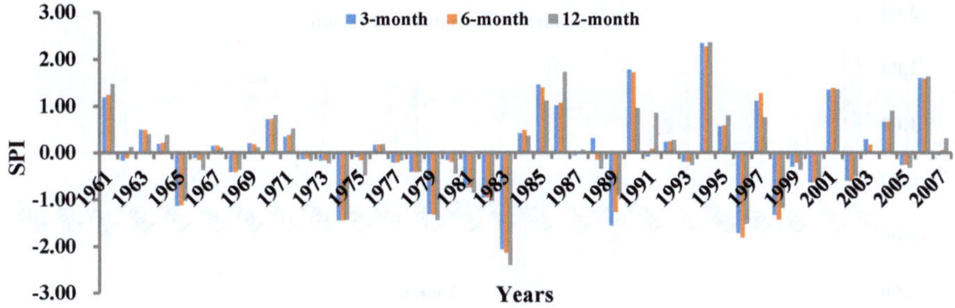

Fig. 13 SPI for 3, 6, and 12-month time scales during the period 1961–2007 in Tentulikhunti

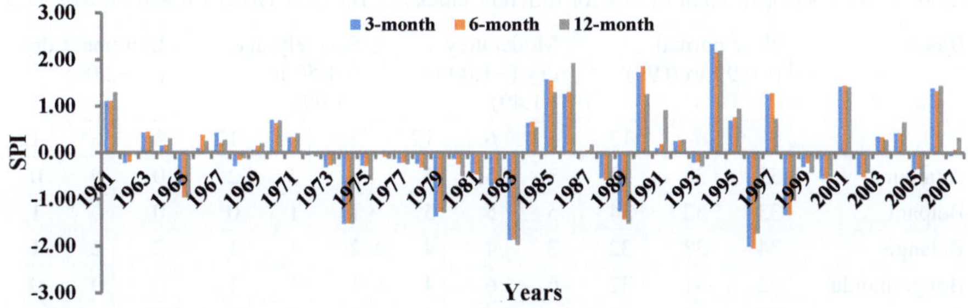

Fig. 14 SPI for 3, 6, and 12-month time scales during the period 1961–2007 in Titlagarh

and 2000. Extremely wet situation observed by the block in 2001. Moderately dry situation occurred in Tentulikhunti block in the years 1965, 1974, 1979, and 1998 and extremely dry condition was observed in 1983 (Fig. 13). In 1989, severely dry, moderately dry and near normal condition experienced as per 3, 6, and 12 months' SPI value respectively. Severely dry situation was detected in 1996. Extremely wet situation observed in 1994 whereas very wet condition faced by the block in 2006.

In the years 1974, 1979, and 1998, Titlagarh block experienced moderately dry condition whereas severely dry situation was observed in 1983 (Fig. 14). Moderately dry condition as per 3 and 6 months' SPI and severely dry according to 12 months' SPI were observed in 1989. In 1996, extremely dry situation occurred according to 3 and 6 months' SPI whereas severely dry condition occurred based on 12 months' SPI value. Extremely wet condition occurred in 1994 and moderately wet situation experienced in 2001. In the years 1974, 2002 and 2004, Tureikela block faced moderately dry situation (Fig. 15). Near normal condition as per 3 and 6 months' SPI and moderately dry based on 12 months' SPI occurred in the years 1978, 1980, and 1983. In 1979, severely dry condition was detected based on 3 and 6 months' SPI whereas moderately dry condition occurred according to 12 months' SPI value. Moderately dry condition was experienced by the block according to 3 and 6 months' SPI and near normal based on 12 months' SPI. In the years 1961, 1970, and 1994, extremely wet condition was observed in the blocks.

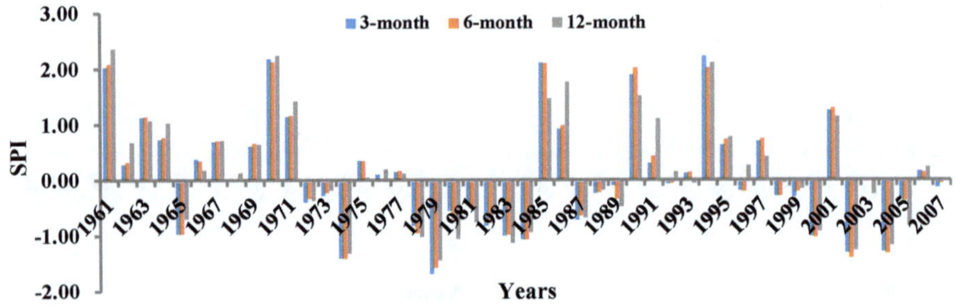

Fig. 15 SPI for 3, 6, and 12-month time scales during the period 1961–2007 in Tureikela

Table 5 Number of drought events for different blocks of Balangir District based on SPI value

Block	Near normal (−0.99 to 0.99)			Moderately dry (−1.00 to −1.49)			Severely dry (−1.50 to −1.99)			Extremely dry (≤−2.00)		
	3	6	12	3	6	12	3	6	12	3	6	12
Agalpur	35	33	33	3	5	5	3	3	2	0	0	0
Belpara	33	33	34	5	5	5	1	1	0	0	0	1
Balangir	34	33	32	3	4	4	2	1	1	2	2	1
Bongomunda	32	31	32	6	6	4	1	1	1	0	0	0
Deogaon	35	35	35	3	3	1	2	2	2	0	0	0
Khaprakhol	34	34	34	5	5	4	1	1	2	0	0	0
Loisinga	35	35	32	3	3	4	1	1	1	1	1	1
Muribahal	34	36	33	5	5	6	1	1	1	0	0	0
Patnagarh	35	35	32	3	3	4	1	2	2	1	0	1
Puintala	35	35	35	4	5	2	1	1	2	1	1	1
Saintala	37	36	35	4	4	4	0	0	0	0	0	0
Tentulikhunti	32	32	34	4	5	5	2	1	1	1	1	1
Titlagarh	33	33	34	3	3	2	2	2	4	1	1	0
Tureikela	33	33	29	5	5	7	1	1	0	0	0	0

The pattern of drought events was not similar for 14 blocks of the Balangir district based on the SPI value for 3, 6, and 12 months' time scale due to spatial and temporal variation of rainfall. Therefore different management strategy has to be adopted in order to tackle the different types of drought situation experienced by different blocks. The blocks, which were facing frequent dry situations, should have provision for storage of water in order to overcome the agricultural as well as hydrological drought. For the blocks experiencing frequent agricultural droughts, cropping pattern has to be changed. Small duration crops or water-resistant crop varieties may be planted for these types of blocks.

Different types of drought events were experienced by the 14 blocks of Balangir district based on 3, 6, and 12 months' SPI value as presented in Table 5. Extremely drought events were experienced by Balangir, Loisinga, Patnagarh, Puintala, Tentulikhunti, and Titlagarh as per 3 months' SPI scale, therefore these blocks faced extreme agricultural drought conditions. Extreme hydrological drought conditions were faced by Belpara, Balangir, Loisinga, Patnagarh, Puintala, and Tentulikhunti blocks according to 12 months' SPI value. Highest number of severely drought conditions were detected in Titlagarh block based on 12 months' SPI value whereas maximum number of moderately drought situations faced by Tureikela block. Agalpur, Bongomunda, Deogaon, Khaprakhol, Muribahal, Saintala, and Tureikela blocks did not experience extreme dry events based on 3, 6, and 12 months' SPI value. Both Titlagarh and Agalpur block experienced highest number (i.e., 8) of severely drought events. Seventeen moderately drought events occurred altogether according to 3, 6, and 12 months' SPI in Tureikela.

Conclusion

In this study, the capabilities of different drought indicators, including SPI, DI, MI, and HTC, in detecting the severity of drought across 14 blocks of Balangir district using precipitation and temperature data were evaluated. SPI index was found to be suitable among all the drought indices, as through this index, the severity of drought can be easily identified. Belpara, Saintala, and Tureikela blocks did not face any severe drought events during 1961–2007. Maximum number of marginal drought events (i.e., 23) were experienced by Deogaon and Titlagarh blocks whereas Muribahal faced maximum number of moderate drought events (i.e., 11). Deogaon, Loisinga, and Titlagarh are chronically drought-prone areas. Marginal droughts return period varies from 2–4 years, 4–16 years for moderate droughts and 24–48 years in case of severe droughts for all the blocks of Balangir district. Since precipitation is fundamental for rain-fed crops in these drought-prone regions, improvements in drought monitoring and early warning will improve the capacity to detect, anticipate, and mitigate famine. Agalpur, Bongomunda, Deogaon, Khaprakhol, Muribahal, Saintala, and Tureikela blocks did not experience any extreme dry events based on 3, 6, and 12 months' SPI value. Balangir, Loisinga, Patnagarh, Puintala, Tentulikhunti, and Titlagarh blocks faced extreme agricultural drought conditions. Extreme hydrological drought conditions were faced by Belpara, Balangir, Loisinga, Patnagarh, Puintala, and Tentulikhunti blocks. Early warning systems to predict possible likelihood of droughts by developing the drought indices will help people to perceive the possible occurrence of droughts and prepare themselves to adapt to drought conditions by adopting alternative cropping patterns to reduce the risk. It will also help the policy makers to take early measures to review the situation such as the progress of agricultural operations; condition of the standing crops; adequacy of stock of food grains available in the district and adequacy of water for kharif through Canals/Minor Irrigation Project/Lift Irrigation points, etc. The analysis of these drought indices led to several useful and practicable inferences for better understanding the drought attributes of the study area. This study will help in planning drought preparedness and its mitigation in a realistic and appropriate manner.

References

Alley WM (1984) The palmer drought severity index: limitation and assumptions. J Climate Appl Meteorol 23:1100–1109

Appa Rao G (1991) Drought and south west monsoon. Training course, Meteorology 3rd WMO Asian/African Monsoon Workshop, Pune, India

Benson C, Clay EJ (1998) The impact of drought on Sub-Saharan African economies: a preliminary examination. World Bank Technical Paper 401, Washington, D.C.: World Bank, p 80

Benson C, Clay EJ (2000) Developing countries and the economic impacts of natural disasters. In: Kreimer A, Arnold M (eds.) Managing disaster risk in emerging economies. Disaster Risk Management Series 2, Washington, D.C.: World Bank, pp 11–21

Budyko MI (1958) The heat balance of the earth's surface. Translated by Nina A. Stepanova, US Department of Commerce, Washington, D.C., 259 p

Gibbs WJ, Maher JV (1967) Rainfall deciles as drought indicators. In: Bureau of meteorology bulletin, no. 48, Commonwealth of Australia, Melbourne

Gore PG, Sinha Ray KC (2002) Drought and aridity over districts of Gujarat. J Agrometeorol 4:75–85

Hagman G (1984) Prevention better than cure: report on human and natural disasters in the third world. Swedish Red Cross, Stockholm

Hayes M, Wilhite DA, Svoboda M, Vanyarkho O (1999) Monitoring the 1996 drought using the standardized precipitation index. Bull Am Meteor Soc 80:429–438

Heim RR Jr (2000) Drought indices: a review. In: Wilhite (ed) Drought: a global assessment, hazards disasters ser., vol I, pp 159–167. Routledge, New York

Hounam CE, Burgos JJ, Kalik MS, Palmer WC, Rodda J (1975) Drought and agriculture. Report of the CAgM working group on assessment of drought. Technical Note No. 138, WMO Publication No. 392, Geneva, Switzerland, 127 pp

Ji L, Peters AJ (2003) Assessing vegetation response to drought in the northern great plains using vegetation and drought indices. Remote Sens Environ 87:85–98

Lloyd-Hughes B, Saunders MA (2002) A drought climatology for Europe. Int J Climatol 22 (13):1571–1592

McKee TB, Doesken NJ, Kleist J (1993) The relationship of drought frequency and duration to time scales. In: Proceedings of the 8th conference on applied climatology, AMS, Boston, MA, pp 179–184

McKee TB, Doesken NJ, Kleist J (1995) Drought monitoring with multiple time scales. In: Proceedings of the 9th conference on applied climatology, AMS, Boston, MA, pp 233–236

Mishra AK, Singh VP (2010) A review of drought concepts. J Hydrol 391:202–216

Narain P, Rao AS, Abrol IP (2000) Managing droughts and Desertification in India—lessons from the past and future strategies. Central Arid zone research Institute, Jodhpur and Center for Advancement of Sustainable Agriculture, New Delhi

Palmer WC (1965) Meteorological drought. Research Paper No. 45, US Department of Commerce Weather Bureau, Washington, DC

Pandey RP, Dash BB, Mishra SK, Singh R (2008) Study of indices for drought characterization in KBK districts in Orissa (India). Hydrol Process 22(12):1895–1907

Patel NR, Chopra P, Dadhwal VK (2007) Analyzing spatial patterns of meteorological drought using standardized precipitation index. Meteorol Appl 14:329–336

Rao AS (1997) Impact of droughts on Indian arid ecosystem. Desertification control in the arid ecosystem of India for sustainable development. Agro Botanical Publishers, Bikaner, pp 120–130
Rathore MS (2005) State level analysis of drought policies and impacts in Rajasthan, India. Colombo, Sri Lanka: IWMI, 40 p (Working paper 93: Drought Series Paper No. 6)
Sahoo BB (2015) Critical appraisal of different drought indices of drought prediction and their application in KBK Districts of Odisha. Master dissertation, National Institute of Technology Rourkela, Odisha, India
Selyaninov GT (1930) Methods of agricultural climatology (in Russian). In Agricultural Meteorology, No. 22 L
Sinha Ray KC, Shewale MP (2000) Probability of occurrence of drought in various sub-divisions of India. Mausam 52(3):541–546
Subbiah AR (1993) Indian drought management: from vulnerability to resilience. In: White DW (ed) Drought assessment, management and planning: theory and case studies. Kluwer Academic Publishers, Boston, pp 157–179
Subramanya K (2013) Engineering hydrology. McGraw Hill Education (India) Private Limited, New Delhi, India, pp 534
Wang L, Qu JJ (2009) Satellite remote sensing applications for surface soil moisture monitoring: a review. Front Earth Sci China 3:237–247
Wilhite DA (2000) Drought as a natural hazard: concepts and definitions. In: Wilhite DA (ed) Drought: a global assessment, vol 1. Routledge, London, pp 3–18
Wilhite DA, Vanyarkho O (2000) Drought: pervasive impacts of a creeping phenomenon. In: Wilhite DA (ed) Drought: a global assessment, vol 1. Routledge Publishers, London, England, pp 245–255
Zargar A, Sadiq R, Naser B, Khan F (2011) A review of drought indices. Environ Rev 19:333–349
Zhang J, Xu Y, Yao F, Wang P, Guo W, Li L, Yang L (2010) Advances in estimation methods of vegetation water content based on optical remote sensing techniques. Sci China-Technol Sci 53:1159–1167

Impact of HFC Fire Extinguishing Clean Agents on Climate Change and Its System Design Requirements for Fire Hazards in India—A Brief Study

R.S. Chimote

Abstract This paper discusses the impact on climate change of hydrofluorocarbons (HFCs) in fire suppression applications. Alternatives and substitutes for HFCs, perfluorocarbons (PFCs), and ozone depleting substances (ODSs) have recently been extensively evaluated. NFPA 2001 defines a clean fire extinguishing agent as an electrically non-conducting, volatile, or gaseous fire suppressant that does not leave a residue upon evaporation. A clean agent must have no known effect on the ozone layer and also, no effect on any human survival within an enclosure protected by a clean agent, and in normally occupied areas must be used in a concentration that is less than "no observed adverse effect level (NOAEL)". NOAEL is a measure of clean agent toxicity to humans under test conditions. The HFCs that are projected for large volume use have global warming potentials (GWPs) lower than the replacing ODSs. GWPs of HFCs replacing ODSs ranges from 120 to 12,000 as per the year 2000 data of Intergovernmental Panel on Climate Change (IPCC). HFC-23 with a GWP of 12,000 is used as a replacement for ODSs to a very limited extent. However, there are relatively large emissions of HFC-23 from the HCFC-22 manufacturing process. However, the majority of HFCs have GWPs much lower than that of HFC-23. NFPA 2001 standard demonstrates the fact that the GWP value considered by itself does not provide an indication of the impact of fire extinguishing clean agent on climate change. Further, the paper briefly describes the clean agent fire extinguishing system design considerations to extinguish fires either by flame extinguishment or by inerting in accordance with the changing characteristics of fire hazard scenarios in building and industrial occupancies. An important finding of this brief study is that the value of 0.4858 kg/m^3 is a total flooding factor for HFC-227ea fire extinguishing agent representing the quantity of halocarbon clean agent required to achieve a selected design fire extinguishing concentration of 6% at a specified ambient temperature of 21 °C. It is further important to understand that the impact of a fire extinguishing clean agent on

R.S. Chimote (✉)
Fire Research Laboratory, CSIR-Central Building Research Institute,
Roorkee 247667, India
e-mail: ratnakar1961@yahoo.com

V.P. Singh et al. (eds.), *Climate Change Impacts*, Water Science
and Technology Library 82, https://doi.org/10.1007/978-981-10-5714-4_22

climate change is a function of both the GWP of the gas and the amount of gas emitted. For example, carbon dioxide has one of the lowest GWP values of all greenhouse gas emissions (GWP = 1), yet emissions of CO_2 account for approximately 85% of the impact of all greenhouse gas (GHG) emissions. The characteristics of fire hazard scenarios with respect to anticipated fires have been continuously changing in India due to emerging trends in the up gradation/modern furnishing and interior design considerations/requirements in almost all the urban, semi-urban, and rural occupancies. The data from IPCC and Asia Pacific Fire Magazine, October 25, 2011 showed that if nothing changes, the HFC emissions are likely to be equivalent to between 9 and 19% of global greenhouse gas emissions by 2050, which indicates that the impact of HFC fire extinguishing clean agents on climate change is minuscule. As a result, HFCs are expected to remain viable, sustainable, and environmentally acceptable replacements for Halon 1301, which was phased out due to ozone depletion potential problems under Montreal and other protocols.

Keywords Climate change · Hydro fluorocarbons (HFCs) · Fire extinguishing agents · Fire hazards · System design

Introduction

The results of a study of the effect of chlorofluorocarbons (CFCs) on the ozone layer won the Nobel Prize for two chemists (1–38) at the University of California Irvine, Frank Sherwood Rowland and Mario Mocina which resulted in a landmark international agreement, the Montreal Protocol, signed by the United States and 24 other countries in 1987, with significant amendments in 1990 and 1992. Alternatives and substitutes for HFCs, perfluorocarbons (PFCs), and ozone depleting substances (ODSs) have recently been extensively evaluated. The Montreal Protocol Technology and Economic Assessment Panel (TEAP) and its technical committees published a comprehensive assessment. Furthermore, reports were published within the framework of the joint IPCC/TEAP workshop and the second non-CO_2 greenhouse gases conference.

Kofi Annan, former Secretary General of the United Nations, said "perhaps the single most successful international agreement to date was the Montreal Protocol of 1987." The agreement was intended to sharply restrict the production of chemicals that had been identified as contributing to depletion of the stratospheric ozone layer. The ozone layer is a protective layer of our stratosphere that helps to filter the ultraviolet rays of the sun before they reach Earth. In the absence of the ozone layer, the incidence of skin cancer and melanoma increase. An ozone molecule consists of three oxygen atoms (O_3). Freon, released from air conditioners, and halogenated extinguishing agents rise to the stratosphere. Bromine and chlorine molecules from these agents break up the O_3 molecules and attach themselves to one of the free oxygen molecules. These gases, therefore, were included in the list of ozone depleting agents.

In advance of the Montreal Protocol, the Vienna Convention for the Protection of the Ozone Layer provided the framework for negotiations in 1985 (1–38). Immediately subsequent to the initial signing of the Montreal Protocol, evidence continued to mount that the ozone layer was continuing to shrink at a frightening rate. Numerous additional countries signed the Montreal Protocol, and the target date for ceasing production of halogenated hydrocarbons was advanced to January 1994. At present, 191 nations have signed the Montreal Protocol, making it one of the planet's most successful international agreements.

The cessation of halon production rapidly rendered existing halon systems (1–38) which are to be gradually phased out of the fire protection industry, has placed owners of halon systems and the companies that insure the hazards protected by halon systems, in an extremely uncomfortable position. Although the Montreal Protocol did not call for removing all existing halon systems, it prohibited the manufacture of new halon—making it impractical to legally purchase new halons. Owners and insurers of halon systems were faced with the prospect of a total loss of fire protection pursuant to an accidental or purposeful halon system discharge.

Figure 1 shows (1–9) (a) Arctic map from world atlas (b) Artic Ozone Hole, yellow area within red circle, could expose millions of people, wildlife and plants to

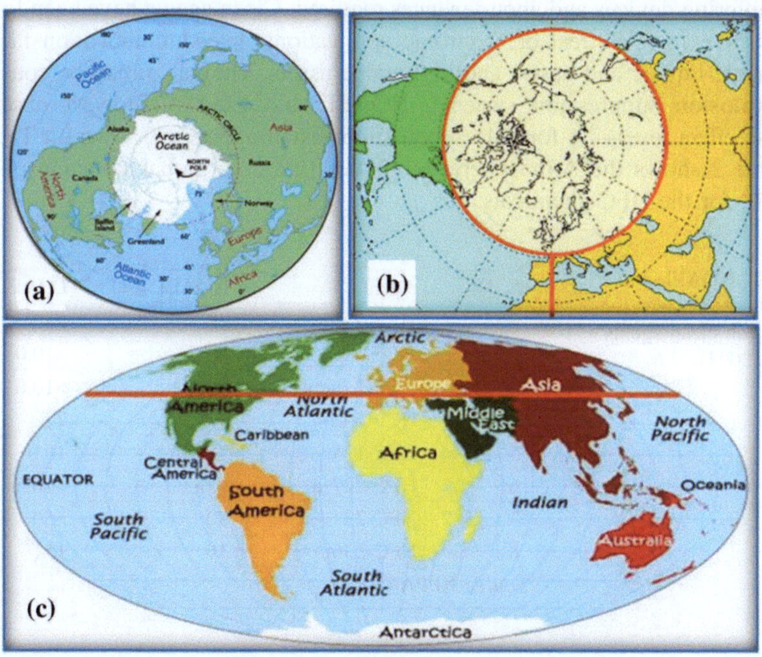

Fig. 1 It shows **a** arctic map from world atlas **b** artic ozone hole, *yellow area* within *red circle*, could expose millions of people, wildlife and plants to dangerous UV ray levels **c** the area that will be affected by formation of arctic ozone hole above the *redline*

dangerous UV ray levels; and (c) the area that will be affected by formation of Arctic Ozone Hole above the redline. To protect against this eventuality, many system owners opted to replace their halon systems with either a substitute gaseous system replacement or a water-based system replacement. Although introductory background of halon's replacements is covered in this paper, the primary focus is on clean agents and halon-alternative fire extinguishing agents.

Clean Fire Extinguishing Agents

NFPA 2001 (3–7) defines a clean agent as an electrically non-conducting, volatile, or gaseous fire extinguishant that does not leave a residue upon evaporation. For an agent to qualify as a clean agent, it must have no known effect on the ozone layer, it must have no effect on human survival within an enclosure protected by a clean agent, and in normally occupied areas must be used in a concentration that is less than the NOAEL—an abbreviation for "no observed adverse effect level." NOAEL is a measure of clean agent toxicity to humans, under test conditions. At present time, no drop-in agent is available that would allow Halon 1301 to be removed and an equivalent amount of replacement agent inserted. Systems with gaseous halon replacement agents require that more gas than halon be stored on a volumetric basis, with differing devices and appurtenances required. Clean agents have been found to be effective for electrical or electronic applications, telecommunication facilities, flammable liquids and gases, and high-value assets. They also may be considered for explosion suppression systems. NOAEL and LOAEL—lowest observable adverse effect level—% for halocarbon suppression agents as shown on Table 1.

Table 2 shows the Atmospheric properties [lifetime, global warming potential (GWP)] for the HFC chemicals.

Table 1 NOAEL and LOAEL percentages for halocarbon clean agents (NFPA 2004)

S. No.	Agent	NOAEL (%)	LOAEL (%)
1	FC-3-1-10	40	>40
2	FK-5-1-12	10.0	>10.0
3	HCFC Blend A	10.0	>10.0
4	HCFC-124	1.0	2.5
5	HFC-125	7.5	10.0
6	HFC-227ea	9.0	>10.5
7	HFC-23	30	>50
8	HFC-236fa	10	15

Source NFPA (2004)

Table 2 Atmospheric properties [lifetime, global warming potential (GWP)] for the HFC chemicals (1–38)

S. No.	Sub-sector	Chemical formula	Lifetime (year) (IPCC 1996)	GWP (100 year) (IPCC 1996)	Lifetime (year) (IPCC 2000)	GWP (100 year) (IPCC 2000)
1	HFC-23	CHF_3	264	11,700	260	12,000
2	HFC-32	CH_2F_2	5.6	650	5.0	550
3	HFC-125	CHF_2CF_3	32.6	2800	29	3400
4	HFC-134a	CH_2FCF_3	14.6	1300	13.8	1300
5	HFC-143a	CH_3CF_3	48.3	3800	52	4300
6	HFC-152a	–	1.5	140	1.4	120
7	HFC-227ea	CF_3CHFCF_3	36.5	2900	33	3500
8	HFC-245fa[a]	$CF_3CH_2CHF_2$	–	–	7.2	950
9	HFC-365mfc[a]	$CF_3CH_2CF_2CH_3$	–	–	9.9	890
10	HFC-4310mee	$CF_3CHFCHFCF_2CF_3$	17.1	1300	15	1500
11	R-404A					
	44%	HFC-125		3260		
	4%	HFC-134a				
	52%	HFC143a				
12	R-407C					
	23%	HFC-32		1525		
	25%	HFC-125				
	52%	HFC-134a				
13	R-410A					
	50%	HFC-32		1725		
	50%	HFC-125				
14	R-507					
	50%	HFC-125		3300		
	50%	HFC-143a				

[a]Atoms of middle Carbon-Cl_2

Clean Agent Classification

The two basic classifications of clean agents are halocarbon agents and inert gas agents. Agents addressed by NFPA 2001 are listed on Table 3 (1–38). Halocarbon agents consist of hydrofluorocarbons (HFCs), hydro-chlorofluorocarbons (HCFCs), and perfluorocarbons (PFCs) and are given numerical descriptions as shown in Table 3 in accordance with ANSI (American National Standards Institute) and ASHRAE (American Society of Heating, Refrigerating and Air Conditioning Engineers) standards. Halocarbons are stored as a liquid and distributed to the hazard as a gas, and extinguish fires by chemical and physical mechanisms, as opposed to oxygen deprivation. The extinguishment mechanism is breaking the combustion chain. Inert gas agents contain one or more non-reactive gases, such as helium, neon, and argon, mixed with nitrogen or carbon dioxide. They extinguish

Table 3 Clean fire extinguishing agents (1–38)

1	FC-3-1-10	Perfluorobutane	C_4F_{10}
2	FK-5-1-12	Dodecafluoro-2-methylpentan-3-one	$CF_2CF_2C(O)CF(CF_3)_2$
3	HCFC blend A	Dichlorotrifluoroethane HCFC-123 (4.75%) Chlorodifluoromethane HCFC-22 (82%) Chlorotetrafluoroethane HCFC-124 (9.5%) Isopropenyl-1-methylcyclohexene (3.75%)	$CHCl_2CF_3$ $CHClF_2$ $CHClFCF_3$
4	HCFC-124	Chlorotetrafluoroethane	$CHClFCF_3$
5	HFC-125	Pentafluoroethane	CHF_2CF_3
6	HFC-227ea	Heptafluoropropane	CF_3CHFCF_3
7	HFC-23	Trifluoromethane	CHF_3
8	HFC-236fa	Hexafluoropropane	$CF_3CH_2CF_3$
9	FIC-1311	Trifluoroiodide	CF_3I
10	IG-01	Argon	Ar
11	IG-100	Nitrogen	N_2
12	IG-541	Nitrogen (52%)	N_2
		Argon (40%)	Ar
		Carbon dioxide (8%)	CO_2
13	IG-55	Nitrogen (50%)	N_2
		Argon (50%)	Ar

Notes Other agents could be added via the NFPA process in future editions or amendments of the standard. Composition of inert gas agents are given in vol.%. Composition of HCFC blend A is given in wt%. The full analogous ASHRAE nomenclature for FK-5-1-12 is FK-5-1-12 mmy2. *Source* NFPA 2001 (2004)

fires by lowering the oxygen concentration within a room from normal condition of 21% to a level below 15%—usually 12–13%, less than the level required to sustain combustion for most combustibles. Inert gas agents are approximately the same density as air and, therefore, mix better and display less settling to the floor than other gaseous agents such as carbon dioxide.

Discussion

In 1986, less than half of total ODS use was in insulating foams, fire protection, refrigeration, air conditioning, and heat pumps, with more than half as aerosol product propellants, non-insulating foam, solvent, and specialized applications (1–38). Alternatives and substitutes for HFCs, perfluorocarbons (PFCs), and ozone depleting substances (ODSs) have recently been extensively evaluated. The Montreal Protocol

Technology and Economic Assessment Panel (TEAP) and its technical committees published a comprehensive assessment (UNEP 1999b). Furthermore, reports were published within the framework of the joint IPCC/TEAP workshop and the second non-CO_2 greenhouse gases conference.

The HFCs that are projected for large volume use have global warming potentials (GWPs) which are generally lower than those of the ODSs they replace. The GWP of HFCs replacing ODSs range from 140 to 11,700. HFC-23 with a GWP of 11,700 is used as a replacement for ODSs to only a very minor extent. However, there are relatively large emissions of HFC-23 from the HCFC-22 manufacturing process. The majority of HFCs have GWPs much lower than that of HFC-23. PFCs have GWPs that are generally higher than those of the ODSs they replace, ranging from 7000 to 9200 as shown in Table 2 (1–38).

However, by 1997 (1–9), the global consumption of fluorocarbons (CFCs, HCFCs, and HFCs) had decreased by about 50% as solvent, aerosol product, and non-insulating foam applications switched to alternatives other than fluorocarbons. Refrigeration, air conditioning, and insulating foam accounted for about 85% of the remaining total fluorocarbon use. Eighty percent of projected chlorofluorocarbon demand was avoided by reducing emissions, redesign, and use of non-fluorocarbon technologies.

As CFCs, halons, and HCFCs are phased out globally, the quantities of fluorocarbons are expected to continue to decline in the short term, but are expected to grow in the longer term. Most HFCs are used for energy-consuming applications such as refrigeration, air conditioning and heat pumps, and building and appliance insulation. Life cycle climate performance (LCCP) analysis is being used to estimate the net contribution to climate change. The LCCP is a very system specific parameter that can be used to make relative rankings. However, LCCP analysis involves regional differences—including different fuel sources—and the related equipment operating conditions; the results can therefore not be generalized in order to make globally valid comparisons.

The energy efficiency of equipment and products can be expressed in at least three ways: theoretical maximum efficiency, maximum efficiency achievable with current technology, and actual efficiency for commercial scale production (often expressed as a range of values). Unlike anthropogenic greenhouse gases emitted as an immediate consequence of the burning of fossil fuels to generate energy, most HFCs and PFCs are contained within equipment or products for periods ranging from a few months (e.g., in aerosol propellants) to years (e.g., in refrigeration equipment) to decades (e.g., in insulating foams). Thus, emissions significantly lag consumption and, because HFC systems are relatively new, emissions will continue to grow after 2010. Systems optimized for a new refrigerant have been compared to sub-optimum systems with other refrigerants.

Furthermore, appliance sizes and features that influence energy performance vary between studies and test conditions, and methodologies are often significantly different. These factors have led to a wide range of energy efficiency claims in technical reports and commercial publications. Ultimately, the performance and cost effectiveness of specific products from commercial scale production must be directly compared. Furthermore, costs reported might not always be comparable because of differing estimation methods, including estimates based on both consumer and producer costs.

Future global HFC and PFC consumption and/or emissions as substitutes for ODSs have been estimated by IPCC (1995). Projected consumption data for 2000 and 2010 are primarily based on UNEP reports (1–9).

Considering that emissions lag consumption by many years, none of the scenarios have considered the implications of new uses of HFCs or PFCs other than as substitutes for ODSs (1–9).

With regard to the impact of HFCs in fire suppression on climate change, the International Fire Protection Magazine article "Climate Change and the HFC-Based Clean Extinguishing Agents" referenced in the original post for a factual discussion is worth reading (8–9). Based on US EPA data, the contribution of HFCs in fire suppression applications to climate change represents approximately 0.01% of the contribution of all greenhouse gases. With regard to regulations, HFCs in fire suppression applications are being treated differently than HFCs employed in other applications, and as a result it is important when encountered with an assertion of impending legislative or regulatory action related to HFCs in fire suppression applications to always request two items: (i) a copy of the legislation and (ii) the location of the text that is specifically related to HFCs in fire suppression applications (1–9).

FM-200 (HFC-227ea) and Novec 1230 are fire extinguishing agents characterized by zero ODP and whose use in fire suppression applications results in a negligible contribution to climate change (global warming). There are three major differences between Novec 1230 and all of the other clean agents, including FM-200. As regard to Chemical reactivity (8–9), unlike the HFC and inert gas clean agents, which are characterized by very low chemical reactivity, Novec 1230 is characterized by high chemical reactivity. The HFC and inert gas clean agents are all unreactive with water, alcohols, amines, and solvents. Novec 1230, on the other hand, is characterized by high chemical reactivity.

For example, Novec 1230 design manuals indicate (8–9) the following: (i) Contact of Novec 1230 with water or solvents either polar or hydrocarbon could render Novec 1230 fluid ineffective, (ii) the transfer of Novec 1230 requires the use of a drier because humid air may cause the agent to convert to acid. It is reported (8–9) that Novec 1230 is chemically reactive with nucleophiles such as alcohols. Novec 1230 is also chemically reactive with other fire extinguishing agents (8–9), e.g., it has been reported that Novec 1230 undergoes reaction with sodium bicarbonate. The reaction of Novec 1230 with water produces HFC-227ea and Perfluoropropionic acid, a strong, corrosive organic acid. Due to its high reactivity,

Novec 1230 is the only clean agent that is classified as a volatile organic compound (VOC) (8–9).

As regard to the interaction in the body (8–9), unlike the HFC and inert gas agents, Novec 1230 undergoes reaction in the lungs. Novec 1230 reacts to form HFC-227ea and Perfluoropropionic acid when it crosses the lung-air interface. In contrast, FM-200 does not react to form potentially hazardous products; the toxicity of FM-200 is so low that it is approved for use as a propellant in metered dose inhalers (MDIs), where it is employed to propel a medicament down the throat of the patient into his/her lungs. As regard to the physical state, unlike the HFC and inert gas clean agents, which are all gaseous at room temperature, Novec 1230 is a high boiling liquid which increases the possibility of a liquid discharge with Novec 1230 compared to the other clean agents and also affects its performance. For example, recent studies within the aviation industry have indicated that Novec 1230 is ineffective in several civil aviation applications (8–9).

Human Safety Concerns with Clean Fire Extinguishing Agents (1–9)

NFPA 2001 does not recommend exposure to halon-carbon clean agents for more than 300 s with less exposure in higher concentrations, as shown on Tables 4, 5, 6 and 7. Designers of fire protection systems need to specially exercise care in the

Table 4 Time for safe egress for halocarbon clean agents (NFPA 2001 2004)

Time for safe human exposure at stated concentrations for HFC-125			
S. No.	HFC-125 concentration		Maximum permitted human exposure time (min)
	%v/v	ppm	
1	7.5	75,000	5.00
2	8.0	80,000	5.00
3	8.5	85,000	5.00
4	9.0	90,000	5.00
5	9.5	95,000	5.00
6	10.0	100,000	5.00
7	10.5	105,000	5.00
8	11.0	110,000	5.00
9	11.5	115,000	5.00
10	12.0	120,000	1.67
11	12.5	125,000	0.59
12	13.0	130,000	0.54
13	13.5	135,000	0.49

Notes Data derived from the EPA-approved and peer-reviewed physiologically based pharmacokinetic (PBPK) model or its equivalent. Based on LOAEL of 10.0% in dogs

Table 5 Time for safe human exposure at stated concentrations for HFC-236fa (NFPA 2001 2004)

Time for safe human exposure at stated concentrations for HFC-236fa			
S. No.	HFC-236fa concentration		Maximum permitted human exposure time (min)
	%v/v	ppm	
1	10.0	100,000	5.00
2	10.5	105,000	5.00
3	11.0	110,000	5.00
4	11.5	115,000	5.00
5	12.0	120,000	5.00
6	12.5	125,000	5.00
7	13.0	130,000	1.65
8	13.5	135,000	0.92
9	14.0	140,000	0.79
10	14.5	145,000	0.64
11	15.0	150,000	0.49

Note Data derived from EPA-approved/peer-reviewed PBPK model on LOAEL of 15% in dogs

Table 6 Time for safe human exposure at concentrations for HFC-227ea (NFPA 2001 2004)

Time for safe human exposure at stated concentrations for HFC-227ea			
S. No.	HFC-227ea concentration		Maximum permitted human exposure time (min)
	%v/v	ppm	
1	9.0	90,000	5.00
2	9.5	95,000	5.00
3	10.0	100,000	5.00
4	10.5	105,000	5.00
5	11.0	110,000	1.13
6	11.5	115,000	0.60
7	12.0	120,000	0.49

Note Data derived from EPA-approved/peer-reviewed PBPK model on LOAEL of 10.5% in dogs

Table 7 Time for safe human exposure at stated concentrations for FIC-1311 (NFPA 2001 2004)

Time for safe human exposure at stated concentrations for FIC-1311			
S. No.	FIC-1311 concentration		Maximum permitted human exposure time (min)
	%v/v	ppm	
1	0.20	2000	5.00
2	0.25	2500	5.00
3	0.30	3000	5.00
4	0.35	3500	4.30
5	0.40	4000	0.85
6	0.45	4500	0.49
7	0.50	5000	0.35

Note Data derived from EPA-approved/peer-reviewed PBPK model on LOAEL of 0.4% in dogs design of clean agent systems for enclosures where human exposure to the agent is possible.

Of particular concern is human exposure to the decomposition byproducts formed by breakdown of the extinguishant when exposed to high temperatures or an open flame. For example, halocarbon agents containing fluorine have the potential to form toxic hydrogen fluoride. Inert gas agents do not create decomposition products, but care must be taken to avoid high application concentrations. Table 8 lists the toxicity of clean agents. Inert gas agents contain about 8% carbon dioxide, but the CO_2 is not a concern at normal inert gas concentrations. Care must be taken to avoid over-design which could result in excessive inert gas concentrations and reduce oxygen concentrations below 10%. NFPA 2001 (2004) prohibits the application of halocarbon agents into occupied rooms at concentrations greater than 24% and requires that the NOAEL limits listed in Table 1 not be exceeded for any clean agent.

Table 8 provides information for designers relative to NOAEL and LOAEL % ages for halocarbon clear agents and time for safe exposure for HFC-125, HFC-236fa, HFC-277ea, and FIC-1311.

The ALC is the approximate lethal concentration. The cardiac sensitization levels are based on the observance or non-observance of serious heart arrhythmias in a dog. The usual protocol is a 5-min exposure followed by a challenge with epinephrine. High concentration values are determined with addition of oxygen to prevent asphyxiation.

These tables enable clean agent system designers to consider methodologies for keeping human exposure to a minimum. Methods to protect personnel exposed to clean agents include

- Ensuring that exits are well situated, well marked, and well lighted, adequate number and width to allow rapid egress of all occupants, and are readily accessible with clear and unobstructed aisles or passageways to the exits.
- Consider the provision of extra egress doors; specifying that doors are required to swing in the direction of egress travel and to reclose automatically.

Table 8 Toxicity information for halocarbon clean agents (NFPA 2001 2004)

S. No.	Agent	LC_{50} or ALC (%)	NOAEL (%)	LOAEL (%)
1	FC-3-1-10	>80	40	>40
2	FIC-1311	>12.8	0.2	0.4
3	FK-5-1-12	>10.0	10	>10.0
4	HCFC Blend A	64	10	>10.0
5	HCFC-124	23–29	1	2.5
6	HFC-125	>70	7.5	10
7	HFC-227ea	>80	9	10.5
8	HFC-23	>65	50	>50
9	HFC-236fa	>18.9	10	15

Notes LC_{50} is the concentration lethal to 50% of a rat population during a 4-h exposure

- Providing adequate alarm notification before clean agent discharge.
- Providing training of personnel to ensure proper identification and response to an alarm; and providing continuous alarms during discharge and agent containment.
- Providing alarms, locks, signs and methods to prevent reentry to a room during agent containment.
- Specifying placement of breathing apparatus and room ventilation requirements.
- Specifying a plan for rescue of anyone who may become trapped within the room or otherwise overexposed to the suppressing agent and its combustion byproducts.

Consideration also must be given to the possibility of confusion or disorientation of occupants during discharge. Clean agent discharge may be noisy, and the force of discharge may create reduced visibility. It may produce a swirl of dislodged papers or other loose materials. The low temperature of discharging gas may be a shock to a personnel. Training of responsible personnel is a necessity, and the specification of clean agent systems should not be considered for "at risk" persons, such as in public and patient areas in hospitals and nursing homes.

Clean Agent System Design Consideration for Urban/Semi-urban Occupancies in India

The clean agent system design considerations for the halon-alternative fire extinguishing agents that must be planned and designed for urban and semi-urban occupancies in India such that the fire extinguishant containers should not be in the hazardous area, and it shall suitably be in a protected location as close as possible to the hazard. Piping and fittings must be of a pressure rating commensurate with expected system pressures, and must be corrosion-resistant. Piping and fittings must be metallic, and the fittings cannot be of cast iron and it may be of welded, brazed, or malleable iron. Fire suppression and detection shall be selected/designed to be appropriate for the anticipated class of fires and emerging fire load density pattern/layout with appropriately designed discharge flow rate, particle/droplet size distribution with respect to fire extinguishing efficiency parameters.

An existing detection system may possibly be reused when designing a clean agent system, provided that the characteristics of the anticipated fire have not been changed because local application has not been found to be effective by the committee responsible for the technical content of NFPA 2001 (2004), clean agent systems are to be specified and designed for total flooding of enclosures. The enclosure protected by a clean agent must be rendered amenable to the application and retention of agent by

- Arranging for the automatic closing of doors
- Sealing opening and cracks around doors and windows

- Clipping down and restraining ceiling tiles and sealing them where necessary
- Shutting down supply and return air to a room with dampers in ducts to prevent loss of clean agent
- Attempting to limit loss of clean agent through floor drains, trenches, pipe penetrations through walls, and other wall and floor penetrations
- Shutting down gas or other flammable supplies
- Shutting down electrical power to energized electrical components where necessary

Systems must be designed such that the agent containers are not in the hazard area, and are in a protected location as close as possible to the hazard. Piping and fittings must be of a pressure rating commensurate with expected system pressures, and must be corrosion-resistant. Piping and fittings must be metallic, and fittings cannot be cast iron. They can be welded, brazed, or malleable iron. Detection shall be selected to be appropriate for the anticipated fire, as discussed. An existing detection system possibly may be reused when designing a clean agent system for a room currently protected by halon, provided that the characteristics of the anticipated fire have not changed.

Designing Halocarbon Clean Agents Total Fire Suppression Quantitative Requirement (1–9)

Clean agent systems are designed to extinguish fires either by flame extinguishment or by inerting. For reduction of flammable concentration in an atmosphere, inerting is required to be done below one-half of its lower flammable limit. Flame extinguishment is designed to cease combustion of a combustible solid or a flammable liquid. Halogenated clean agents are required to possess the properties listed in Table 9, and their systems are required to be designed to operate within the working pressure as shown in Table 10. Halocarbon agent total flooding quantity, assuming normal leakage from a tight enclosure, is calculated by using the same formula being used for halon

$$W = \frac{(V).(C).(A)}{(s).(100 - C)}$$

Table 9 Halogenated agent quality requirements (NFPA 2001 2004)

S. No.	Property	Specification
1	Agent purity, mole%, minimum	99.0
2	Acidity, ppm (by weight HCl equivalent), maximum	3.0
3	Water content, % by weight, maximum	0.001
4	Nonvolatile residues, g/100 ml maximum	0.05

Table 10 Minimum design working pressure for halocarbon clean agent system piping (1–9)

S. No.	Agent	Agent container maximum fill density (kg/m^3)	Agent container charging pressure at 21 °C (kPa)	Agent container pressure at 55 °C (kPa)	Minimum piping design pressure at 21 °C (kPa)
1	HFC-227ea	1260.75	1029	1708.14	1372
		1210.32	2469.6	3567.2	2853.76
		1210.32	4116	7031.5	5625.2
2	FC-3-1-10	1344.8	2469.6	3087	2469.6
3	HCFC blend A	944.722	4116	5831	4664.8
		944.722	2469.6	3704.4	2963.52
4	HFC 23	806.88	4177.05[a]	11751.18	9405.06
		756.45	4177.05[a]	10701.6	8561.28
		672.4	4177.05[a]	9480.52	7587.16
		588.35	4177.05[a]	8629.88	6908.02
		504.3	4177.05[a]	7943.88	6359.22
5	HCFC-124	1243.94	1646.4	2428.44	1941.38
6	HCFC-124	1243.94	2469.6	3978.8	3183.04
7	HFC-125	907.74	2469.6	4218.9	3375.12
8	HFC-125	941.36	4116	7168.7	5734.96
9	HFC-236fa	1243.94	1646.4	2469.6	1920.8
10	HFC-236fa	1260.75	2469.6	4116	3292.8
11	HFC-236fa	1243.94	4116	7546	6036.8
12	FK-5-1-12	1512.9	2469.6	2833.18	2469.6

[a]Not super-pressurized with nitrogen

$$s = (K1) + (K2.T),$$

where W = weight of halocarbon clean agent (kg); V = net volume of protected enclosure (m^3); S = specific volume (m^3/kg); A = altitude correction factor as per Table 11 (NFPA 2001 2004).

C = halocarbon clean agent design concentration that represents percentage of clean agent per volume. For example, if the halocarbon clean agent concentration is 6%, C = 6, not 0.06, and if the concentration is 6.5%, C = 6.5, not 0.065. k_1 and k_2 = constants that relate to the specific volume of hydrocarbon agent used; these constants are listed in Table 12 (NFPA 2001 2004).

Alternatively, the required agent quantity can be determined using the flooding factors found in Annexure A of NFPA 2001. For each agent, the flooding factor multiplied by the room volume gives the agent quantity, which is multiplied by the altitude correction factor. This can be better illustrated from following examples for the general understanding of the readers for the clean agents: HFC-227ea and IG-541.

Table 11 Atmospheric correction factors (NFPA 2001 2004)

S. No.	Equivalent altitude (km)	Enclosure pressure (mm Hg)	Atmospheric correction factor
1	−0.92	840	1.11
2	−0.61	812	1.07
3	−0.30	787	1.04
4	0.00	760	1.00
5	0.30	733	0.96
6	0.61	705	0.93
7	0.91	678	0.89
8	1.22	650	0.86
9	1.52	622	0.82
10	1.83	596	0.78
11	2.13	570	0.75
12	2.45	550	0.72
13	2.74	528	0.69
14	3.05	505	0.66

Case 1: Design of HFC-227ea Halocarbon Clean Agent Concentration

A sophisticated instrumentation room of size: 3 m wide × 6 m long × 3 m height with an ambient temperature of 21 °C is protected by a halocarbon clean agent, HFC-227ea, at a 6% design concentration. Design HFC-227ea requirement by weight to protect the room, assuming an elevation at sea level?

Solution: As specific volume of superheated HFC-227ea vapor can be approximated by the formula:

$$s = (k1) + (k2.t),$$

where t = temp. (°C) given as 21 °C, $k_1 = 0.1209$ and $k_2 = 0.00049$ for HFC-227ea.

Therefore

$$\begin{aligned} s &= 0.1209 + 0.00049.t = 0.1209 + (0.00049 \times 21) \\ &= 0.1313\,\text{m}^3/\text{kg} \end{aligned}$$

Refer Table 11, approx. same value at 21 °C.

$$\begin{aligned} V &= (3\ \text{m}) \times (6\ \text{m}) \times (3\ \text{m}) = 54\ \text{m}^3 \\ C &= 6\ (\text{given}) \end{aligned}$$

Table 12 HFC-227ea total flooding quantity (NFPA 2001 2004)

S. No.	Temperature, *t*	Specific vapor vol., s	Weight requirements of hazard volume, *W*/*V* (kg/m^3)[b] Design concentration (% by volume)[e]									
	(°C)[c]	(m^3/kg)[d]	6%	7%	8%	9%	10%	11%	12%	13%	14%	15%
1	−12	0.1146	0.556411	0.657271	0.758131	0.862353	0.95817	1.079202	1.190148	1.304456	1.420445	1.539796
2	−6.6	0.1173	0.542963	0.640461	0.741321	0.842181	0.946403	1.052306	1.161571	1.272517	1.386825	1.502814
3	−1	0.1202	0.531196	0.625332	0.72283	0.822009	0.92455	1.028772	1.134675	1.242259	1.353205	1.467513
4	4.5	0.1230	0.519429	0.611884	0.707701	0.803518	0.902697	1.005238	1.107779	1.215363	1.322947	1.433893
5	10	0.1258	0.507662	0.598436	0.690891	0.786708	0.882525	0.981704	1.084245	1.188467	1.29437	1.403635
6	15.5	0.1285	0.495895	0.584988	0.675762	0.769898	0.864034	0.961532	1.060711	1.161571	1.265793	1.373377
7	21	0.1313	0.485809	0.573221	0.662314	0.753088	0.845543	0.94136	1.038858	1.138037	1.238897	1.343119
8	26.6	0.1340	0.475723	0.561454	0.648866	0.737959	0.828733	0.921188	1.017005	1.114503	1.213682	1.316223
9	32	0.1367	0.467318	0.549687	0.635418	0.72283	0.811923	0.904378	0.996833	1.09265	1.190148	1.289327
10	38	0.1395	0.457232	0.539601	0.623651	0.709382	0.796794	0.885887	0.976661	1.070797	1.166614	1.264112
11	43	0.1422	0.448827	0.529515	0.611884	0.695934	0.781665	0.869077	0.95817	1.050625	1.144761	1.240578
12	49	0.1449	0.440422	0.519429	0.600117	0.682486	0.766536	0.852267	0.94136	1.030453	1.122908	1.217044
13	54	0.1476	0.432017	0.509343	0.58835	0.669038	0.753088	0.837138	0.922869	1.011962	1.102736	1.195191
14	60	0.1503	0.425293	0.500938	0.578264	0.657271	0.73964	0.822009	0.90774	0.993471	1.082564	1.173338
15	65.5	0.1530	0.416888	0.492533	0.568178	0.645504	0.726192	0.80688	0.89093	0.976661	1.064073	1.153166
16	71	0.1556	0.410164	0.484128	0.558092	0.635418	0.714425	0.793432	0.875801	0.959851	1.045582	1.132994
17	76.6	0.1583	0.40344	0.475723	0.549687	0.623651	0.700977	0.779984	0.860672	0.943041	1.027091	1.114503
18	82	0.1610	0.396716	0.467318	0.539601	0.613565	0.68921	0.768217	0.847224	0.927912	1.010281	1.096012

(continued)

Table 12 (continued)

S. No.	Temperature, t	Specific vapor vol., s	Weight requirements of hazard volume, W/V (kg/m^3)[b] Design concentration (% by volume)[e]									
	(°C)[c]	(m^3/kg)[d]	6%	7%	8%	9%	10%	11%	12%	13%	14%	15%
19	88	0.1637	0.389992	0.460594	0.531196	0.603479	0.679124	0.754769	0.833776	0.912783	0.995152	1.077521
20	93	0.1662	0.383268	0.452189	0.522791	0.595074	0.667357	0.743002	0.820328	0.899335	0.978342	1.060711

[a]The manufacturer's listing specifies the temperature range for operation
[b]W/V [agent weight requirements (kg/m^3)] = kilograms of agent required per cubic meter of protected volume to produce indicated concentration at temperature specified: $W = \frac{V}{s}\left(\frac{C}{100-C}\right)$
[c]t [temperature (°C)] = the design temperature in the hazard area
[d]s [specific volume (m^3/kg)] = specific volume of superheated HFC-227ea vapor can be approximated by the formula: $s = 0.1209 + 0.00049.t$, where t = temp. (°C)
[e]C [conc. (%)] = Volumetric conc. of HFC-227ea in air at the temperature indicated
[f]For other halocarbon clean agents, refer to NFPA 2001

A = sea level, or 0 feet elevation, and the correction factor is therefore 1, per Table 10.

$$W = \frac{(V).(C).(A)}{s.(100 - C)} = \frac{(54).(6).(1)}{(0.1312).(100 - 6)}$$
$$= 26.23 \approx 27 \text{ kg}$$

An alternative method of determining that design flooding quantity is to use the tables contained in the Annex of NFPA 2001. The table for HFC-227ea is included as Table 12.

The weight requirement corresponding to 21 °C and 6% is shown as 0.4858. Multiplying this factor times the room volume gives

$$W = \left(0.4858 \frac{\text{kg}}{\text{m}^3}\right) \times \left(54 \text{ m}^3\right) = 26.23 \approx 27 \text{ kg}$$

It may be noted that for this Case-1, the results are identical whether using the formula or the table. NFPA 2001 (2004) includes a table for each clean agent recognized by the standard.

The value 0.4858 kg/m^3 is a flooding factor, representing the quantity of halocarbon clean agent required to achieve a selected design concentration (6%) at a specified temperature (21 °C).

Inert Gas Clean Agent Total Flooding Quantity

Inert gas agents are required to be used at the minimum working pressures shown in Table 13 (NFPA 2001 2004) and must possess the quality shown in Table 14 (NFPA 2001 2004).

Inert gas quantity is based on finding the volume of gas needed, as opposed to finding the weight, as we did with halocarbons.

The formula for determining the volume of gas required is

$$V_g = (2.303).(V).\left(\frac{V_S}{s}\right). \log 10\left(\frac{100}{100 - C}\right).(A)$$

where

s	$(k_1) + (k_2.T)$
V	net volume of protected enclosure (m^3)
S	specific volume of inert gas (m^3/kg) at specified temperature
C	inert gas clean agent design concentration (v%)
V_s	specific volume of inert gas at 21 °C
A	altitude correction factor, refer Table 11.

Table 13 Minimum design working pressure for inert gas clean agent system piping (NFPA 2001 2004)

S. No.	Agent	Agent container pressure at 21 °C (kPa)	Agent Container Pressure at 55°C (kPa)	Minimum design pressure at 21 °C of piping upstream of pressure reducer (kPa)
1	IG-01	16,341	18,271	16,341
		20,436	22,781	20,436
2	IG-541	14,997	17,755	14,997
		19,996	23,671	19,996
3	IG-55	15,320	17,065	15,320
		20,423	22,753	20,423
		30,634	34,130	30,634
4	IG-100	16,575	19,299	16,575
		22,312	26,015	22,312
		28,000	32,778	28,000

Table 14 Inert gas agent quality requirements (NFPA 2001 2004)

S. No.			IG-01	IG-100	IG-541	IG-55
1	Composition, % by volume	N_2 Ar CO_2	Minimum 99.9%	Minimum 99.9%	52% ± 4% 40% ± 4% 8% + 1% − 0.0%	50% ± 5% 50% ± 5%
2	Water content, % by weight		Maximum 0.005%	Maximum 0.005%	Maximum 0.005%	Maximum 0.005%

$k_1 + k_2$ constants that relate to the specific volume of inert gas clean agent used, as listed on Table 14.

Note that Table 14 provides constants for IG-541 only, and that NFPA 2001 should be consulted for other inert gas agents.

V_g = volume of inert gas added at standard conditions per volume of hazard space.

Case 2: Design of IG-541 Inert Gas Clean Agent Concentration

A compartment/room of size: 6.06 m wide by 15.15 m long by 2.42 m height with an ambient temperature of 21 °C, is protected by inert gas clean agent IG-541, with a concentration of 34%. Design the minimum volume requirement of IG-541 to protect the compartment/room, assuming an elevation at sea level? (Given: V_s for IG-541 at 21 °C is 0.675 m^3/kg.)

Solution: As specific volume of superheated IG-541 vapor can be approximated by the formula

Table 15 IG-541 total flooding quantity (NFPA 2001 2004)

S. No.	Temp, t	Specific vapor volume s	Volume requirements of agent per unit volume of hazard, $V_{agent}/V^{b}_{enclosure}$ Design concentration (% by volume)[e]							
	(°C)[c]	(m^3/kg)[d]	34	38	42	46	50	54	58	62
1	−40	0.5353	0.524	0.603	0.686	0.802	0.873	0.977	1.096	1.218
2	−34.4	0.5480	0.513	0.590	0.672	0.760	0.855	0.958	1.070	1.194
3	−29	0.5608	0.501	0.576	0.657	0.743	0.836	0.936	1.046	1.166
4	−23	0.5735	0.490	0.563	0.642	0.726	0.817	0.915	1.022	1.140
5	−17.7	0.5863	0.479	0.551	0.628	0.710	0.799	0.895	1.000	1.116
6	−12	0.5990	0.469	0.539	0.615	0.695	0.782	0.876	0.979	1.092
7	−6.6	0.6117	0.459	0.528	0.602	0.681	0.766	0.858	0.958	1.069
8	−1	0.6245	0.450	0.517	0.590	0.667	0.750	0.840	0.939	1.047
9	4.44	0.6372	0.441	0.507	0.578	0.653	0.735	0.824	0.920	1.026
10	10	0.6499	0.432	0.497	0.566	0.641	0.721	0.807	0.902	1.006
11	15.55	0.6627	0.424	0.487	0.555	0.628	0.707	0.792	0.885	0.987
12	21	0.6755	0.416	0.478	0.545	0.616	0.693	0.777	0.868	0.968
13	26.6	0.6882	0.408	0.469	0.535	0.605	0.681	0.762	0.852	0.950
14	32	0.7009	0.401	0.461	0.525	0.594	0.668	0.749	0.836	0.933
15	37.7	0.7137	0.393	0.453	0.516	0.583	0.656	0.735	0.821	0.916
16	43	0.7264	0.386	0.445	0.507	0.573	0.645	0.722	0.807	0.900
17	49	0.7392	0.380	0.437	0.498	0.563	0.634	0.710	0.793	0.884
18	54.4	0.7519	0.373	0.430	0.489	0.554	0.623	0.698	0.779	0.869
19	60	0.7647	0.367	0.422	0.481	0.544	0.612	0.686	0.766	0.855
20	65.5	0.7774	0.361	0.415	0.473	0.535	0.602	0.675	0.754	0.841
21	71	0.7902	0.355	0.409	0.466	0.527	0.593	0.664	0.742	0.827
22	76.6	0.8029	0.350	0.402	0.458	0.518	0.583	0.653	0.730	0.814
23	82	0.8157	0.344	0.396	0.451	0.510	0.574	0.643	0.718	0.801
24	87.7	0.8284	0.339	0.390	0.444	0.502	0.565	0.633	0.707	0.789
25	93	0.8412	0.334	0.384	0.437	0.495	0.557	0.624	0.697	0.777

Note The manufacturer's listing specifies the temperature range for operation
[a]Atoms of middle Carbon-Cl_2
[b]Atoms of middle Carbon-Cl, F
[c]Atoms of middle Carbon-F_2
[d]Atoms of middle Carbon-Cl, H
[e]Atoms of middle Carbon-H, F
[f]Atoms of middle Carbon-H_2

$$s = k_1 + k_2.t,$$

where t = temperature (°C); k_1 = 0.6271 and k_2 = 0.00229 for IG-541.

Therefore,

$$\begin{aligned} s &= (0.6271) + (0.00229 \times 21) \\ &= 0.675\ \mathrm{m^3/kg} \end{aligned}$$

Refer Table 15, approx. same value at 21 °C.

$$V = (6.06) \times (15.15) \times (2.42) = 222.17\ \text{m}^3$$

$$\text{Concentration } (\%), C = 34 \text{ (given)}$$
$$V_s = 0.675\ \text{m}^3/\text{kg}, \text{as per NFPA 2001}$$

A = sea level elevation, and the correction factor is, therefore 1, as per Table 11.

$$\begin{aligned} V_g &= (2.303).(222.17).\left(\frac{0.675}{0.675}\right).\log_{10}\left(\frac{100}{100-34}\right).(1) \\ &= (2.303) \times (222.17) \times (1) \times (0.1805) \times (1) = 92.35\ \text{m}^3 \end{aligned}$$

In a manner analogous to the tabular solution presented in Case-1, designers can use Table 15 to obtain a solution. It may be very important to note that the flooding factor corresponding to 21 °C and 34% concentration is 0.416 m^3/m^3 which when multiplied with 222.17 m^3 gives rise to 92.42 m^3 of IG-541, which is slightly more than what was obtained using the design calculation method, which could be attributed to rounding of logarithmic functions.

- V_g [agent volume requirements (kg/m^3)] = Kilogram of agent required per cubic meter of protected volume to produce indicated concentration at temperature specified.

$$V_g = 2.303 \times \left(\frac{V_S}{s}\right).\log_{10}\left(\frac{100}{100-C}\right) = \left(\frac{V_S}{s}\right) \times \ln\left(\frac{100}{100-C}\right)$$

- t [temperature (°C)] = the design temperature in the hazard area.
- s [specific volume (m^3/kg)] = specific volume of superheated IG-541 vapor can be approximated by the formula: $s = 0.6271 + 0.00229t$, where t = temperature (°C).
- C [concentration (%)] = Vol. Conc. of IG-541 in air at the temperature indicated.

Note: V_s = the term $V_g = \ln\ [100/(100 - C)]$ gives volume at a rated concentration (%) and temperature to reach air-agent mixture at the end of flooding time in a volume of 1 m^3.

Halocarbon and Inert Gas Discharge Time

Halocarbon clean agents must be discharged within 10 s. Inert gas agents that do not create decomposition products may be discharged within one minute. The room must hold the gas for a time sufficient to extinguish a deep-seated fire without re-ignition.

Fig. 2 A clean agent storage arrangement and discharge from nozzles

Clean Agent Storage and Nozzle Discharge Arrangement

A clean agent storage arrangement, clean agent nozzles and clean agent discharge are shown in Fig. 2 for suppression of fires in a room or compartment.

Pressure Relief Venting for Clean Agent Systems

NFPA 2001 (2004) requires that where clean agent valving arrangements on the pilot piping or on the discharge piping create closed piping arrangements where pressure could increase beyond the pressure rating of the piping, fittings, and nozzles, pressure relief devices are to be installed. The pressure relief devices are required to discharge in such a manner as not to be hazardous to personnel. The NFPA 2001 Annex describes pressure relief isometric diagrams for clean agent cylinders, showing pressure compatibilities for a variety of clean agent storage conditions. The Annex further recommends that pressure relief venting for closed piping sections follow the FSSA Pipe Design Handbook.

Novel and New Water-Based Options for Halon Replacement Options

Three water-based options for Halon replacement are

1. Water mist system
2. Double-interlocked pre-action Water Mist systems
3. Standard automatic water mist systems

Concluding Remarks

1. The design of new halon systems has been essentially halted as the result of cessation of production of halon in accordance with the Montreal Protocol, which prohibits the manufacture of halogenated agents in countries participating in the agreement. Although pure halon in concentrations between 5 and 10% is considered nontoxic to humans during brief exposure, the products of decomposition can be dangerous if breathed.
2. Clean agent systems may be considered as halon system replacements when designed in accordance with NFPA 2001.
3. From Case-1 study, it may be important to note that the value of 0.4858 kg/m^3 is a total flooding factor of HFC-227ea fire extinguishing agent representing the quantity of halocarbon clean agent required to achieve a selected design fire extinguishing concentration of 6% at a specified ambient temperature of 21 °C.
4. From Case-2 Study, it may be very important to note that the flooding factor for IG-541 clean agent fire extinguishing agent corresponding to ambient temperature of 21 °C and minimum fire extinguishing 34% concentration is 0.416 m^3/m^3, which is slightly more than that was obtained using the design calculation method, which could be attributed to rounding of logarithmic functions.
5. The clean agent containers should not be kept/mounted in the hazardous area, and it shall suitably be installed in a protected location as close as possible to the hazard. Piping and fittings must be of a pressure rating commensurate with expected system pressures, and must be corrosion-resistant. Piping and fittings must be metallic, and the fittings cannot be of cast iron and it may be of welded, brazed, or malleable iron.
6. Fire suppression and detection shall be selected/designed to be appropriate for the anticipated class of fires and emerging fire load density pattern/layout with appropriately designed discharge flow rate, particle/droplet size distribution with respect to fire extinguishing efficiency parameters.
7. An existing detection system may possibly be reused when designing a clean agent system, provided that the characteristics of the anticipated fire have not been changed
8. Clean agents include halocarbon and inert gas agents that are in conformance with NOAEL and EPA guidelines.
9. Halocarbon agents develop products of decomposition that may be harmful to personnel, hence such agents may be strategically used as per EPA and local government guidelines, rules, and regulations prevailing under Protocol Agreement.

Acknowledgements The authors are thankful to the Director, CSIR-Central Building Research Institute, Roorkee for according his permission to submit the paper in the International Conference on Water, Environment, Energy and Society is being organized jointly by the Texas A & M University, Texas, USA and AISECT University, Bhopal, India from March 15–18, 2016 in Bhopal, M.P., India in association with ICE WaRM, Australia, Environment Management & Human Welfare Council, Water and Land Management Institute, MANIT Bhopal, M.P. Council of

Science and Technology, International Institute of Solid Waste Management, CIPET, IISS, WALMI, Institute of Disaster Management, University of Western Sydney, etc.

References

Climate Change Assessments (2010) Review of the processes and procedures of the IPCC. Committee to Review the Intergovernmental Panel on Climate Change, Oct 2010

IPCC Second Assessment Climate Change (1995) A report of the intergovernmental panel on climate change

NFPA 2001 (2004) Clean agent fire extinguishing systems. National Fire Protection Association, Quincy, MA 02169

NFPA 12A. Standard on Halon 1301 fire extinguishing systems

NFPA 12B. Standard on Halon 1211 fire extinguishing systems

NFPA 12C-T. Standard on Halon 2402 extinguishing systems

NFPA 75. Standard for the protection of electronic computer data processing equipment

Robin ML (2012) Ph.D., DuPont chemicals & fluoroproducts, 28 June 2012

http://www2.dupont.com/FE/en_US/assets/downloads/pdf/201202_IFP_article_Climate_Change_and_the_HFC-Based_Clean_Extinguishing_Agents.pdf

Ford CL (1970) Halon 1301 fire extinguishing agent: properties and applications. Fire J 64(6)

Wickham RT (1972) Engineering and economic aspects of Halon extinguishing equipment. An Appraisal of Halogenated Fire Extinguishing Agents, National Academy of Sciences, Washington, DC

Jensen R (1972) Halogenated extinguishing agent systems. Fire J 66(3):37–39

Ford C (1962) Overview of Halon 1301 systems. In: Symposium on the mechanism of halogenated extinguishing agents. ACS Symposia Series

Bischoff BG (1978) Gaseous extinguishing agents. Heating, ventilation and air conditioning

Clarke DG (1970) The toxicity of bromotrifluoromethane (FE1301) in animals and man. Industrial Hygiene Research Laboratory, Imperial Chemical Industries, Alderley Park, Chesire, UK

Franck TE (1971) Clean room protection using Halon 1301. Fire J 65(2):77–79

Steinberg M (1972) Toxic hazards from extinguishing gasoline fires using Halon 1301 extinguishers in armored personnel carriers. An Appraisal of Halogenated Fire Extinguishing Agents, National Academy of Sciences, Washington, DC

Williamson HV (1972) Halon 1301—minimum concentrations for extinguishing deep-seated fires. Fire Technol 8(4)

National Academy of Sciences (1972) An appraisal of halogenated fire extinguishing agents. Proceedings of a symposium

McDaniel DE (1972) Evaluation of Halon 1301 for shipboard use. An Appraisal of Halogenated Fire Extinguishing Agents, National Academy of Sciences, Washington, DC

Wiersman SJ (1978) Flow characteristics of Halon 1301 in pipelines. Fire Technol 14(1):5–14

Strasser A, Liebman I, Kuchta JM (1974) Methane flame extinguishment with layered Halon or carbon dioxide. Fire Technol 10(1):25–34

Robinson VB (1978) Partial flooding of volumes with Halon 1301. Fire Technol 14(2):97–109

Edmonds Albert (1972) Use of Halon 1211 in hand extinguishers and local application systems. An Appraisal of Halogenated Fire Extinguishing Agents, National Academy of Sciences, Washington, DC

Carbon Dioxide Fire Extinguishing Systems (1984) Walter Kidde and Co., Wake Forest, NC, 1981, revised Aug 1984

Beiggs AA (1977) Use of nitrogen-filled high expansion foam to protect a 500—ton fuel tank. Fire Research Note 1074, Fire Research Station, Boreham Wood, Herts, England, Aug 1977

NFPA 11A. Standard for medium and high expansion foam systems

NFPA 403. Recommended practice for aircraft rescue and fire fighting services at airports and heliports
NFPA 17. Standard for dry chemical extinguishing systems
Guise AB (1967) Extinguishment of natural gas pressure fires. Fire Technol 3(3):175–193
Lee TG, Robertson AF (1960) Extinguishing effectiveness of some powdered materials on hydrocarbon fires. Fire Res Abstr Rev 2(1)
Meldrum DN (1962) Combined use of foam and dry chemical. NFPA Q 56(1):28–34
Underwriters Laboratories Inc. (1963) The compatibility relationship between mechanical foam and dry chemical fire extinguishing agents. UL Bull Res 54 (Chicago, IL)
Woolhouse RA, Sayers DR (1973) Monnex compared with other potassium-based dry chemicals. Fire J 671(1):85–88
Wesson HR (1972) Studies of the effects of particle size on the flow characteristics of dry chemical. Fire Technol 8(3):173–180
Tuve RL (1964) Light water and potassium bicarbonate dry chemical—a new two agent extinguishing system. NFPA Q 58(1):64–69
Haessler WM (1974) The extinguishment of fire, revised edition. National Fire Protection Association, Quincy, MA

MIX
Papier aus verantwortungsvollen Quellen
Paper from responsible sources
FSC® C105338

If you have any concerns about our products,
you can contact us on
ProductSafety@springernature.com

In case Publisher is established outside the EU,
the EU authorized representative is:
Springer Nature Customer Service Center GmbH
Europaplatz 3, 69115 Heidelberg, Germany

Printed by Libri Plureos GmbH
in Hamburg, Germany